THEORY OF DISTRIBUTIONS

MONOGRAPHS AND TEXTBOOKS IN PURE AND APPLIED MATHEMATICS

1. *K. Yano,* Integral Formulas in Riemannian Geometry (1970)
2. *S. Kobayashi,* Hyperbolic Manifolds and Holomorphic Mappings (1970)
3. *V. S. Vladimirov,* Equations of Mathematical Physics (A. Jeffrey, editor; A. Littlewood, translator) (1970)
4. *B. N. Pshenichnyi,* Necessary Conditions for an Extremum (L. Neustadt, translation editor; K. Makowski, translator) (1971)
5. *L. Narici, E. Beckenstein, and G. Bachman,* Functional Analysis and Valuation Theory (1971)
6. *S. S. Passman,* Infinite Group Rings (1971)
7. *L. Dornhoff,* Group Representation Theory (in two parts). Part A: Ordinary Representation Theory. Part B: Modular Representation Theory (1971, 1972)
8. *W. Boothby and G. L. Weiss (eds.),* Symmetric Spaces: Short Courses Presented at Washington University (1972)
9. *Y. Matsushima,* Differentiable Manifolds (E. T. Kobayashi, translator) (1972)
10. *L. E. Ward, Jr.,* Topology: An Outline for a First Course (1972)
11. *A. Babakhanian,* Cohomological Methods in Group Theory (1972)
12. *R. Gilmer,* Multiplicative Ideal Theory (1972)
13. *J. Yeh,* Stochastic Processes and the Wiener Integral (1973)
14. *J. Barros-Neto,* Introduction to the Theory of Distributions (1973)
15. *R. Larsen,* Functional Analysis: An Introduction (1973)
16. *K. Yano and S. Ishihara,* Tangent and Cotangent Bundles: Differential Geometry (1973)
17. *C. Procesi,* Rings with Polynomial Identities (1973)
18. *R. Hermann,* Geometry, Physics, and Systems (1973)
19. *N. R. Wallach,* Harmonic Analysis on Homogeneous Spaces (1973)
20. *J. Dieudonné,* Introduction to the Theory of Formal Groups (1973)
21. *I. Vaisman,* Cohomology and Differential Forms (1973)
22. *B.-Y. Chen,* Geometry of Submanifolds (1973)
23. *M. Marcus,* Finite Dimensional Multilinear Algebra (in two parts) (1973, 1975)
24. *R. Larsen,* Banach Algebras: An Introduction (1973)
25. *R. O. Kujala and A. L. Vitter (eds.),* Value Distribution Theory: Part A; Part B: Deficit and Bezout Estimates by Wilhelm Stoll (1973)
26. *K. B. Stolarsky,* Algebraic Numbers and Diophantine Approximation (1974)
27. *A. R. Magid,* The Separable Galois Theory of Commutative Rings (1974)
28. *B. R. McDonald,* Finite Rings with Identity (1974)
29. *J. Satake,* Linear Algebra (S. Koh, T. A. Akiba, and S. Ihara, translators) (1975)
30. *J. S. Golan,* Localization of Noncommutative Rings (1975)
31. *G. Klambauer,* Mathematical Analysis (1975)
32. *M. K. Agoston,* Algebraic Topology: A First Course (1976)
33. *K. R. Goodearl,* Ring Theory: Nonsingular Rings and Modules (1976)
34. *L. E. Mansfield,* Linear Algebra with Geometric Applications: Selected Topics (1976)
35. *N. J. Pullman,* Matrix Theory and Its Applications (1976)
36. *B. R. McDonald,* Geometric Algebra Over Local Rings (1976)
37. *C. W. Groetsch,* Generalized Inverses of Linear Operators: Representation and Approximation (1977)
38. *J. E. Kuczkowski and J. L. Gersting,* Abstract Algebra: A First Look (1977)
39. *C. O. Christenson and W. L. Voxman,* Aspects of Topology (1977)

40. *M. Nagata,* Field Theory (1977)
41. *R. L. Long,* Algebraic Number Theory (1977)
42. *W. F. Pfeffer,* Integrals and Measures (1977)
43. *R. L. Wheeden and A. Zygmund,* Measure and Integral: An Introduction to Real Analysis (1977)
44. *J. H. Curtiss,* Introduction to Functions of a Complex Variable (1978)
45. *K. Hrbacek and T. Jech,* Introduction to Set Theory (1978)
46. *W. S. Massey,* Homology and Cohomology Theory (1978)
47. *M. Marcus,* Introduction to Modern Algebra (1978)
48. *E. C. Young,* Vector and Tensor Analysis (1978)
49. *S. B. Nadler, Jr.,* Hyperspaces of Sets (1978)
50. *S. K. Segal,* Topics in Group Kings (1978)
51. *A. C. M. van Rooij,* Non-Archimedean Functional Analysis (1978)
52. *L. Corwin and R. Szczarba,* Calculus in Vector Spaces (1979)
53. *C. Sadosky,* Interpolation of Operators and Singular Integrals: An Introduction to Harmonic Analysis (1979)
54. *J. Cronin,* Differential Equations: Introduction and Quantitative Theory (1980)
55. *C. W. Groetsch,* Elements of Applicable Functional Analysis (1980)
56. *I. Vaisman,* Foundations of Three-Dimensional Euclidean Geometry (1980)
57. *H. I. Freedan,* Deterministic Mathematical Models in Population Ecology (1980)
58. *S. B. Chae,* Lebesgue Integration (1980)
59. *C. S. Rees, S. M. Shah, and C. V. Stanojević,* Theory and Applications of Fourier Analysis (1981)
60. *L. Nachbin,* Introduction to Functional Analysis: Banach Spaces and Differential Calculus (R. M. Aron, translator) (1981)
61. *G. Orzech and M. Orzech,* Plane Algebraic Curves: An Introduction Via Valuations (1981)
62. *R. Johnsonbaugh and W. E. Pfaffenberger,* Foundations of Mathematical Analysis (1981)
63. *W. L. Voxman and R. H. Goetschel,* Advanced Calculus: An Introduction to Modern Analysis (1981)
64. *L. J. Corwin and R. H. Szcarba,* Multivariable Calculus (1982)
65. *V. I. Istrăţescu,* Introduction to Linear Operator Theory (1981)
66. *R. D. Järvinen,* Finite and Infinite Dimensional Linear Spaces: A Comparative Study in Algebraic and Analytic Settings (1981)
67. *J. K. Beem and P. E. Ehrlich,* Global Lorentzian Geometry (1981)
68. *D. L. Armacost,* The Structure of Locally Compact Abelian Groups (1981)
69. *J. W. Brewer and M. K. Smith, eds.,* Emily Noether: A Tribute to Her Life and Work (1981)
70. *K. H. Kim,* Boolean Matrix Theory and Applications (1982)
71. *T. W. Wieting,* The Mathematical Theory of Chromatic Plane Ornaments (1982)
72. *D. B. Gauld,* Differential Topology: An Introduction (1982)
73. *R. L. Faber,* Foundations of Euclidean and Non-Euclidean Geometry (1983)
74. *M. Carmeli,* Statistical Theory and Random Matrices (1983)
75. *J. H. Carruth, J. A. Hildebrant, and R. J. Koch,* The Theory of Topological Semigroups (1983)
76. *R. L. Faber,* Differential Geometry and Relativity Theory: An Introduction (1983)
77. *S. Barnett,* Polynomials and Linear Control Systems (1983)
78. *G. Karpilovsky,* Commutative Group Algebras (1983)
79. *F. Van Oystaeyen and A. Verschoren,* Relative Invariants of Rings: The Commutative Theory (1983)
80. *I. Vaisman,* A First Course in Differential Geometry (1984)
81. *G. W. Swan,* Applications of Optimal Control Theory in Biomedicine (1984)

82. *T. Petrie and J. D. Randall,* Transformation Groups on Manifolds (1984)
83. *K. Goebel and S. Reich,* Uniform Convexity, Hyperbolic Geometry, and Non-expansive Mappings (1984)
84. *T. Albu and C. Năstăsescu,* Relative Finiteness in Module Theory (1984)
85. *K. Hrbacek and T. Jech,* Introduction to Set Theory: Second Edition, Revised and Expanded (1984)
86. *F. Van Oystaeyen and A. Verschoren,* Relative Invariants of Rings: The Non-commutative Theory (1984)
87. *B. R. McDonald,* Linear Algebra Over Commutative Rings (1984)
88. *M. Namba,* Geometry of Projective Algebraic Curves (1984)
89. *G. F. Webb,* Theory of Nonlinear Age-Dependent Population Dynamics (1985)
90. *M. R. Bremner, R. V. Moody, and J. Patera,* Tables of Dominant Weight Multiplicities for Representations of Simple Lie Algebras (1985)
91. *A. E. Fekete,* Real Linear Algebra (1985)
92. *S. B. Chae,* Holomorphy and Calculus in Normed Spaces (1985)
93. *A. J. Jerri,* Introduction to Integral Equations with Applications (1985)
94. *G. Karpilovsky,* Projective Representations of Finite Groups (1985)
95. *L. Narici and E. Beckenstein,* Topological Vector Spaces (1985)
96. *J. Weeks,* The Shape of Space: How to Visualize Surfaces and Three-Dimensional Manifolds (1985)
97. *P. R. Gribik and K. O. Kortanek,* Extremal Methods of Operations Research (1985)
98. *J.-A. Chao and W. A. Woyczynski, eds.,* Probability Theory and Harmonic Analysis (1986)
99. *G. D. Crown, M. H. Fenrick, and R. J. Valenza,* Abstract Algebra (1986)
100. *J. H. Carruth, J. A. Hildebrant, and R. J. Koch,* The Theory of Topological Semigroups, Volume 2 (1986)
101. *R. S. Doran and V. A. Belfi,* Characterizations of C*-Algebras: The Gelfand-Naimark Theorems (1986)
102. *M. W. Jeter,* Mathematical Programming: An Introduction to Optimization (1986)
103. *M. Altman,* A Unified Theory of Nonlinear Operator and Evolution Equations with Applications: A New Approach to Nonlinear Partial Differential Equations (1986)
104. *A. Verschoren,* Relative Invariants of Sheaves (1987)
105. *R. A. Usmani,* Applied Linear Algebra (1987)
106. *P. Blass and J. Lang,* Zariski Surfaces and Differential Equations in Characteristic $p > 0$ (1987)
107. *J. A. Reneke, R. E. Fennell, and R. B. Minton,* Structured Hereditary Systems (1987)
108. *H. Busemann and B. B. Phadke,* Spaces with Distinguished Geodesics (1987)
109. *R. Harte,* Invertibility and Singularity for Bounded Linear Operators (1988)
110. *G. S. Ladde, V. Lakshmikantham, and B. G. Zhang,* Oscillation Theory of Differential Equations with Deviating Arguments (1987)
111. *L. Dudkin, I. Rabinovich, and I. Vakhutinsky,* Iterative Aggregation Theory: Mathematical Methods of Coordinating Detailed and Aggregate Problems in Large Control Systems (1987)
112. T. Okubo, *Differential Geometry* (1987)
113. *D. L. Stancl and M. L. Stancl,* Real Analysis with Point-Set Topology (1987)
114. *T. C. Gard,* Introduction to Stochastic Differential Equations (1988)
115. *S. S. Abhyankar,* Enumerative Combinatorics of Young Tableaux (1988)
116. *H. Strade and R. Farnsteiner,* Modular Lie Algebras and Their Representations (1988)
117. *J. A. Huckaba,* Commutative Rings with Zero Divisors (1988)
118. *W. D. Wallis,* Combinatorial Designs (1988)
119. *W. Więsław,* Topological Fields (1988)

120. *G. Karpilovsky,* Field Theory: Classical Foundations and Multiplicative Groups (1988)
121. *S. Caenepeel and F. Van Oystaeyen,* Brauer Groups and the Cohomology of Graded Rings (1989)
122. *W. Kozlowski,* Modular Function Spaces (1988)
123. *E. Lowen-Colebunders,* Function Classes of Cauchy Continuous Maps (1989)
124. *M. Pavel,* Fundamentals of Pattern Recognition (1989)
125. *V. Lakshmikantham, S. Leela, and A. A. Martynyuk,* Stability Analysis of Nonlinear Systems (1989)
126. *R. Sivaramakrishnan,* The Classical Theory of Arithmetic Functions (1989)
127. *N. A. Watson,* Parabolic Equations on an Infinite Strip (1989)
128. *K. J. Hastings,* Introduction to the Mathematics of Operations Research (1989)
129. *B. Fine,* Algebraic Theory of the Bianchi Groups (1989)
130. *D. N. Dikranjan, I. R. Prodanov, and L. N. Stoyanov,* Topological Groups: Characters, Dualities, and Minimal Group Topologies (1989)
131. *J. C. Morgan II,* Point Set Theory (1990)
132. *P. Biler and A. Witkowski,* Problems in Mathematical Analysis (1990)
133. *H. J. Sussmann,* Nonlinear Controllability and Optimal Control (1990)
134. *J.-P. Florens, M. Mouchart, and J. M. Rolin,* Elements of Bayesian Statistics (1990)
135. *N. Shell,* Topological Fields and Near Valuations (1990)
136. *B. F. Doolin and C. F. Martin,* Introduction to Differential Geometry for Engineers (1990)
137. *S. S. Holland, Jr.,* Applied Analysis by the Hilbert Space Method (1990)
138. *J. Okniński,* Semigroup Algebras (1990)
139. *K. Zhu,* Operator Theory in Function Spaces (1990)
140. *G. B. Price,* An Introduction to Multicomplex Spaces and Functions (1991)
141. *R. B. Darst,* Introduction to Linear Programming: Applications and Extensions (1991)
142. *P. L. Sachdev,* Nonlinear Ordinary Differential Equations and Their Applications (1991)
143. *T. Husain,* Orthogonal Schauder Bases (1991)
144. *J. Foran,* Fundamentals of Real Analysis (1991)
145. *W. C. Brown,* Matrices and Vector Spaces (1991)
146. *M. M. Rao and Z. D. Ren,* Theory of Orlicz Spaces (1991)
147. *J. S. Golan and T. Head,* Modules and the Structures of Rings: A Primer (1991)
148. *C. Small,* Arithmetic of Finite Fields (1991)
149. *K. Yang,* Complex Algebraic Geometry: An Introduction to Curves and Surfaces (1991)
150. *D. G. Hoffman, D. A. Leonard, C. C. Lindner, K. T. Phelps, C. A. Rodger, and J. R. Wall,* Coding Theory: The Essentials (1991)
151. *M. O. González,* Classical Complex Analysis (1992)
152. *M. O. González,* Complex Analysis: Selected Topics (1992)
153. *L. W. Baggett,* Functional Analysis: A Primer (1992)
154. *M. Sniedovich,* Dynamic Programming (1992)
155. *R. P. Agarwal,* Difference Equations and Inequalities: Theory, Methods, and Applications (1992)
156. *C. Brezinski,* Biorthogonality and Its Applications to Numerical Analysis (1992)
157. *C. Swartz,* An Introduction to Functional Analysis (1992)
158. *S. B. Nadler, Jr.,* Continuous Theory: An Introduction (1992)
159. *M. A. Al-Gwaiz,* Theory of Distributions (1992)

Additional Volumes in Preparation

THEORY OF DISTRIBUTIONS

M. A. Al-Gwaiz

King Saud University
Riyadh, Saudi Arabia

CRC Press
Taylor & Francis Group
Boca Raton London New York

CRC Press is an imprint of the
Taylor & Francis Group, an **informa** business
A TAYLOR & FRANCIS BOOK

First published 1992 by Marcel Dekker, Inc.

Published 2019 by CRC Press
Taylor & Francis Group
6000 Broken Sound Parkway NW, Suite 300
Boca Raton, FL 33487-2742

First issued in paperback 2020

ISBN 13: 978-0-367-57996-8 (pbk)
ISBN 13: 978-0-8247-8672-4 (hbk)

Library of Congress Cataloging-in-Publication Data

Al-Gwaiz, M. A.
Theory of distributions / M.A. Al-Gwaiz.
p. cm.
Includes bibliographical references and index.
ISBN 0-8247-8672-6
1. Theory of distributions (Functional analysis) I. Title.
QA324.A52 1992 92-1351
515'.782--dc20 CIP

In memory of Omar (1948-1988)

Preface

The need to "differentiate" nondifferentiable functions was recognized early in the treatment of differential equations. It was also realized that the formula for integration by parts,

$$\int_a^b f'(x)g(x)\,dx = -\int_a^b f(x)g'(x)\,dx$$

provided a sort of *generalized derivative* f' in such case, g being an arbitrary smooth (test) function that vanishes at a and b, for the right-hand side of the equation is well defined if f is merely continuous. This led to Heaviside's operational calculus at the turn of the century, which was widely used by applied mathematicians, physicists, and engineers because it worked, though it was not clear why. In particular, the "δ-function," which had become an indispensable tool in these disciplines, was differentiated and transformed quite freely according to the rules of this calculus. What was lacking was a clear mathematical framework for justifying these procedures and interpreting the generalized functions they produced.

In the 1930s, S. L. Sobolev, in his investigations of the Cauchy problem for a hyperbolic partial differential equation, came a long way toward constructing a mathematical theory for generalized functions when he realized that they were continuous linear functionals over a certain space of test functions. This point of view was developed and articulated in the language of modern functional analysis with the publication of L. Schwartz' *Theorie des Distributions* in the early 1950s. The resulting mathematical structure, based on the duality of certain topological vector spaces, is as striking in its power and sweep as it is in its simplicity and beauty.

This book attempts to provide a concise and motivated introduction to the theory of distributions, or generalized functions, along the lines of L. Schwartz. It is based on lecture notes I have used for a short graduate course on the subject at King Saud University. The book has been written mainly for graduate students of mathematics, both pure and applied, and it assumes the standard undergraduate level of knowledge of topology and analysis, with special emphasis on real analysis. A working knowledge of Lebesgue integration and Hilbert space theory, including the Riesz representation theorem, is also assumed. Many examples have been included to show some of the applications of distributions to differential equations and boundary value problems.

Chapter 1 presents the general background in topology and functional analysis that is needed for defining the space of test functions $\mathfrak{D}$ as a locally convex topological vector space. In Chapter 2 distributions are defined as continuous linear functionals on $\mathfrak{D}$, and the derivative of a distribution is defined by extending the formula for integration by parts. The properties of distributions, which take up the rest of the chapter, all follow from these two definitions.

By enlarging the space of test functions, we use duality to generate special subspaces of distributions, namely those of compact support, finite order, and the so-called tempered distributions. Duality is also used to define the convolution product of two distributions and the Fourier transform of a distribution, the definition in each case being limited to the appropriate subspace. This forms the subject of Chapters 3 and 4. With the Fourier transformation the theory of distributions reaches its climax and becomes a very effective tool for handling linear partial differential equations. Chapter 5 covers the Hilbert space aspect of the theory, which makes it especially suited to the treatment of boundary value problems. These are discussed in Chapter 6 through some representative examples.

Several colleagues at King Saud University were kind enough to read and comment on parts of the manuscript. I am especially grateful to Professor V. Anandam, who has taken the time to read the entire manuscript, for suggesting a number of improvements and providing me with a collection of worked-out examples, most of which have been added to the text. Dr. Sharief Deshmukh was most helpful in proofreading, and Mr. Dilip Thomas prepared the drawings for the figures.

M. A. Al-Gwaiz

Contents

THEORY OF DISTRIBUTIONS

1
Locally Convex Spaces

1.1 PRELIMINARY CONCEPTS

Linear Spaces

We shall denote by **R** and **C** the fields of real and complex numbers, respectively, and we shall use Φ to denote either of these two fields. A *linear space*, or a *vector space*, over Φ is a nonempty set X on which two operations, *addition* and *scalar multiplication*, are defined such that

(a) X is an abelian group under addition, i.e., to every pair $x,y \in X$ the sum $x + y$ is also in X, and we have for all $x,y,z \in X$

- (i) $x + y = y + x$
- (ii) $x + (y + z) = (x + y) + z$
- (iii) There is a zero element $0 \in X$ such that $x + 0 = x$ for all x
- (iv) For each $x \in X$ there is an element $-x \in X$ such that $x + (-x) = 0$.

(b) For every pair a,x with $a \in \Phi$ and $x \in X$ the scalar product $a \cdot x$ is an element in X, and we have for all $a,b \in \Phi$ and $x,y \in X$

- (i) $1 \cdot x = x$
- (ii) $a \cdot (b \cdot x) = (a \cdot b) \cdot x$

(iii) $a \cdot (x + y) = a \cdot x + a \cdot y$
(iv) $(a + b) \cdot x = a \cdot x + b \cdot x$.

From these properties it can be shown that the zero element is unique and that every $x \in X$ has a unique additive inverse $-x$. Furthermore, it follows that $0 \cdot x = 0$ and $(-1) \cdot x = -x$ for every $x \in X$, and that $a \cdot 0 = 0$ for every $a \in \Phi$. In this notation we have used the same symbol 0 to denote the zeros of both Φ and X. As usual, the dot symbol for the product will be dropped.

We call X a real or a complex linear space depending on whether $\Phi = \mathbf{R}$ or $\Phi = \mathbf{C}$. The elements of X are called *vectors*. The vectors $x_1, \ldots, x_n$ are *linearly independent* if the equation $a_1x_1 + \cdots + a_nx_n = 0$, with $a_k \in \Phi$, implies $a_k = 0$ for all k; otherwise they are *linearly dependent*. The set $\{x_1, \ldots, x_n\}$ of vectors in X is said to *span* the space X if any $x \in X$ can be represented by a linear combination of the form

$$x = a_1x_1 + \cdots + a_nx_n$$

where $a_k \in \Phi$. Any (finite) set of linearly independent vectors $\{x_1, \ldots, x_n\}$ which spans X is called a *basis* of X. The *dimension* of X is then n, the number of elements in its basis. If no such basis exists, X is said to be *infinite dimensional*. In a linear space X with basis $\{x_1, \ldots, x_n\}$, any vector $x \in X$ has a unique representation of the form $x = a_1x_1 + \cdots + a_nx_n$, in the sense that the scalar coefficients a_k are uniquely determined by x.

A nonempty subset M of a linear space X is called a (linear) *subspace* of X if $x + y \in M$ and $ax \in M$ whenever $x,y \in M$ and $a \in \Phi$; for M is then a linear space in its own right. $\{0\}$ is a subspace of every linear space.

With $x \in X$, $\lambda \in \Phi$ and $A \subset X$ we shall use the notation

$$x + A = \{x + y : y \in A\} \qquad \lambda A = \{\lambda y : y \in A\}$$

The "sum" $A + B$ of two subsets of X denotes the set $\{x + y : x \in A, y \in B\}$. But the "difference" $A - B$ will be used to denote the set $\{x \in A : x \notin B\}$ for *any* pair of sets A and B, i.e. the complement of B in A.

We now define three types of subsets of the linear space X:

(1) $E \subset X$ is *convex* if, whenever $x, y \in E$ and $0 \leq \lambda \leq 1$, then $\lambda x + (1 - \lambda)y \in E$. Thus a convex set contains the "line segment" joining x and y whenever it contains x and y.
(2) $E \subset X$ is *balanced* if, whenever $x \in E$ and $|\lambda| \leq 1$, then $\lambda x \in E$. By choosing $\lambda = 0$ we see that every balanced set in X contains 0.
(3) $E \subset X$ is *absorbing* if for every $x \in X$ there is a $\lambda > 0$ such that $x \in \lambda E$. Here again it is obvious that 0 is contained in every absorbing set.

The set $\mathbf{R}^n$ whose elements are the n-tuples $(x_1, \ldots, x_n)$ with $x_k \in \mathbf{R}$, $1 \leq k \leq n$, is an n-dimensional linear space over $\mathbf{R}$ under the operations

$$(x_1, \ldots, x_n) + (y_1, \ldots, y_n) = (x_1 + y_1, \ldots, x_n + y_n)$$

$$a(x_1, \ldots, x_n) = (ax_1, \ldots, ax_n)$$

If we define the distance between any two vectors (points) $x = (x_1, \ldots, x_n)$ and $y = (y_1, \ldots, y_n)$ in $\mathbf{R}^n$ by

$$|x - y| = \left[\sum_{k=1}^{n} (x_k - y_k)^2 \right]^{1/2}$$

then $\mathbf{R}^n$ is called *n-dimensional Euclidean space*. In this space the set of points x such that $|x| \leq r$, for some positive number r, defines a ball of radius r and center 0. Such balls are convex, balanced, and absorbing. Obviously they have other features as well, but it turns out that the properties of being convex, balanced, and absorbing are the basic features which we need to abstract and build upon.

Topological Spaces

A *topological space* is a nonempty set X in which a collection τ of subsets is defined such that τ contains X and the empty set $\emptyset$, the intersection of any pair in τ, and the union of any subcollection of τ. The members of τ are known as *open sets* and τ is said to define a *topology* on X. Since different topologies may be defined on the same set X, the topological space should properly be denoted by the pair (X,τ), but we shall often use X to denote the topological space when it is clear, or unimportant, what topology is being considered.

In what follows we give the standard terminology used in connection with the topological space (X,τ):

(1) A *neighborhood* of $x \in X$ is any subset of X which contains an open set containing x.

(2) (X,τ) is a *Hausdorff space* if distinct points of X have disjoint neighborhoods.

(3) If τ and σ are two topologies on X, we say that τ is *stronger (finer)* than σ, or that σ is *weaker (coarser)* than τ, if $\sigma \subset \tau$, i.e. if every open set in (X,σ) is open in (X,τ).

(4) A collection $\sigma \subset \tau$ of open sets is a *base* for τ if every member of τ is a union of sets of σ.

(5) The *product topology* on $X \times Y$, where Y is another topological space,

is the topology which has as a base the collection of all sets of the form $U \times V$, where U is an open set in X and V is an open set in Y.

(6) A collection σ of neighborhoods of $x \in X$ is a *local base* at x if every neighborhood of x contains a member of σ.

(7) A sequence $(x_n : n \in \mathbf{N})$ in the topological space X *converges* to a *limit* $x \in X$, written $\lim x_n = x$ or $x_n \to x$, if every neighborhood of x contains all but finitely many elements of the sequence. A weaker requirement is to have an element of (x_n), different from x, in every neighborhood of x, in which case x is called a *cluster point* of (x_n). In a Hausdorff space the limit of a sequence, if it exists, is unique.

(8) The *interior* of a set $E \subset X$ is the union of all open subsets of E. It is denoted by $\mathring{E}$ and is clearly an open set.

(9) A subset E of X is *closed* if its complement in X, $E^c = X - E$, is open. The *closure* of $E \subset X$ is the intersection of all closed sets which contain E. The closure of E, denoted by $\overline{E}$, is always closed since its complement, by De Morgan's law, is the union of open subsets of E^c, and is therefore open. In other words $(\overline{E})^c = (E^c)^\circ$.

(10) A subset E of X is *dense* in X if $\overline{E} = X$. Even when E is not a subset of X, we still say that E is dense in X if $E \cap X$ is dense in X.

(11) The *boundary* of $E \subset X$ is the set $\partial E = \overline{E} - \mathring{E}$, which is closed since it is the intersection of the two closed sets $\overline{E}$ and $(\mathring{E})^c$.

(12) $E \subset X$ is *compact* if every collection of open sets of X whose union contains E has a finite subcollection whose union contains E.

(13) If $E \subset X$ and σ is the collection of sets $E \cap U$ where U runs through the open sets in τ, then σ is a topology on E. With this *inherited* topology any subset of X becomes a topological space in its own right. This topology is also referred to as the *subspace* topology of E in X.

Mappings

Here we recall the basic features and nomenclature of a *mapping*, or a *map*, from a nonempty set X to a nonempty set Y, written symbolically as

$$T : X \to Y$$

When $Y = \Phi$ the mapping T is usually referred to as a *function* from X to Y.

(i) The *image* of any $x \in X$ is denoted by $T(x) \in Y$. If $A \subset X$ and $B \subset Y$, the set

$$T(A) = \{T(x) : x \in A\} \subset Y$$

is the image, under T, of A, and

$$T^{-1}(B) = \{x \in X : T(x) \in B\} \subset X$$

is the *preimage*, under T, of B. T is *injective* (or *one-to-one*) if $T(x_1) = T(x_2)$ implies $x_1 = x_2$ for any pair $x_1, x_2 \in X$, and *surjective* (or *onto*) if $T(X) = Y$. When T is both injective and surjective it is called *bijective*, and in this case the inverse mapping $T^{-1} : Y \to X$ is defined by $T^{-1}(y) = x$ if and only if $T(x) = y$. A bijective map from X to Y is also referred to as a *bijection* or a *one-to-one correspondence* from X to Y.

(ii) If X and Y are linear spaces over Φ then T is *linear* if $T(ax + by) = aT(x) + bT(y)$ for every $a,b \in \Phi$ and $x,y \in X$. When T is linear it follows that $T(0) = 0$, $T(A)$ is a subspace of Y whenever A is a subspace of X, and $T^{-1}(B)$ is a subspace of X whenever B is a subspace of Y. In particular the subspace $T^{-1}(\{0\}) \subset X$ is called the *null space* of T and is denoted by $N(T) = \{x \in X : T(x) = 0\}$. When X is a linear space over Φ and $Y = \Phi$ the linear function T is called a *linear functional*.

(iii) If X and Y are topological spaces then T is *continuous* at $x \in X$ if for every neighborhood V of $T(x)$ the set $T^{-1}(V)$ is a neighborhood of x. T is continuous on X, or simply continuous, if it is continuous at every point in X. Equivalently, T is continuous on X if and only if $T^{-1}(V)$ is an open set in X whenever V is an open set in Y. By taking complements we can also state that T is continuous if and only if $T^{-1}(V)$ is a closed subset of X whenever V is a closed subset of Y. Consequently, if $E \subset X$, then the identity mapping from E into X is continuous on E provided the topology of E is either the topology inherited from X or a stronger topology; for then the inverse image of an open subset $U \subset X$ is simply $U \cap E$. A *homeomorphism* from X to Y is a continuous bijection from X to Y whose inverse is continuous. Thus, when there is a homeomorphism from X to Y, the image of an open set in X is an open set in Y, and the inverse image of an open set in Y is an open set in X. The topologies on X and Y are therefore in a one-to-one correspondence, and the two spaces are said to be *homeomorphic*.

(iv) If X and Y are topological spaces and $T : X \to Y$ is an injective continuous mapping, then the mapping $S : X \to Z = T(X)$, defined by $S(x) = T(x)$ for all $x \in X$, is clearly bijective. If S is a homeomorphism from X to Z, T is called a *(topological) imbedding* of X in Y. When $X \subset Y$ the identity mapping from X to Y is always an imbedding whenever the topology of X

coincides with its subspace topology in Y. If X carries a stronger topology then we merely have a continuous injection of X into Y.

Metric Spaces

A *metric space* X is a topological space in which the topology is generated by a metric, or distance, function $d : X \times X \to \mathbf{R}$ satisfying

(i) $0 \leq d(x,y) < \infty$
(ii) $d(x,y) = 0$ if and only if $x = y$
(iii) $d(x,y) = d(y,x)$
(iv) $d(x,y) \leq d(x,z) + d(z,y)$, for all $x,y,z \in X$.

For any $x \in X$ and $r > 0$, the set $B(x,r) = \{y \in X : d(x,y) < r\}$ is called an *open ball* with center at x and radius r. By defining a subset of X to be open if and only if it is a (possibly empty) union of open balls, the axioms of open sets are satisfied and X becomes a topological space. For any distinct pair $x,y \in X$ the open balls $B(x,r)$ and $B(y,r)$ are disjoint if we choose $r < \frac{1}{2}d(x,y)$. Thus every metric space is Hausdorff.

Every point of a metric space has a countable base of neighborhoods. One such choice is $\{B(x, 1/n) : n \in \mathbf{N}\}$. Using this property it can be shown (see [1]) that if f maps the metric space X into the topological space Y, then f is continuous at $x \in X$ if and only if for every sequence (x_n) in X which converges to x, the sequence $(f(x_n))$ converges to $f(x)$ in Y. If E is a subset of X and x is a cluster point of E then there is a sequence in E which converges to x. The set of cluster points of E is then contained in its closure $\overline{E}$, and E will be dense in X if every $x \in X$ is the limit of a sequence (x_n) in E.

In a metric space X the sequence (x_n) is called a *Cauchy sequence* if for every $\varepsilon > 0$ there is a positive integer N such that $d(x_n,x_m) < \varepsilon$ for all $n \geq N$ and $m \geq N$. X is said to be *(sequentially) complete* if every Cauchy sequence in X converges to a point in X.

Not every topological space (X,τ) is a metric space because it is not always possible to define a metric on X, with the above properties, which will generate the topology τ. When this is possible, we say that the topological space is *metrizable*, and that its topology can be *induced* or *generated* by a metric.

A homeomorphism h from a metric space (X,d_1) to another (Y,d_2) is called an *isometry* if it preserves distances, in the sense that $d_2(h(x_1),h(x_2)) = d_1(x_1,x_2)$ for any pair $x_1,x_2 \in X$. In this connection, two metrics d_1 and d_2 on the same set X are said to be *equivalent* if the identity map from (X,d_1) onto (X,d_2) is a homeomorphism, i.e. if a set is open with respect to one metric whenever it is open with respect to the other.

1.2 TOPOLOGICAL VECTOR SPACES

A *topological vector space* is a linear space on which a topology is defined in such a way as to preserve the continuity of the operations of the underlying linear space. More precisely, a topological vector space is a linear space X on which a topology τ is defined so that the operations of addition from $X \times X$ to X and scalar multiplication from $\Phi \times X$ to X are continuous. One way we can achieve this is to define a *norm* on the space X, so that X becomes a *normed linear space*.

A linear space X is a normed (linear) space if to every $x \in X$ corresponds a real number $\|x\|$, called the norm of x, such that

(i) $\|x + y\| \leqslant \|x\| + \|y\|$ for all $x, y \in X$
(ii) $\|cx\| = |c|\,\|x\|$ for all $c \in \Phi$ and $x \in X$
(iii) $\|x\| \neq 0$ whenever $x \neq 0$.

Since property (i) leads to the inequality

$$\big|\|x\| - \|y\|\big| \leqslant \|x - y\|$$

the norm of a vector is never negative. Furthermore, property (ii) implies that the norm of the zero vector is zero.

Clearly a normed space is also a metric space if we define $d(x,y) = \|x - y\|$ as the distance from x to y, and when it is complete in this metric the normed space is called a *Banach space*. The continuity of the operations $(x,y) \mapsto x + y$ and $(c,x) \mapsto cx$ is then a direct consequence of the above properties of the norm; for if $x_n \to x$ and $y_n \to y$ in X and $c_n \to c$ in Φ, then the inequalities

$$\|(x_n + y_n) - (x + y)\| \leqslant \|x_n - x\| + \|y_n - y\|$$

$$\begin{aligned}\|c_n x_n - cx\| &= \|c_n(x_n - x) + (c_n - c)x\| \\ &\leqslant |c_n|\,\|x_n - x\| + |c_n - c|\,\|x\|\end{aligned}$$

imply the $x_n + y_n \to x + y$ and $c_n x_n \to cx$ in X.

The n-dimensional Euclidean space $\mathbf{R}^n$, with the usual Euclidean distance, is the simplest example of a finite dimensional Banach space. But the significance of the theory of Banach spaces is that it provides a suitable framework for studying infinite dimensional spaces, including some of the important function spaces, such as the continuous functions on compact sets, L^p spaces and Hilbert spaces. This approach has provided much insight into the structure of such spaces, and there is no doubt today that Banach space theory is one of the outstanding achievements of classical functional analysis.

There are, however, important function spaces which are not Banach spaces, such as the various spaces of differentiable functions on $\mathbf{R}^n$, which will be discussed later in this chapter, and the space of test functions to be defined in Chapter 2. These are all linear spaces where the natural topology that makes them into topological vector spaces cannot be induced by a norm.

We say that a subset E of a topological vector space X is *bounded* if, with a suitable contraction, it can be contained in any neighborhood of 0. More precisely, $E \subset X$ is bounded if for every neighborhood U of $0 \in X$ there is a number $\lambda > 0$ such that $E \subset \lambda U$.

For any point x_0 in the topological vector space X, the translation from X onto X defined by $x \to x + x_0$ is injective and, by assumption, continuous. Its inverse $x \to x - x_0$ is also continuous. Therefore translation by x_0 is a homeomorphism in X, and the set

$$U + x_0 = \{x + x_0 : x \in U\}$$

is open whenever U is open. Thus τ is completely determined by any local base, which will always be taken at 0.

Similarly, for any nonzero $\lambda \in \Phi$, the mapping from X onto itself defined by $x \to \lambda x$ is a homeomorphism, and we can use the continuity of the mapping $(\lambda,x) \to \lambda x$ to conclude that every neighborhood of 0 is absorbing and contains a balanced neighborhood of 0. For if U is a neighborhood of $0 \in X$ and x is any (nonzero) point in X, then the mapping $\lambda \to \lambda x$ is continuous at $\lambda = 0$, so there is a neighborhood $\{\lambda \in \Phi : |\lambda| < \varepsilon\}$ of $0 \in \Phi$ which is mapped into U; hence $\lambda x \in U$ for all $|\lambda| < \varepsilon$ and $x \in \mu U$ whenever $|\mu| > 1/\varepsilon$. On the other hand, the continuity of the mapping $(\lambda,x) \to \lambda x$ at $(\lambda,x) = (0,0)$ implies that there is a neighborhood V of $0 \in X$ and a positive number ε such that $\lambda V \subset U$ whenever $|\lambda| < \varepsilon$. Now the set $W = \bigcup_{|\lambda|<\varepsilon} \lambda V$ is a balanced neighborhood of 0 which is contained in U.

Thus every topological vector space X has a balanced, absorbing local base. A topological vector space X is called a

(i) *locally convex space* if its topology has a local base whose members are convex sets
(ii) *locally bounded space* if 0 has a bounded neighborhood
(iii) *Fréchet space* if it is locally convex, metrizable and complete
(iv) *normable space* if a norm can be defined on X which is compatible with the topology of X, in the sense that it generates the topology

In a topological vector space X we can define the notion of a Cauchy sequence even in the absence of a metric: If $\mathcal{B}$ is a local base for the topology of X, then the sequence (x_n) in X is a Cauchy sequence if to every $U \in \mathcal{B}$ corresponds an N such that $x_n - x_m \in U$ for all $n \geq N$ and $m \geq N$. But the notion of completeness of a topological vector space X is more general than sequential completeness, and is defined in terms of *Cauchy filters* instead of Cauchy sequences (see [1]). When X is metrizable the two concepts are equivalent. However, when X is a nonmetrizable locally convex space it suffices, for our purposes, to confine ourselves to sequential completeness.

If (x_n) is a Cauchy sequence in the topological vector space X, the sequence (x_n) is bounded in the sense that the set $\{x_n\}$ is bounded. That is because if U is a neighborhood of 0, then there is an N such that $x_n - x_N \in U$ for all $k \geqslant N$. This means that

$$\{x_k : k \geqslant N\} \subset x_N + U$$

and since $\lambda(x_N + U)$ may be contained in any neighborhood of 0 by a suitable choice of $\lambda > 0$, the sequence (x_n) is clearly bounded.

Let X and Y be topological vector spaces over the same field Φ, and let T be a linear map from X to Y. T is said to be *bounded* if $T(A)$ is a bounded subset of Y for every bounded subset A of X. But since every bounded subset of X may be mapped homeomorphically into any neighborhood of $0 \in X$, it is clear that T is bounded if and only if it is bounded on a neighborhood of 0. Similarly T is continuous if and only if it is continuous at 0. For if T is continuous at 0, then for every neighborhood V of $0 \in Y$ there is a neighborhood U of $0 \in X$ such that $T(U) \subset V$. But then

$$T(x_0 + U) = T(x_0) + T(U) \subset T(x_0) + V$$

for any $x_0 \in X$.

An important class of linear mappings consists of those for which $Y = \Phi$, i.e., the linear functionals on X. They are denoted by X^*, the *algebraic dual* of X. With the definition

$$(aT + bS)(x) = aT(x) + bS(x)$$

for any $a,b \in \Phi$ and $x \in X$, X^* is a linear space over Φ. In the usual metric topology of Φ, the continuous linear functionals on X constitute a subspace of X^* which is denoted by X', the *topological dual* of X, or simply the *dual* of X. The following theorem gives some useful characterizations of a continuous linear functional.

Theorem 1.1 If T is a linear functional on a topological vector space X, then the following statements are equivalent:

(i) T is continuous at 0.
(ii) T is continuous.

(iii) $T^{-1}(\{0\})$ is closed.
(iv) T is bounded.

Proof. The equivalence between (i) and (ii) has already been proved. To show that (ii) implies (iii), we note that if T is continuous then $T^{-1}(\{0\}) = N(T)$ is closed in X because $\{0\}$ is closed in Φ.

To show that (iii) implies (iv), suppose $N(T)$ is closed. If T is not identically zero, in which case it is bounded, then there is a point $x_0 \in X - N(T)$ with $T(x_0) = 1$. Since $N(T)$ is closed, its complement $X - N(T)$ is open and there is a balanced neighborhood U of 0 such that $x_0 + U \subset X - N(T)$, which implies that $(x_0 + U) \cap N(T) = \emptyset$. If $|T(x)| \geq 1$ for some $x \in U$, then $y = -x/T(x) \in U$ and $T(x_0 + y) = 1 - 1 = 0$, so that $(x_0 + U) \cap N(T) \neq \emptyset$. This contradiction implies that $|T(x)| < 1$ for all $x \in U$. Thus T is bounded.

To show that (iv) implies (i), let T be bounded on some neighborhood U of 0; then there is a number $M > 0$ such that $|T(x)| < M$ for every $x \in U$. For any $\varepsilon > 0$ we therefore have $|T(x)| < \varepsilon$ whenever x is in $(\varepsilon/M)U$. Since $(\varepsilon/M)U$ is also a neighborhood of 0, T is continuous at 0. □

1.3 SEMINORMS AND LOCALLY CONVEX SPACES

Our purpose in this section is to provide a framework for defining a natural topology on each of the various linear spaces of functions that we shall have occasion to deal with, i.e. a topology which is compatible with the algebraic structure of the linear space, in the sense defined in Section 1.2, and resulting in a topological vector space. This is necessary for any analytical study of such spaces.

Since we cannot always hope to generate such topologies through the concept of a norm, as pointed out earlier, it is natural to seek a less restrictive concept which will play a similar role. Thus we are led to the *seminorm* as a means for defining a local base for the topology of a linear space. It turns out that the local base defined in this fashion is made up of convex open sets, and so we arrive at the notion of a locally convex space as the proper framework we seek. It is broad enough to cover the function spaces of interest, providing at the same time a concrete method for defining each topology through a system of seminorms.

Let X be a linear space over Φ. A seminorm on X is a real-valued function p satisfying

(i) $p(x + y) \leq p(x) + p(y)$
(ii) $p(\lambda x) = |\lambda|\, p(x)$

for all $x, y \in X$ and $\lambda \in \Phi$. The subadditive property (i) implies $p(x - y) + p(y) \geq p(x)$. Interchanging x and y and using property (ii) we obtain $p(x - y) + p(x) \geq p(y)$. Therefore we always have $p(x - y) \geq |p(x) - p(y)|$ and hence $p(x) \geq 0$ for all $x \in X$. The equality $p(0) = 0$ follows directly from (ii) but it may happen that $p(x) = 0$ for some $x \neq 0$. When $p(x) = 0$ implies $x = 0$ then p is a norm on X. For any linear functional T on X the function $p(x) = |T(x)|$ is an example of a seminorm on X.

The set $\{x \in X : p(x) < r\}$ for $r > 0$ corresponds to the ball $B(0,r)$ in a metric space with center 0 and radius r, and will be denoted by $B_p(r)$.

Theorem 1.2 In a linear space X equipped with a seminorm p, the p-ball $B_p(r) = \{x \in X : p(x) < r\}$ is convex, balanced, and absorbing.

Proof. If $|\lambda| \leq 1$ and $x \in B_p(r)$ then $p(\lambda x) = |\lambda|\, p(x) < r$, and therefore $\lambda x \in B_p(r)$. Thus $B_p(r)$ is balanced.

If $0 \leq \lambda \leq 1$ and $x, y \in B_p(r)$, then

$$p(\lambda x + (1 - \lambda)\, y) \leq \lambda p(x) + (1 - \lambda)\, p(y) < r$$

and so $B_p(r)$ is convex.

Finally, if $x \in X$ and $\lambda > p(x)$ then $p(r/\lambda\, x) = [p(x)/\lambda]\, r < r$ and hence $x \in (\lambda/r)\, B_p(r)$, which means that $B_p(r)$ is absorbing. □

Let E be an absorbing subset of the linear space X and let x be any point of X, then there is always a finite positive number λ such that $\lambda^{-1} x \in E$. We define the *Minkowski functional* μ_E of E by

$$\mu_E(x) = \inf\{\lambda > 0 : \lambda^{-1} x \in E\}$$

for any $x \in X$. $\mu_E(0) = 0$ since every absorbing subset of X contains 0, and we clearly have $\mu_E : X \to [0,\infty)$. If, in addition, E is convex then for each $x \in X$ the set

$$M_E(x) = \{\lambda > 0 : \lambda^{-1} x \in E\} = \{\lambda > 0 : x \in \lambda E\}$$

is convex and unbounded. In fact, $M_E(x)$ is the semi-infinite interval whose left endpoint is $\mu_E(x)$.

Theorem 1.3 In a linear space X the Minkowski functional of a convex, balanced and absorbing set is a seminorm on X.

Proof. Let E be a convex and absorbing subset of X. For any $x, y \in X$ we choose λ_1 and λ_2 so that

$$\mu_E(x) < \lambda_1 \qquad \mu_E(y) < \lambda_2$$

Since E is convex, it then follows that

$$\lambda_1^{-1}\, x \in E \qquad \lambda_2^{-1}\, y \in E$$

and that

$$\frac{1}{\lambda_1 + \lambda_2}(x + y) = \frac{\lambda_1}{\lambda_1 + \lambda_2}\lambda_1^{-1}\, x + \frac{\lambda_2}{\lambda_1 + \lambda_2}\lambda_2^{-1}\, y$$

is in E. Thus $\mu_E(x + y) \leq \lambda_1 + \lambda_2$. But since λ_1 and λ_2 can be taken arbitrarily close to $\mu_E(x)$ and $\mu_E(y)$, respectively, we conclude that $\mu_E(x + y) \leq \mu_E(x) + \mu_E(y)$. the relation

$$\mu_E(cx) = |c|\mu_E(x)$$

is always true for $c > 0$. When E is balanced it is also true for $|c| = 1$. Thus $\mu_E\,(cx) = |c|\mu_E(x)$ for any $c \in \Phi$. □

Note that the definition of μ_E implies that

$$\{x \in X : \mu_E(x) < 1\} \subset E \subset \{x \in X : \mu_E(x) \leq 1\}$$

for any convex (and absorbing) subset E of the linear space X. If, moreover, X is a topological vector space then E is a neighborhood of 0 if and only if μ_E is continuous.

To see this, suppose E is a neighborhood of 0; then the inequality

$$|\mu_E(x) - \mu_E(y)| \leq \mu_E(x - y)$$

which follows from the subadditive property of μ_E, shows that it suffices to prove continuity at 0. But for any $\varepsilon > 0$, if $x \in \varepsilon E$, then $\mu_E(x) \leq \varepsilon$. Conversely, if μ_E is continuous, then $\{x \in X : \mu_E(x) < 1\}$ is an open set which contains 0 and is contained in E. Indeed it is clear, in this case, that $\{x \in X\colon \mu_E(x) < 1\}$ is the interior $\mathring{E}$ of E and that $\{x \in X : \mu_E(x) \leq 1\}$ is its closure $\bar{E}$.

In what follows, we shall exploit the connection expressed in Theorem 1.3 between the convex, balanced, and absorbing subsets of a linear space and their corresponding seminorms in order to build a topology on the linear space which makes it into a locally convex topological space.

Theorem 1.4 Given any set $\{p_i : i \in I\}$ of seminorms on a linear space X, there is a topology on X, compatible with its algebraic structure, in which every seminorm p_i is continuous. Under this topology X is a locally convex topological space.

Proof. For each seminorm in the set $\mathcal{P} = \{p_i : i \in I\}$, the p_i-ball

$$B_i(r) = \{x \in X : p_i(x) < r\} \qquad i \in I,\ r > 0$$

is convex, balanced, and absorbing, according to Theorem 1.2. We shall take $B_i(r)$ to be an open neighborhood of 0. For any finite subset I' of I, let

$$\mathcal{P}' = \{p_i : i \in I'\}$$

be the corresponding finite subset of $\mathcal{P}$, and define

$$B' = \bigcap_{i \in I'} B_i(1)$$

Clearly B' is a convex, balanced, and absorbing set. The collection

$$\mathcal{B} = \{rB' : \mathcal{P}' \subset \mathcal{P},\ r > 0\}$$

where $\mathcal{P}'$ runs through the finite subsets of $\mathcal{P}$, satisfies the properties of a base of neighborhoods of the origin, and therefore $(X,\mathcal{B})$ is a locally convex space. In this topology every p_i is continuous because every $B_i(r)$ is a neighborhood of 0.

It remains to show that the algebraic operations on X are also continuous. For any pair $x,y \in X$ and any B in $\mathcal{B}$, we have

$$(x + \tfrac{1}{2}B) + (y + \tfrac{1}{2}B) = (x + y) + B$$

since B is convex, so that addition is continuous on $(X,\mathcal{B})$. To show that scalar multiplication is continuous, suppose that $x \in X$, $\lambda \in \Phi$ and $B \in \mathcal{B}$. Since

$$\mu y - \lambda x = \mu(y - x) + (\mu - \lambda)\, x$$

we see that $\mu y - \lambda x$ is in B if $\mu(y - x) \in \frac{1}{2}B$ and $(\mu - \lambda)\, x \in \frac{1}{2}B$. The second condition is satisfied by taking μ close enough to λ, say $|\mu - \lambda| < \varepsilon$, and the first by choosing

$$y \in x + \frac{1}{2(|\lambda| + \varepsilon)} B$$ □

It is worth noting that the topology defined on X in this proof is the weakest topology in which every seminorm p_i is continuous. It will henceforth be referred to as the topology generated by the family of semi-norms $\{p_i\}$. The reason we need to resort to a *family* of seminorms in order to define the topology of X is that a single seminorm does not convey enough topological information. We have already observed that $p(x) = 0$ does not guarantee that $x = 0$. However, if *enough* seminorms vanished at x then, presumably, we may safely conclude that $x = 0$.

Definition A family $\mathcal{P}$ of seminorms on the linear space X satisfies the *separation axiom*, or is *separating*, if to each $x \neq 0$ in X there is a seminorm $p \in \mathcal{P}$ such that $p(x) \neq 0$.

In any locally convex topological vector space X, let $\mathcal{B}$ be a convex and balanced local base in X. For each $B \in \mathcal{B}$, which is always absorbing, Theorem 1.3 shows that the Minkowski functional μ_B is a seminorm on X. If X is Hausdorff then for any nonzero vector x in X there is a $B \in \mathcal{B}$ such that $x \not\in B$ and consequently $\mu_B(x) \geqslant 1$. Thus $\{\mu_B : B \in \mathcal{B}\}$ is a separating family of seminorms in X. Conversely, if the seminorms $\{p_i\}$ on the linear space X are separating, then the topology which they generate on X, according to Theorem 1.4, is Hausdorff. This follows from the observation that if $x - y \neq 0$ then there is a p_i such that $p_i(x - y) = r > 0$, and the two neighborhoods $x + B_i(\frac{1}{2}r)$ and $y + B_i(\frac{1}{2}r)$ are disjoint.

We therefore arrive at the important result that a linear space with a separating family of seminorms may be topologized through the procedure of Theorem 1.4 to produce a locally convex Hausdorff space in which each seminorm is continuous. And, conversely, any locally convex Hausdorff space X is a topological vector space in which the topolology is generated by a separating family of continuous seminorms defined by the Minkowski functionals of the convex local base of X.

Theorem 1.5 A locally convex space X is metrizable if and only if it is Hausdorff and has a countable local base.

Proof. If X is metrizable, with metric d, then the balls

$$\{x \in X : d(0,x) < 1/n,\ n \in \mathbf{N}\}$$

are convex, balanced and absorbing and form a countable base at 0 for a Hausdorff topology.

If, on the other hand, X is Hausdorff and has a countable local base $\mathcal{B} = \{B_i\}$ then its topology is generated by the countable, separating family of seminorms $\mathcal{P} = \{p_i\}$, where p_i is the Minkowski functional of B_i. We define

$$d(x,y) = \sum_{i=1}^{\infty} \frac{2^{-i} p_i(x - y)}{1 + p_i(x - y)}$$

for any $x,y \in X$. In view of the inequality

$$\frac{a}{1+a} \leqslant \frac{b}{1+b}$$

which is true when $0 \leqslant a \leqslant b$, we have

$$\begin{aligned}
\frac{p_i(x-y)}{1+p_i(x-y)} &= \frac{p_i[(x-z)+(z-y)]}{1+p_i[(x-z)+(z-y)]} \\
&\leqslant \frac{p_i(x-z)+p_i(z-y)}{1+p_i(x-z)+p_i(z-y)} \\
&\leqslant \frac{p_i(x-z)}{1+p_i(x-z)} + \frac{p_i(z-y)}{1+p_i(z-y)}
\end{aligned}$$

Therefore d is subadditive, and $d(x,y) = 0$ implies $x = y$ because $\mathcal{P}$ is separating; so d is clearly a metric on X.

Since $d(x + z, y + z) = d(x,y)$ for any $x \in X$, the sets

$$U_n = \{x \in X : d(0,x) < 2^{-n}\}$$

form a base of neighborhoods at 0 for the topology of (X,d). Now the series which defines d converges uniformly on $X \times X$ and p_i is continuous on X. Hence d is continuous on $X \times X$ and U_n is open in $(X,\mathcal{B})$. Moreover $U_{n+1} \subset B_n$ because if $x \notin B_n$ then $p_n(x) \geqslant 1$ and therefore

$$d(0,x) \geqslant 2^{-n} \frac{p_n(x)}{1+p_n(x)} \geqslant 2^{-n-1}$$

Thus $\{U_n\}$ forms a local base for the topology of $(X,\mathcal{B})$. □

Corollary A countable, separating family of seminorms on a linear space X generates a locally convex, metrizable topology on X.

Theorem 1.6 A locally convex, Hausdorff space X is normable if and only if its zero vector has a bounded neighborhood.

Proof. If X is normable then the open unit ball $\{x \in X : \|x\| < 1\}$ is a bounded neighborhood of 0. If U, on the other hand, is a bounded neighborhood of 0 in the locally convex space (X,τ), then it contains a convex, balanced and absorbing open set U_0 which is also bounded. Let p_0 be the Minkowski functional of U_0. If $p_0(x) = 0$ then $x \in \lambda U_0$ for any $\lambda > 0$. But, since U_0 is bounded, every neighborhood of 0 contains λU_0 for some $\lambda >$

0. Therefore x is in every neighborhood of 0 and, since X is Hausdorff, $x = 0$. Therefore p_0 is a norm on X.

The normed space (X,p_0) has a local base given by $\{\lambda U_0 : \lambda > 0\}$. But each λU_0 is an open set in the original topology τ of X, and every neighborhood of 0 in (X,τ) contains λU_0 for some $\lambda > 0$. Thus p_0 generates τ. □

In practice we shall have no occasion to deal with locally convex spaces which are not Hausdorff; so we shall tacitly assume from now on that all locally convex spaces are Hausdorff in order to avoid unnecessary technical details.

When the topological vector space X is normable then a subset E of X is bounded, in the sense that it is absorbable by any neighborhood of 0, if and only if it is bounded in X as a normed linear space, i.e. there exists a positive constant M such that $\|x\| \leq M$ for all $x \in E$. But in general these two notions of boundedness are not equivalent.

1.4 EXAMPLES OF LOCALLY CONVEX SPACES

In the calculus of n variables we shall find it convenient to use the n-tuple $\alpha = (\alpha_1, \ldots, \alpha_n)$ of nonnegative integers α_i as a multi-index and define $|\alpha| = \alpha_1 + \cdots + \alpha_n$ and $\alpha! = \alpha_1! \cdots \alpha_n!$. With $x = (x_1, \ldots, x_n) \in \mathbf{R}^n$ we shall use the notation

$$x^\alpha = x_1^{\alpha_1} \cdots x_n^{\alpha_n}$$

$$\partial = (\partial_1, \ldots, \partial_n) = \left(\frac{\partial}{\partial x_1}, \ldots, \frac{\partial}{\partial x_n}\right)$$

$$\partial^\alpha = \partial_1^{\alpha_1} \cdots \partial_n^{\alpha_n} = \frac{\partial^{|\alpha|}}{\partial^{\alpha_1} x_1 \cdots \partial^{\alpha_n} x_n}$$

The functions we shall be interested in will, in general, be complex-valued and defined on an open subset Ω of $\mathbf{R}^n$, with the usual Euclidean topology on $\mathbf{R}^n$. The *support* of a function $\phi : \Omega \to \mathbf{C}$, denoted by supp ϕ, is defined to be the closure of the set $\{x \in \Omega : \phi(x) \neq 0\}$ in the topological space Ω, i.e. the smallest closed set containing $\{x \in \Omega : \phi(x) \neq 0\}$.

The following examples of function spaces illustrate the ideas we have discussed so far, and they also serve to prepare the grounds for the developments of the next chapter:

(i) $C^m(\Omega)$ denotes the set of (complex-valued) functions defined on Ω with continuous derivatives of order m, where $m < \infty$, i.e. $\partial^\alpha\phi$ is continuous on Ω for every α with $|\alpha| \leqslant m$. When $m = 0$ we have the set $C^0(\Omega)$ of continuous functions on Ω. Clearly, $C^m(\Omega) \subset C^{m-1}(\Omega) \subset \cdots \subset C^0(\Omega)$.

(ii) $C^\infty(\Omega) = \bigcap_{m \geqslant 0} C^m(\Omega)$ is the set of functions on Ω with continuous derivatives of all orders.

(iii) $C_K^m(\Omega)$ is the set of functions in $C^m(\Omega)$ with support in K, where K will always denote a compact subset of Ω.

(iv) $C_K^\infty(\Omega)$ is the set of functions in $C^\infty(\Omega)$ with support in K.

Clearly $C^m(\Omega)$ is a linear space over $\mathbf{C}$, for $m \leqslant \infty$, by the usual definition of addition of functions and multiplication by complex numbers

$$(\phi + \psi)(x) = \phi(x) + \psi(x) \qquad (c\phi)(x) = c\phi(x)$$

and $C_K^m(\Omega)$ is a subspace of $C^m(\Omega)$ for every m. A well-known example of a $C^\infty(\mathbf{R}^n)$ function of compact support is given by

$$\alpha(x) = \begin{cases} \exp\left(-\dfrac{1}{1 - |x|^2}\right) & \text{on } |x| < 1 \\ 0 & \text{on } |x| \geqslant 1 \end{cases} \tag{1.1}$$

which has support in the unit ball $\{x \in \mathbf{R}^n : |x| \leqslant 1\}$.

We now use the results of Section 1.3 to define suitable topologies for these spaces.

(i) $C^0(\Omega)$

Since any open subset of $\mathbf{R}^n$ may be expressed as a countable union of compact sets in $\mathbf{R}^n$, we can write $\Omega = \cup K_i$, where K_i is a compact subset of $\mathbf{R}^n$ for all $i \in \mathbf{N}$. Without loss of generality, we may choose $K_1 \subset K_2 \subset K_3 \subset \cdots$. For any $\phi \in C^0(\Omega)$ we define the seminorm

$$p_i(\phi) = \sup\{|\phi(x)| : x \in K_i\} \qquad i \in \mathbf{N}$$

and note that the increasing sequence (p_i) is clearly separating.

The sets

$$B_i(r) = \{\phi \in C^0(\Omega) : p_i(\phi) < r\} \qquad i \in \mathbf{N},\ r > 0$$

form a convex local topological base for $C^0(\Omega)$, and the resulting topology is compatible with the metric

$$d(\phi,\psi) = \sum_{i=1}^{\infty} 2^{-i} \frac{p_i(\phi - \psi)}{1 + p_i(\phi - \psi)}$$

Since convergence in this metric is uniform on every compact subset of Ω, the limit of every Cauchy sequence is always a continuous function on Ω. Thus the metric space $C^0(\Omega)$ is complete and is therefore a Fréchet space. But $C^0(\Omega)$ is not normable because in every $B_i(r)$ we can always find a function ϕ for which $p_{i+1}(\phi)$ is as large as we please, so that no $B_i(r)$ can be bounded. Note, however, that every $B_i(r)$ is bounded in the metric d. In fact the whole space $C^0(\Omega)$ is bounded in this metric.

(ii) $C^m(\Omega)$

Again we assume that Ω is the union of a sequence of compact sets $K_1 \subset K_2 \subset \cdots$ and, for any $\phi \in C^m(\Omega)$ with $1 \leq m < \infty$, we define the separating countable family of seminorms

$$P_{i,m}(\phi) = \sup\{|\partial^\alpha \phi(x)| : x \in K_i, |\alpha| \leq m\}$$

The corresponding balls

$$B_{i,m}(r) = \{\phi \in C^m(\Omega) : p_{i,m}(\phi) < r\}$$

provide a base for a topology on $C^m(\Omega)$ which makes it into a locally convex, metrizable space.

The convergence of (ϕ_k) in $C^m(\Omega)$ is equivalent to the uniform convergence of $(\partial^\alpha\phi_k)$ on every compact subset of Ω for all $|\alpha| \leq m$. It is clear that the topology of $C^m(\Omega)$ is the weakest in which the linear map

$$\partial^\alpha : C^m(\Omega) \to C^0(\Omega) \qquad |\alpha| \leq m$$

is continuous, where $C^0(\Omega)$ carries its natural topology of uniform convergence. We therefore conclude that a sequence (ϕ_k) converges to ϕ in $C^m(\Omega)$ if and only if the sequence $(\partial^\alpha\phi_k)$ converges to $\partial^\alpha\phi$ in $C^0(\Omega)$ for all $|\alpha| \leq m$, which is equivalent to the uniform convergence of $\partial^\alpha\phi_k$ to $\partial^\alpha\phi$ on every compact subset of Ω. This implies the uniform convergence of (ϕ_k) to ϕ in $C^0(\Omega)$, so the topology of $C^m(\Omega)$ is clearly stronger than its subspace topology in $C^0(\Omega)$. More generally, the topology of $C^l(\Omega)$ is stronger than its subspace topology in $C^m(\Omega)$ whenever $l \geq m \geq 0$, and consequently the identity map from $C^l(\Omega)$ into $C^m(\Omega)$ is continuous.

Theorem 1.7 The locally convex space $C^m(\Omega)$ is complete.

Proof. Let (ϕ_k) be a Cauchy sequence in $C^m(\Omega)$, which necessarily makes it a Cauchy sequence in $C^0(\Omega)$. Since $C^0(\Omega)$ is complete

$$\phi_k \to \phi \in C^0(\Omega)$$

The sequence $(\partial^\alpha \phi_k)$ is also a Cauchy sequence in $C^0(\Omega)$ for every α satisfying $|\alpha| \leqslant m$, and therefore

$$\partial^\alpha \phi_k \to \phi_\alpha \in C^0(\Omega)$$

Since the operator $\partial_\alpha : C^m(\Omega) \to C^0(\Omega)$ is continuous, we conclude that $\partial^\alpha \phi = \lim \partial^\alpha \phi_k = \phi_\alpha$ is in $C^0(\Omega)$ and ϕ is therefore in $C^m(\Omega)$. □

This theorem shows that $C^m(\Omega)$ is a Fréchet space. It is not normable because here also every neighborhood of 0 is unbouded.

(iii) $C^\infty(\Omega)$
We write Ω as the union of an increasing sequence of compact sets (K_i) and we define the seminorms

$$p_i(\phi) = \sup\{|\partial^\alpha \phi(x)| : x \in K_i, |\alpha| \leqslant i\} \qquad \phi \in C^\infty(\Omega) \tag{1.2}$$

The balls

$$B_i(r) = \{\phi \in C^\infty(\Omega) : p_i(\phi) < r\}$$

form a local base for the topology of $C^\infty(\Omega)$, and the same argument as above shows that, with this topology, $C^\infty(\Omega)$ is a Fréchet space which is not normable. It is the weakest topology which makes the linear map

$$\partial^\alpha : C^\infty(\Omega) \to C^m(\Omega) \qquad \text{for all } m \geqslant 0$$

continuous, where $C^m(\Omega)$ carries its natural topology defined in (ii). The next theorem characterizes the bounded subsets of $C^\infty(\Omega)$.

Theorem 1.8 A subset E of $C^\infty(\Omega)$ is bounded if and only if for every $m \in \mathbf{N}_0 = \{0,1,2, \ldots\}$ and every compact set $K \subset \Omega$ there is a positive constant M, which depends on m and K, such that

$$|\partial^\alpha \phi(x)| \leqslant M \tag{1.3}$$

whenever $|\alpha| \leqslant m$, $x \in K$ and $\phi \in E$.

Proof. Suppose the elements of E satisfy the inequality (1.3) and let U be a neighborhood of $0 \in C^\infty(\Omega)$. Any such neighborhood contains the balls

$$B_i = \{\phi \in C^\infty(\Omega) : p_i(\phi) < 1/i\}$$

for all values of i greater than some positive integer. We can choose i large enough so that $i \geq m$ and $K_i \supset K$. If we now choose λ such that $0 < \lambda < 1/M(i + 1)$, we obtain

$$\lambda E = \left\{\phi \in E : |\partial^\alpha \phi(x)| \leq \frac{1}{i+1}, x \in K, |\alpha| \leq m\right\} \subset B_{i+1} \subset U$$

which means that E is bounded.

Conversely, if E is bounded, then by a suitable choice of the positive number λ the set λE may be contained in any neighborhood of 0. In particular, for every B_i there is a $\lambda_i > 0$ such that

$$\lambda_i E \subset B_i$$

which means that

$$p_i(\phi) < 1/i\lambda_i \qquad \phi \in E$$

Given any $m \in \mathbf{N}_0$ and any compact $K \subset \Omega$ we can choose i so that $i \geq m$ and $K_i \supset K$, and inequality (1.3) follows by choosing $M = 1/i\lambda_i$. □

Note that, according to this theorem, no B_i can be bounded for any finite integer i.

The system of seminorms that we have used to define the topology of $C^\infty(\Omega)$ is equivalently given by

$$p_{i,K}(\phi) = \sup\{|\partial^\alpha \phi(x)| : x \in K, |\alpha| \leq i\}$$

as i runs through the nonnegative integers and K through the compact subsets of Ω. Theorem 1.8 may then be restated as follows: $E \subset C^\infty(\Omega)$ is bounded if and only if for every $m \in \mathbf{N}_0$ and every compact $K \subset \Omega$ there is an $M > 0$ such that $p_{m,K}(\phi) \leq M$ for every $\phi \in E$. Furthermore, the set

$$\{\phi \in C^\infty(\Omega) : p_{m,K}(\phi) < r\}$$

is a neighborhood of 0 for every $m \in \mathbf{N}_0$, $K \subset \Omega$, and $r > 0$.

Here again the convergence of (ϕ_k) in $C^\infty(\Omega)$ is equivalent to the uniform convergence of $(\partial^\alpha \phi_k)$, for every multi-index $\alpha = (\alpha_1, \ldots, \alpha_n)$, on every compact subset of Ω.

(iv) $C_K^m(\Omega)$, $m \leq \infty$

To show that $C_K^m(\Omega)$ is a closed subspace of $C^m(\Omega)$ for $m \leq \infty$ we note that, for any $x \in \Omega$, the linear mapping T_x from $C^m(\Omega)$ to $\mathbf{C}$ defined by $T_x(\phi) = \phi(x)$ is continuous and so its null space

$$N(T_x) = \{\phi \in C^m(\Omega) : \phi(x) = 0\}$$

is closed, by Theorem 1.1, for every $x \in \Omega$. Since

$$C_K^m(\Omega) = \bigcap_{x \in \Omega - K} N(T_x)$$

we see that C_K^m is closed in $C^m(\Omega)$. Thus $C_K^m(\Omega)$ is also a Fréchet space.

For $m \leqslant \infty$ the seminorms on $C_K^m(\Omega)$ are given by

$$p_i(\phi) = \sup\{|\partial^\alpha \phi(x)| : x \in K, |\alpha| \leqslant i\} \qquad 0 \leqslant i \leqslant m \tag{1.4}$$

and the local base they define is the collection of balls

$$B_i(r) = \{\phi \in C_K^\infty(\Omega) : p_i(\phi) < r\}$$

These seminorms, in contrast to those of the previous examples, actually define norms on $C_K^m(\Omega)$, since $p_i(\phi) = 0$ for any $i \in \mathbf{N}_0$, $i \leqslant m$, implies $\phi = 0$.

If E is a bounded subset of $C_K^\infty(\Omega)$ then, by definition, for every B_i there is a positive number λ_i such that $E \subset \lambda_i B_i$. This is equivalent to saying that, for every nonnegative integer i, there is a constant M_i such that

$$\sup\{|\partial^\alpha \phi(x)| : \phi \in E, x \in K, |\alpha| \leqslant i\} \leqslant M_i$$

Every B_i is therefore unbounded because it contains a ϕ for which $p_{i+1}(\phi)$ is arbitrarily large. Hence $C_K^\infty(\Omega)$ is not normable. But when $m < \infty$ the largest of the seminorms, i.e., p_m, which is actually a norm, makes $C_K^m(\Omega)$ into a Banach space.

If $K_1 \subset K_2 \subset \Omega$, then $C_{K_1}^m(\Omega)$ is a closed subspace of $C_{K_2}^m(\Omega)$, where $m \leqslant \infty$, and the topology on $C_{K_1}^m(\Omega)$ is the topology it inherits as a subspace of $C_{K_2}^m(\Omega)$.

EXERCISES

1.1 If $\{\tau_\lambda\}$ is a collection of topologies on X, show that $\bigcap \tau_\lambda$ and $\bigcup \tau_\lambda$ are also topologies on X.

1.2 If the mapping $f : X \to Y$ is continuous and surjective, and E is a dense set in X, show that $f(E)$ is dense in Y.

1.3 If (X,d) is a metric space, show that the topology on X induced by the metric d is the weakest topology in which the function $d : X \times X \to \mathbf{R}$ is continuous.

1.4 If Ω is an open set in $\mathbf{R}^n$, prove that there is an increasing sequence of compact sets (K_i) such that $K_i \subset \Omega$ and $\cup K_i = \Omega$.

1.5 Let K_1 and K_2 be compact sets in Ω such that $K_1 \subset K_2$. Show that the identity mapping from $C^m_{K_1}(\Omega)$ to $C^m_{K_2}(\Omega)$ is continuous and that $C^m_{K_1}(\Omega)$ is a closed subspace of $C^m_{K_2}(\Omega)$.

1.6 Let E be a convex set in the topological vector space X, $x \in \mathring{E}$, and $y \in \overline{E}$. Show that $\lambda x + (1 - \lambda) y \in \mathring{E}$ for all $\lambda \in (0,1)$.

1.7 Prove that the intersection of any collection of convex sets is a convex set.

1.8 Let $\mathcal{P}$ be a nonempty family of seminorms on a linear space X. For every finite subset $\mathcal{P}'$ of $\mathcal{P}$, let

$$p(x) = \max\{p_k(x) : p_k \in \mathcal{P}'\} \qquad x \in X$$

Prove that p is a seminorm on X, and that the topology generated by these seminorms on X, for different choices of $\mathcal{P}'$, is the same as the topology generated by $\mathcal{P}$.

1.9 If E is a set in a locally convex space X, prove that E is bounded if and only if each continuous seminorm on E is bounded. Show that it is sufficient to consider those seminorms which generate the topology of X.

1.10 Show that, if X is a linear space and $E \subset X$ is convex and absorbing, then $\mathring{E} = \{x \in X : \mu_E(x) < 1\}$ and $\overline{E} = \{x \in X : \mu_E(x) \leq 1\}$.

1.11 Let $C^0(a,b)$ denote the linear space of continuous functions on the bounded open interval $(a,b) \subset \mathbf{R}$. For any positive number p, consider the functional from $C^0(a,b)$ to $[0,\infty)$ defined by

$$u \mapsto \|u\|_p = \left[\int_a^b |u(x)|^p \, dx\right]^{1/p}$$

Prove that this functional defines a norm when $1 \leq p < \infty$. Is the resulting normed linear space complete?

1.12 Let $L^p(a,b)$ denote the closure, or *completion*, of $C^0(a,b)$ under the norm $\|\cdot\|_p$, so that $L^p(a,b)$ is a Banach space when $1 \leq p < \infty$. Prove that if p and q are positive numbers related by

$$\frac{1}{p} + \frac{1}{q} = 1$$

and $u \in L^p(a,b)$, $v \in L^q(a,b)$ then $uv \in L^1(a,b)$ and

$$\|uv\|_1 \leq \|u\|_p \, \|v\|_q \qquad \text{(Hölder's inequality)}$$

1.13 In a metric space X a set A is said to be *dense with respect to a set B* if $B \subset \bar{A}$, i.e., if every point of B is a cluster point of A. If A is dense with respect to B, and B is dense with respect to C, show that A is dense with respect to C.

1.14 Given the three metric spaces X_1, X_2, and X_3, show that if X_1 is densely imbedded in X_2 and X_2 is densely imbedded in X_3, then X_1 is densely imbedded in X_3.

1.15 Let $\alpha : \mathbf{R} \rightarrow \mathbf{R}$ be the function defined by equation (1.1). Verify that α is a C^∞ function which vanishes with all its derivatives on $(-\infty, -1] \cup [1, \infty)$.

1.16 Construct a $C^\infty(\mathbf{R})$ function whose support is the finite interval $[a,b]$.

2

Test Functions and Distributions

2.1 THE SPACE OF TEST FUNCTIONS $\mathscr{D}$

As in Section 1.4, and throughout this book, Ω will denote a nonempty open subset of $\mathbf{R}^n$. For every compact subset K of Ω we have already defined the linear space $C_K^\infty(\Omega)$ and the topology which makes it into a Fréchet space. The union of the spaces $C_K^\infty(\Omega)$ as K ranges over all compact subsets of Ω is denoted by $C_0^\infty(\Omega)$, so that every function in $C_0^\infty(\Omega)$ is infinitely differentiable on Ω and its support is a compact subset of Ω. The topology of $C_K^\infty(\Omega)$ as a closed subspace of $C^\infty(\Omega)$ was defined in Section 1.4 by the seminorms

$$p_m(\phi) = \sup \{|\partial^\alpha \phi(x)| : x \in K, |\alpha| \leq m\} \qquad m \in \mathbf{N}_0$$

with the sets

$$B_m(r) = \{\phi \in C_K^\infty(\Omega) : p_m(\phi) < r\}$$

as a local base.

But $C_K^\infty(\Omega)$ is also a closed subspace of $C_0^\infty(\Omega)$, and we define the topology of

$$C_0^\infty(\Omega) = \bigcup_{K \subset \Omega} C_K^\infty(\Omega)$$

to be the finest locally convex topology for which the identity map $C_K^\infty(\Omega) \to C_0^\infty(\Omega)$ is continuous for every $K \subset \Omega$. This means that a convex, balanced set $U \subset C_0^\infty(\Omega)$ is a neighborhood of 0 in $C_0^\infty(\Omega)$ if and only if $U \cap C_K^\infty(\Omega)$ is a neighborhood of 0 in $C_K^\infty(\Omega)$ for every $K \subset \Omega$. The collection of all such neighborhoods U constitutes a local base for the topology we have defined on $C_0^\infty(\Omega)$, which is known as the *inductive limit* of the topologies on $C_K^\infty(\Omega)$. In a sense, the topologies of $C_K^\infty(\Omega)$ are "pieced together" to form the topology of the union $C_0^\infty(\Omega)$ (see [2]).

With this topology $C_0^\infty(\Omega)$ is a locally convex space and the original topology on $C_K^\infty(\Omega)$, for any $K \subset \Omega$, is clearly the topology that $C_K^\infty(\Omega)$ inherits as a subspace of $C_0^\infty(\Omega)$. It is also clear that if Ω_1 is an open subset of Ω then $C_0^\infty(\Omega_1)$ is a subspace of $C_0^\infty(\Omega)$, because every function in $C_0^\infty(\Omega_1)$ may be extended as a C_0^∞ function into Ω by defining it to be 0 on $\Omega - \Omega_1$.

Theorem 2.1 A linear functional on $C_0^\infty(\Omega)$ is continuous if and only if its restriction to $C_K^\infty(\Omega)$ is continuous for every compact subset K of Ω.

Proof. Let T be a linear functional on $C_0^\infty(\Omega)$. By Theorem 1.1, T is continuous if and only if it is continuous at $0 \in C_0^\infty(\Omega)$.

Let K be any compact set in Ω, and T_K the restriction of T to $C_K^\infty(\Omega)$. If V is any neighborhood of $0 \in \mathbf{C}$, and T is continuous at 0, then $T^{-1}(V)$ is a neighborhood of 0 in $C_0^\infty(\Omega)$ and therefore $T^{-1}(V) \cap C_K^\infty(\Omega) = T_K^{-1}(V)$ is also a neighborhood of 0 in $C_K^\infty(\Omega)$.

Conversely, if T_K is continuous at 0 for every K then $T_K^{-1}(V) = T^{-1}(V) \cap C_K^\infty(\Omega)$ is a neighborhood of 0 in $C_K^\infty(\Omega)$ for every $K \subset \Omega$, and consequently $T^{-1}(V)$ is a neighborhood of 0 in $C_0^\infty(\Omega)$. □

The locally convex space $C_0^\infty(\Omega)$, endowed with the inductive limit topology, is called the *space of test functions* and is commonly denoted by $\mathfrak{D}(\Omega)$, in accordance with Schwartz' notation in [3]. We shall use $\mathfrak{D}_K$ to denote the locally convex space $C_K^\infty(\Omega)$, where K is a compact subset of Ω.

For any $\phi \in \mathfrak{D}(\Omega)$ we define the norms

$$|\phi|_m = \sup \{|\partial^\alpha \phi(x)| : x \in \Omega, |\alpha| \leq m\} \qquad m \in \mathbf{N}_0$$

and we note that when ϕ is in $\mathfrak{D}_K$, $|\phi|_m$ coincides with the seminorm $p_m(\phi)$ as defined by equation (1.4).

Theorem 2.2 E is a bounded subset of $\mathfrak{D}(\Omega)$ if and only if the following two conditions are satisfied:

(i) $E \subset \mathfrak{D}_K$ for some $K \subset \Omega$.

(ii) E is bounded in $\mathcal{D}_K$, in the sense that for every nonnegative integer m there is a finite constant M_m such that $|\phi|_m \leqslant M_m$ for all $\phi \in E$.

Proof. The sufficiency of (i) and (ii) is clear. To show that they are also necessary, let E be a subset of $\mathcal{D}(\Omega)$ which lies in no $\mathcal{D}_K$. Then there is a sequence of functions $\phi_k \in E$ and a sequence of points $x_k \in \Omega$, with no cluster point in Ω, such that

$$\phi_k(x_k) \neq 0 \qquad k \in \mathbf{N}$$

Let

$$U = \{\phi \in \mathcal{D}(\Omega) : |\phi(x_k)| < k^{-1} |\phi_k(x_k)|, k \in \mathbf{N}\}$$

Since each K contains only a finite number of points of (x_k), the intersection $\mathcal{D}_K \cap U$ is a neighborhood of 0 in $\mathcal{D}_K$ for every K. Hence U is a neighborhood of 0 in $\mathcal{D}(\Omega)$. But since $\phi_k \not\in kU$ for any k, no multiple of U contains E, so E is unbounded.

This proves that if E is bounded in $\mathcal{D}(\Omega)$ then condition (i) must hold. Condition (ii) follows from the fact that the topology of $\mathcal{D}_K$ is the topology it inherits as a subspace of $\mathcal{D}(\Omega)$. □

Theorem 2.3 A sequence of (ϕ_k) in $\mathcal{D}(\Omega)$ converges to 0 if and only if the following two conditions are satisfied:

(i) There is a compact subset K of Ω such that supp $\phi_k \subset K$ for all k.
(ii) $\partial^\alpha \phi_k \to 0$ uniformly on K for all α.

Proof. Conditions (i) and (ii) imply that $\phi_k \to 0$ in $\mathcal{D}_K$, and since the identity map from $\mathcal{D}_K$ to $\mathcal{D}(\Omega)$ is continuous, $\phi_k \to 0$ in $\mathcal{D}(\Omega)$.

Conversely, if $\phi_k \to 0$ in $\mathcal{D}(\Omega)$ then (ϕ_k) is a bounded sequence in $\mathcal{D}(\Omega)$, as we have seen in Section 1.2. From Theorem 2.2 it lies in $\mathcal{D}_K$ for some $K \subset \Omega$ and (i) follows. But then $\phi_k \to 0$ in the subspace topology of $\mathcal{D}_K$ and (ii) follows. □

The disadvantage of the topology that we have defined on $C_0^\infty(\Omega)$ is that it is not metrizable. This may be seen from the following argument: Assuming that d is a metric which defines the topology of $\mathcal{D}(\Omega)$, let $\Omega = \cup K_n$ with K_n compact and $K_n \subset \mathring{K}_{n+1}$ for all n. Choose $\phi_n \in \mathcal{D}(\Omega)$ such that supp $\phi_n \not\subset K_n$. Since multiplication by a constant is a continuous mapping from $\mathcal{D}(\Omega)$ into $\mathcal{D}(\Omega)$, we can find $\lambda_n > 0$ small enough so that $d(0,\lambda_n\phi_n) < 1/n$ for every n. This means that the sequence $\lambda_n\phi_n \to 0$ in $\mathcal{D}(\Omega)$. But this is not possible since supp$(\lambda_n\phi_n)$ cannot be contained in a single compact subset of Ω.

Nevertheless we can still discuss the completeness of $\mathcal{D}(\Omega)$ as a topological vector space, in the sense defined in Section 1.2. If (ϕ_k) is a Cauchy sequence in $\mathcal{D}(\Omega)$ it is bounded and must, according to Theorem 2.1, lie in $\mathcal{D}_K$ for some $K \subset \Omega$. Since $\mathcal{D}_K$ is complete, (ϕ_k) converges in $\mathcal{D}_K$, and consequently in $\mathcal{D}(\Omega)$.

We can also define the topology of $C_0^m(\Omega) = \cup_{K\subset\Omega} C_K^m(\Omega)$ to be the finest locally convex topology in which the identity map from $C_K^m(\Omega)$ to $C_0^m(\Omega)$ is continuous for every compact set $K \subset \Omega$. The resulting topological vector space will be denoted by $\mathcal{D}^m(\Omega)$. This gives the following characterization of convergence in $\mathcal{D}^m(\Omega)$:

Corollary A sequence (ϕ_k) in $\mathcal{D}^m(\Omega)$ converges to 0 if and only if

(i) There is a compact set $K \subset \Omega$ such that supp $\phi_k \subset K$ for all k
(ii) $\partial^\alpha \phi_k \to 0$ uniformly on K for all $|\alpha| \leq m$.

Example 2.1 The function from $\mathbf{R}^n$ to $\mathbf{R}$ defined by

$$\alpha(x) = \begin{cases} \exp\left(-\dfrac{1}{1-|x|^2}\right) & \text{on } |x| < 1 \\ 0 & \text{on } |x| \geq 1 \end{cases}$$

has as support the closed unit ball $\overline{B}(0,1)$ in $\mathbf{R}^n$, and it clearly lies in $\mathcal{D}(\mathbf{R}^n)$. The sequence

$$\alpha_k = \frac{1}{k}\alpha$$

satisfies conditions (i) and (ii) of Theorem 2.3, so it converges to 0 in $\mathcal{D}(\mathbf{R}^n)$. But the sequence

$$\alpha_k = \frac{1}{k}\alpha \circ \frac{1}{k}$$

which is defined at $x \in \mathbf{R}^n$ by $\alpha_k(x) = (1/k)\alpha(x/k)$ does not converge in $\mathcal{D}(\mathbf{R}^n)$ because

$$\text{supp } \alpha_k = \overline{B}(0,k)$$

does not satisfy condition (i). On the other hand, the sequence

$$\alpha_k = \frac{1}{k}\alpha \circ k$$

has a sequence of shrinking supports $\overline{B}(0,1/k)$, but the partial derivatives of α_k do not converge to 0 on any neighborhood of the origin. Condition (ii) of Theorem 2.3 is therefore violated and the sequence diverges.

2.2 DISTRIBUTIONS

Definition A distribution on Ω is a continuous linear functional on $\mathcal{D}(\Omega)$.

In accordance with the terminology of Section 1.2, we shall denote the linear space of all distributions on Ω by $\mathcal{D}'(\Omega)$, the topological dual of $\mathcal{D}(\Omega)$. The following theorem gives a useful characterization of $\mathcal{D}'(\Omega)$.

Theorem 2.4 A linear functional T on $\mathcal{D}(\Omega)$ is a distribution if and only if, for every compact set $K \subset \Omega$, there exists a nonnegative integer m and a finite constant M such that

$$|T(\phi)| \leq M|\phi|_m \tag{2.1}$$

for all $\phi \in \mathcal{D}_K$.

Proof. By Theorems 1.1 and 2.1 T is in $\mathcal{D}'(\Omega)$ if and only if T is bounded on $\mathcal{D}_K$ for every $K \subset \Omega$, which is equivalent, in the topology of $\mathcal{D}_K$, to the inequality (2.1). □

We shall now give some examples of distributions, beginning with those which are defined by Lebesgue integrals, so some familiarity with Lebesgue measure and integration will be assumed (see [4], for example). We shall denote the Lebesgue integral of the measurable function f over the measurable set $E \subset \mathbf{R}^n$ by

$$\int_E f(x)\,dx$$

and this will sometimes be abbreviated to

$$\int_E f\,dx \quad \text{or} \quad \int_E f$$

when the measure function is clear from the context. In this convention E is often dropped when $E = \mathbf{R}^n$. $L^1(\Omega)$ will denote the linear space of complex Lebesgue integrable functions on Ω, i.e. all functions $f: \Omega \rightarrow \mathbf{C}$ whose integral

$$\int_\Omega |f(x)|\,dx$$

is finite. The function f is *locally integrable* on Ω if

$$\int_E |f(x)|\, dx$$

is finite on every compact subset E of Ω, and we shall use $L^1_{\text{loc}}(\Omega)$ to denote the space of locally integrable functions on Ω. All continuous functions on $\mathbf{R}^n$, for example, are locally integrable, although some of them, such as polynomials, are not integrable on $\mathbf{R}^n$. Clearly

$$L^1(\Omega) \subset L^1_{\text{loc}}(\Omega)$$

Example 2.2 If $f \in L^1_{\text{loc}}(\Omega)$ then the linear functional T_f defined on $\mathcal{D}(\Omega)$ by

$$T_f(\phi) = \int_\Omega f(x)\phi(x)\, dx \qquad \phi \in \mathcal{D}(\Omega)$$

is clearly bounded, since

$$|T_f(\phi)| \leq \sup_{x\in\Omega} |\phi(x)| \int_K |f(x)|\, dx = |\phi|_0 \int_K |f(x)|\, dx$$

where $K = \text{supp}\, \phi$, and therefore $T_f \in \mathcal{D}'(\Omega)$.

It is more convenient at times to denote the distribution T_f defined by f in the above example simply by f and to write

$$T_f(\phi) = \langle f, \phi \rangle = \int_\Omega f(x)\phi(x)\, dx \qquad \phi \in \mathcal{D}(\Omega) \tag{2.2}$$

Since continuous functions on Ω are locally integrable, every $f \in C^0(\Omega)$ defines a distribution through equation (2.2).

In Example 2.2 the constant M of the inequality (2.1) is $\int_K |f|$, which clearly depends on K, but the integer $m = 0$ works for all K. T_f is then said to be of *order* 0. The order of the distribution T is the smallest value of m for which the inequality (2.1) holds for all K. If no finite m satisfies the inequality for all K, then T is said to be of *infinite order*.

Example 2.3 Let $f \in L^1_{\text{loc}}(\mathbf{R} - \{0\})$ satisfy $|f(x)| \leq c/|x|^m$ on $|x| \leq 1$ for some positive integer m and a positive constant c. We shall show that there is a distribution $T \in \mathcal{D}'(\mathbf{R})$ of order $\leq m$ such that $T = T_f$ on $\mathcal{D}(\mathbf{R} - \{0\})$.

Let $\phi \in \mathcal{D}(\mathbf{R})$ be arbitrary. Then there is a number $a > 1$ such that $\phi(x) = 0$ on $|x| > a$, and for any x we can use Taylor's formula to write

$$\phi(x) = \phi(0) + x\phi'(0) + \cdots + \frac{x^{m-1}}{(m-1)!}\phi^{(m-1)}(0) + \frac{x^m}{m!}\phi^{(m)}(tx)$$

for some $t \in (0,1)$. Now we define

$$\begin{aligned} T(\phi) &= \int_{|x|>1} f(x)\phi(x)\,dx + \int_{|x|\leq 1} f(x)\left[\phi(x) - \sum_{k=0}^{m-1} \frac{x^k}{k!}\phi^{(k)}(0)\right] dx \\ &= \int_{|x|>1} f(x)\phi(x)\,dx + \int_{|x|\leq 1} f(x)\frac{x^m}{m!}\phi^{(m)}(tx)\,dx \end{aligned}$$

and we obtain

$$\begin{aligned} |T(\phi)| &\leq |\phi|_0 \int_{1<|x|<a} |f(x)|\,dx + \int_{|x|\leq 1} \frac{c}{m!}|\phi^{(m)}(tx)|\,dx \\ &\leq A|\phi|_0 + B|\phi|_m \end{aligned}$$

for some positive constants A and B. Hence T is a distribution on $\mathbf{R}$ of order $\leq m$.

To show that T is represented by f on $\mathbf{R} - \{0\}$, let $\phi \in \mathcal{D}(\mathbf{R})$ with supp $\phi \subset \mathbf{R} - \{0\}$. Then $\phi^{(k)}(0) = 0$ for all k and therefore

$$T(\phi) = \int_{\mathbf{R}} f(x)\phi(x)\,dx = \langle f,\phi\rangle$$

A distribution T is said to be *regular* if there is a locally integrable function f on Ω such that

$$T(\phi) = \langle f,\phi\rangle = \int_{\Omega} f(x)\phi(x)\,dx \qquad \phi \in \mathcal{D}(\Omega)$$

otherwise it is *singular*. The distribution in Example 2.3 corresponding to $f(x) = 1/x^m$, $m \geq 1$, $x \neq 0$, is singular on $\mathbf{R}$ since f is not integrable on a neighborhood of 0. In the next example we give another well-known singular distribution.

Example 2.4 For any fixed point $\xi \in \Omega$ we define

$$T(\phi) = \phi(\xi) \qquad \phi \in \mathcal{D}(\Omega)$$

T is clearly a linear functional on $\mathcal{D}(\Omega)$, which is continuous since $\phi \to 0$ in $\mathcal{D}(\Omega)$ implies that $\phi(\xi) \to 0$ in $\mathbf{C}$. It is known as the *Dirac distribution* and is denoted by δ_ξ, with δ_0 usually abbreviated to δ. Thus

$$\delta(\phi) = \phi(0) \qquad \text{for all } \phi \in \mathcal{D}(\Omega)$$

This distribution obviously has zero order. To show that it is not regular, let

$$\phi_\varepsilon(x) = \alpha\left(\frac{x}{\varepsilon}\right) = \begin{cases} \exp\left(-\dfrac{\varepsilon^2}{\varepsilon^2 - |x|^2}\right) & |x| < \varepsilon \\ 0 & |x| \geqslant \varepsilon \end{cases}$$

where $x \in \mathbf{R}$ and $\varepsilon > 0$. ϕ_ε is clearly in $\mathcal{D}(\mathbf{R})$ and

$$|\phi_\varepsilon(x)| \leqslant \phi_\varepsilon(0) = \frac{1}{e}$$

If δ were regular then we could write

$$\delta(\phi_\varepsilon) = \int f(x)\phi_\varepsilon(x)\,dx = \int_{|x|\leqslant\varepsilon} f(x)\phi_\varepsilon(x)\,dx$$

for some $f \in L^1_{\text{loc}}(\mathbf{R})$. Consequently

$$\frac{1}{e} = \phi_\varepsilon(0) = \delta(\phi_\varepsilon) \leqslant \frac{1}{e}\int_{|x|\leqslant\varepsilon} |f(x)|\,dx \to 0 \qquad \text{as } \varepsilon \to 0$$

which is impossible. Hence δ, and therefore δ_ξ, is singular. Nevertheless we shall often find it convenient to write $\langle\delta_\xi,\phi\rangle$ for $\delta_\xi(\phi) = \phi(\xi)$. In other words, the use of the bracket notation is not restricted to regular distributions.

Example 2.5 If Σ is a hypersurface in $\mathbf{R}^n$ of dimension less than n, then for any locally integrable function f on Σ we can define the distribution

$$T_f(\phi) = \int_\Sigma f\phi\,d\sigma \qquad \phi \in \mathcal{D}(\mathbf{R}^n)$$

which is clearly a generalization of the Dirac distribution from the point 0 to the hypersurface Σ. T_f may be interpreted as a measure on $\mathbf{R}^n$ supported by Σ with density f.

Examples (2.2), (2.4), and (2.5) are typical of a wider class of distributions which are generated by measures on $\mathbf{R}^n$. The Riesz representation theorem [4] asserts that to each continuous linear functional T on $C_0^0(\Omega)$ there corresponds a unique complex, locally finite, regular Borel measure μ on Ω such that

$$T(\phi) = \int_\Omega \phi \, d\mu \tag{2.3}$$

for all $\phi \in C_0^0(\Omega)$. Since such a measure obviously defines a continuous linear functional on $C_0^0(\Omega)$ by the relation (2.3), this correspondence between T and μ is bijective.

The measure function corresponding to the regular distribution T_f in Example 2.2 is given by $\mu(E) = \int_E f$ for any measurable set $E \subset \mathbf{R}^n$. The Dirac distribution δ_ξ which is defined on $\mathfrak{D}(\Omega)$ by $\langle \delta_\xi, \phi \rangle = \phi(\xi)$ is also continuous on $C_0^0(\Omega)$ and corresponds to the measure function

$$\mu(E) = \begin{cases} 1 & \text{if } \xi \in E \\ 0 & \text{if } \xi \notin E \end{cases}$$

However, the space $\mathfrak{D}'(\Omega)$ is wider than the class of Borel measures on Ω, as the next example shows.

Example 2.6 The mapping

$$T(\phi) = \phi'(0)$$

defines a continuous linear functional on $\mathfrak{D}(\mathbf{R})$, in fact on $C_0^m(\mathbf{R})$ for $m \geq 1$, but not on $C_0^0(\mathbf{R})$. Thus T is a distribution which is not a measure.

In higher dimensions, the functional

$$T(\phi) = \partial_k \phi(0) \qquad \phi \in \mathfrak{D}(\mathbf{R}^n)$$

where $1 \leq k \leq n$, is a (sigular) distribution in $\mathbf{R}^n$ of order 1. More generally, the functional $\phi \mapsto \partial^\alpha \phi(0)$, for any $\alpha \in \mathbf{N}_0^n$, is a distribution in $\mathbf{R}^n$ of order $|\alpha|$.

Example 2.7 The function which is defined to be $1/x$ on $(0,\infty)$ and 0 otherwise, is not integrable on any neighborhood of 0, and does not define a distribution on $\mathbf{R}$ by the formula (2.2). Its restriction to $(0,\infty)$, on the other hand, is continuous and therefore defines a regular distribution in $(0,\infty)$.

2.3 DIFFERENTIATION OF DISTRIBUTIONS

When $f \in C^1(\mathbf{R})$ it defines a distribution and has a derivative f' which is also a distribution. To the extent that a distribution is a generalization of a function, it would be desirable to define the distributional derivative of f so that it agrees with f'. We note that integration by parts gives the following result for $f \in C^1(\mathbf{R})$:

$$\langle f',\phi\rangle = \int_{-\infty}^{\infty} f'(x)\phi(x)\,dx$$

$$= f(x)\,\phi(x)\Big|_{-\infty}^{\infty} - \int_{-\infty}^{\infty} f(x)\phi'(x)\,dx$$

$$= -\langle f,\phi'\rangle \qquad \phi \in \mathcal{D}(\mathbf{R})$$

This suggests the following definition:

Definition For any $T \in \mathcal{D}'(\Omega)$ we define

$$\partial_k T(\phi) = -T(\partial_k\phi) \qquad \phi \in \mathcal{D}(\Omega)$$

By using induction we obtain the more general formula

$$\partial^\alpha T(\phi) = (-1)^{|\alpha|}T(\partial^\alpha\phi) \qquad \phi \in \mathcal{D}(\Omega),\ \alpha \in \mathbf{N}_0^n$$

or

$$\langle \partial^\alpha T,\phi\rangle = (-1)^{|\alpha|}\,\langle T,\partial^\alpha\phi\rangle \tag{2.4}$$

Note that the right-hand side of (2.4) is well defined for any multi-index α, because $\phi \in \mathcal{D}(\Omega)$, and represents a continuous linear functional on $\mathcal{D}(\Omega)$. Thus a distribution has derivatives, in the sense of the above definition, of all orders. Furthermore

$$\partial^\alpha\partial^\beta T = \partial^\beta\partial^\alpha T$$

for any $T \in \mathcal{D}'(\Omega)$, because

$$\begin{aligned}\partial^\alpha\partial^\beta\, T(\phi) &= (-1)^{|\alpha|}\,\partial^\beta T(\partial^\alpha\phi)\\ &= (-1)^{|\alpha|+|\beta|}\,T(\partial^\beta\partial^\alpha\phi)\\ &= (-1)^{|\alpha|+|\beta|}\,T(\partial^\alpha\partial^\beta\phi)\\ &= \partial^\beta\partial^\alpha\, T(\phi)\end{aligned}$$

If $f \in C^m(\Omega)$ then the formula for integration by parts can be used to show that the distributional derivative of f coincides with its conventional, or classical, derivative in the sense that

$$\partial^\alpha T_f = T_{\partial^\alpha f} \qquad \text{for all } |\alpha| \leqslant m$$

but in general this relation does not hold, as may be seen from some of the examples which follow.

Example 2.8 Let

$$x_+ = \begin{cases} x & x > 0 \\ 0 & x \leqslant 0 \end{cases}$$

As a function x_+ is not differentiable at $x = 0$ in the classical sense, but as a distribution it can be differentiated by formula (2.4):

$$\begin{aligned}
\langle x_+', \phi \rangle &= -\langle x_+, \phi' \rangle \qquad \phi \in \mathcal{D}(\mathbf{R}) \\
&= -\int_0^\infty x\phi'(x)\, dx \\
&= -x\phi(x)\Big|_0^\infty + \int_0^\infty \phi(x)\, dx \\
&= 0 + \int_{-\infty}^\infty H(x)\phi(x)\, dx \\
&= \langle H, \phi \rangle
\end{aligned}$$

where H is the *Heaviside function* which is defined to be 1 on $x > 0$ and 0 on $x < 0$. Similarly,

$$\begin{aligned}
\langle x_+'', \phi \rangle &= \langle H', \phi \rangle \\
&= -\langle H, \phi' \rangle \\
&= -\int_0^\infty \phi'(x)\, dx \\
&= \phi(0)
\end{aligned}$$

Therefore $x_+'' = H' = \delta$. We can go further:

$$\langle x_+''', \phi \rangle = \langle \delta', \phi \rangle = -\langle \delta, \phi' \rangle = -\phi'(0)$$

$$\vdots$$

$$\langle x_+^{(k+2)}, \phi \rangle = \langle \delta^{(k)}, \phi \rangle = (-1)^k \phi^{(k)}(0)$$

It is important to note in this example that x_+ and H are differentiated as distributions and not as functions. In the case of x_+ it makes no difference, since $x_+' = H$ almost everywhere (a.e.) in the classical sense as well. But the classical derivative of H is 0 a.e., and when we write $H' = \delta$ we really mean $T_H' = \delta$ (see Example 2.10).

The next example shows that, as in the case of x_+, the distributional and the classical derivatives of a function may coincide even when the function is not continuously differentiable.

Example 2.9 Let f be a differentiable function on $I = (a,b)$ whose (classical) derivative f' is integrable on I. Such a function can be expressed as the integral of its derivative (see [4]),

$$f(x) = \int_c^x f'(t)\,dt + f(c) \qquad x,c \in I$$

If $\phi \in \mathcal{D}(I)$, then $(f\phi)' = f'\phi + f\phi'$ and $\int_I (f\phi)' = 0$ because $f\phi$ vanishes outside a closed subinterval of I. Hence

$$\int_I f'\phi + \int_I f\phi' = 0$$

Now if T_f' is the distributional derivative of T_f, then

$$\begin{aligned} T_f'(\phi) &= -T_f(\phi') \\ &= -\int_I f\phi' \\ &= \int_I f'\phi \\ &= T_{f'}(\phi) \end{aligned}$$

Therefore $T_f' = T_{f'}$.

More generally, if f is absolutely continuous on I then f' exists almost everywhere, is integrable on I, and we can write

$$f(x) = \int_c^x f'(t)\,dt + f(c) \qquad x,c \in I$$

The equality $T_f' = T_{f'}$ then follows by the same argument.

Example 2.10 Let $c \in (a,b) = I$ and $f \in C^1(I - \{c\})$. If the left- and right-hand limits at c,

$$f(c^-) = \lim_{\substack{x \to c \\ x<c}} f(x) \qquad f(c^+) = \lim_{\substack{x \to c \\ x>c}} f(x)$$

are finite and f' is bounded in a neighborhood of c, then the distributions $T_{f'}$ and T_f' in $\mathcal{D}'(I)$ are related by

$$T_f' = T_{f'} + [f(c^+) - f(c^-)]\delta_c.$$

For, if ϕ is any function in $\mathcal{D}(I)$, then

$$\begin{aligned}
T_f'(\phi) &= -T_f(\phi') \\
&= -\int_a^b f(x)\phi'(x)\,dx \\
&= -\int_a^c f(x)\phi'(x)\,dx - \int_c^b f(x)\phi'(x)\,dx \\
&= -\lim_{\varepsilon_1 \to 0} \int_a^{c-\varepsilon_1} f(x)\phi'(x)\,dx - \lim_{\varepsilon_2 \to 0} \int_{c+\varepsilon_2}^b f(x)\phi'(x)\,dx \\
&= -\lim_{\varepsilon_1 \to 0} \left[f(x)\phi(x) \Big|_a^{c-\varepsilon_1} - \int_a^{c-\varepsilon_1} f'(x)\phi(x)\,dx \right] \\
&\quad - \lim_{\varepsilon_2 \to 0} \left[f(x)\phi(x) \Big|_{c+\varepsilon_2}^{b} - \int_{c+\varepsilon_2}^b f'(x)\phi(x)\,dx \right] \\
&= -f(c^-)\phi(c) + f(c^+)\phi(c) + \int_a^b f'(x)\phi(x)\,dx \\
&= \langle f',\phi\rangle + [f(c^+) - f(c^-)]\langle \delta_c,\phi\rangle
\end{aligned}$$

In particular, when f is the Heaviside function, $H' = 0$ on $I - \{0\}$ and we obtain the expected result $T_H' = \delta$.

In manipulating distributions it is always helpful to have some clarity and a degree of precision in our notation. Thus x_+ and H are *functions* defined on $\mathbf{R}$ which represent *distributions* on $\mathcal{D}(\mathbf{R})$ by the formula (2.2), since each is locally integrable. The classical derivative of H is the *function* which is 0 almost everywhere, and represents the zero distribution. But the *distributional derivative of H* is δ, which is not a function.

Derivatives will always be taken in the distributional sense, unless otherwise qualified. The pointwise notation $H'(x)$ is therefore meaningful only when it applies to the classical derivative, since we have no way of evaluating a distribution at a point. But if it is interpreted properly, this notation can be useful when we wish to keep track of the point variable, especially when more than one is involved; and we shall find it convenient at times to write the suggestive equalities $H' = \delta$, $H'(x) = \delta(x)$, or $H'_x = \delta_x$, on $\mathbf{R}$, rather than the more accurate $T'_H = \delta$ on $\mathcal{D}(\mathbf{R})$.

Example 2.11 For any subset E of $\mathbf{R}^n$ we define the *characteristic function* of E to be

$$I_E(x) = \begin{cases} 1 & x \in E \\ 0 & x \in \mathbf{R}^n - E \end{cases}$$

If E is a bounded open subset of $\mathbf{R}^n$ with a smooth boundary ∂E, then

$$\begin{aligned} \langle \partial_k I_E, \phi \rangle &= -\langle I_E, \partial_k \phi \rangle \\ &= -\int_E \partial_k \phi(x)\, dx \\ &= -\int_{\partial E} \phi(x) \cos \theta_k \, d\sigma \end{aligned}$$

where θ_k is the angle between the x_k-axis in $\mathbf{R}^n$ and the outward normal to ∂E, $d\sigma$ is the Euclidean measure on ∂E, and the last equality is a consequence of the divergence theorem (see [5]). Thus $\partial_k I_E$ is a measure of density $-\cos \theta_k$ on ∂E.

For the special case when $n = 1$ and $E = (a,b)$ we have

$$I_E(x) = H(x - a) - H(x - b)$$

$$I'_E(x) = \delta(x - a) - \delta(x - b)$$

or $I'_E = \delta_a - \delta_b$

Example 2.12 log $|x|$ is locally integrable on $\mathbf{R}$ so it defines a distribution in $\mathcal{D}'(\mathbf{R})$, but its classical derivative

$$\frac{d}{dx}\log|x| = \frac{1}{x} \qquad x \neq 0$$

does not define a distribution through the formula (2.2), as pointed out in Example 2.7.

Since the distributional derivative of log $|x|$ is also a distribution, it is worth investigating the relationship between this derivative and $1/x$:

$$\left\langle \frac{d}{dx}\log|x|,\phi \right\rangle = -\left\langle \log|x|,\phi' \right\rangle$$

$$= -\int_{-\infty}^{\infty} \log|x|\phi'(x)\,dx$$

since log $|x|$ is locally integrable. Now, with log $|x|\phi'(x)$ integrable in the neighborhood of 0, we can write

$$\left\langle \frac{d}{dx}\log|x|,\phi \right\rangle = -\lim_{\varepsilon\to 0}\int_{|x|\geq\varepsilon} \log|x|\phi'(x)\,dx$$

But since ϕ has compact support,

$$\left\langle \frac{d}{dx}\log|x|,\phi \right\rangle = -\lim_{\varepsilon\to 0}\left[\log|x|\phi(x)\,\Big|_{\varepsilon}^{-\varepsilon} - \int_{|x|\geq\varepsilon}\frac{1}{x}\phi(x)\,dx\right]$$

$$= \lim_{\varepsilon\to 0}\left[2\varepsilon\log\varepsilon\,\frac{\phi(\varepsilon)-\phi(-\varepsilon)}{2\varepsilon} + \int_{|x|\geq\varepsilon}\frac{1}{x}\phi(x)\,dx\right]$$

$$= \lim_{\varepsilon\to 0}\int_{|x|\geq\varepsilon}\frac{1}{x}\phi(x)\,dx$$

ϕ being differentiable at $x = 0$.

The last integral is called the *Cauchy principal value* of the divergent integral

$$\int_{-\infty}^{\infty}\frac{1}{x}\phi(x)\,dx$$

and is denoted by

$$pv \int_{-\infty}^{\infty} \frac{1}{x} \phi(x)\, dx$$

Thus the distributional derivative of log $|x|$, which is not a function and which we shall denote by $pv\ 1/x$, is obtained from the divergent integral

$$\int_{-\infty}^{\infty} \frac{1}{x} \phi(x)\, dx$$

by taking its principal value, a process known as *regularizing* the integral.

More generally, if the function f is locally integrable but $\partial^\alpha f$ is not, then $\partial^\alpha T_f$ is called a *regularization* of $T_{\partial^\alpha f}$. By the same token

$$\begin{aligned}\left\langle \frac{d}{dx} pv \frac{1}{x}, \phi \right\rangle &= -\left\langle pv \frac{1}{x}, \phi' \right\rangle \\ &= -\lim_{\varepsilon \to 0} \int_{|x| \geq \varepsilon} \frac{1}{x} \phi'(x)\, dx \\ &= -\lim_{\varepsilon \to 0} \left[\log |x| \phi'(x) \Big|_{\varepsilon}^{-\varepsilon} - \int_{|x| \geq \varepsilon} \log |x| \phi''(x)\, dx \right] \\ &= \lim_{\varepsilon \to 0} \int_{|x| \geq \varepsilon} \log |x| \phi''(x)\, dx\end{aligned}$$

since ϕ' is differentiable at $x = 0$. Now this last integral is well defined, since $\log |x| \phi''$ is integrable on **R**, and represents the action of the distribution $(d/dx)pv(1/x)$ on ϕ. We shall return to this subject in Section 2.8 when we give more explicit representations of distributions which are defined by powers of x on **R**.

Example 2.13 Let

$$L = \frac{d^2}{dx^2} - 3 \frac{d}{dx} + 2$$

be a differential operator in **R**, and

$$h(x) = \begin{cases} e^x & x \leq 0 \\ e^{2x} & x > 0 \end{cases}$$

If T_h is the distribution defined by the continuous function h, we shall show that $LT_h = \delta$.

For any $\phi \in \mathcal{D}(\mathbf{R})$, we have

$$\begin{aligned} LT_h(\phi) &= \langle h'' - 3h' + 2h, \phi \rangle \\ &= \langle h,\phi'' \rangle + 3\langle h,\phi' \rangle + 2\langle h,\phi \rangle \end{aligned}$$

Now

$$\begin{aligned} \langle h,\phi'' \rangle &= \int_{-\infty}^{0} e^x\phi''(x)\,dx + \int_{0}^{\infty} e^{2x}\phi''(x)\,dx \\ &= \left[\phi'(0) - \int_{-\infty}^{0} e^x\phi'(x)\,dx\right] + \left[-\phi'(0) - 2\int_{0}^{\infty} e^{2x}\phi'(x)\,dx\right] \\ &= -\left[\phi(0) - \int_{-\infty}^{0} e^x\phi(x)\,dx\right] - 2\left[-\phi(0) - 2\int_{0}^{\infty} e^{2x}\phi(x)\,dx\right] \\ &= \phi(0) + \int_{-\infty}^{0} e^x\phi(x)\,dx + 4\int_{0}^{\infty} e^{2x}\phi(x)\,dx \end{aligned}$$

$$\begin{aligned} \langle h,\phi' \rangle &= \int_{-\infty}^{0} e^x\phi'(x)\,dx + \int_{0}^{\infty} e^{2x}\phi'(x)\,dx \\ &= -\int_{-\infty}^{0} e^x\phi(x)\,dx - 2\int_{0}^{\infty} e^{2x}\phi(x)\,dx \end{aligned}$$

$$\langle h,\phi \rangle = \int_{-\infty}^{0} e^x\phi(x)\,dx + \int_{0}^{\infty} e^{2x}\phi(x)\,dx$$

Hence $LT_h(\phi) = \phi(0)$ for every $\phi \in \mathcal{D}(\mathbf{R})$, which means that $LT_h = \delta$.

It is worth noting in this example that the function h, though continuous, has a jump discontinuity in its derivative at $x = 0$ given by

$$h'(0^+) - h'(0^-) = 2e^0 - e^0 = 1$$

and this accounts for the δ distribution when h is differentiated a second time. On $\mathbf{R}-\{0\}$ the function h is twice differentiable and satisfies $Lh = 0$.

We can generalize Example 2.13 in the following way:

Let

$$L = \frac{d^2}{dx^2} + a\frac{d}{dx} + b$$

with $a, b \in \mathbf{C}$, be a differential operator in $\mathbf{R}$, and suppose that f_1 and f_2 are two C^2 solutions in $\mathbf{R}$ of $Lf = 0$, which satisfy

$$f_1(0) = f_2(0)$$

$$f_2'(0) - f_1'(0) = 1$$

If h is the continuous function defined by

$$h(x) = \begin{cases} f_1(x) & x \leqslant 0 \\ f_2(x) & x > 0 \end{cases}$$

and T_h is the distribution defined by h, then it is a simple matter to verify that $LT_h = \delta$. Note that the solution x_+ of $T'' = \delta$ in Example 2.8 is in accordance with this construction.

In $\mathbf{R}^n$ the partial differential operator

$$\sum_{k=1}^{n} \partial_k^2$$

is known as the *Laplacian operator*, and will be denoted by Δ.

Example 2.14 The function $\log |x|$ is locally integrable in $\mathbf{R}^2$. We shall now obtain its (distributional) Laplacian derivative

$$\Delta \log |x| = (\partial_1^2 + \partial_2^2) \log |x|$$

By the differentiation formula (2.4),

$$\begin{aligned} \langle \Delta \log |x|, \phi \rangle &= \langle \log |x|, \Delta\phi \rangle \qquad \phi \in \mathcal{D}(\mathbf{R}^2) \\ &= \int_{\mathbf{R}^2} \log |x| \Delta\phi(x)\, dx \\ &= \lim_{\varepsilon \to 0} \int_{|x| \geqslant \varepsilon} \log |x| \Delta\phi(x)\, dx \qquad (2.5) \end{aligned}$$

For any bounded open set $\Omega \subset \mathbf{R}^n$ with a sufficiently smooth boundary and any pair of functions $u, v \in C^2(\overline{\Omega})$, *Green's first formula*

$$\int_{\Omega} \left[u\, \Delta v + \sum_{k=1}^{n} (\partial_k u)(\partial_k v) \right] = \int_{\partial\Omega} u\, \partial_\eta v$$

follows directly from the divergence theorem [5]. Here ∂_η is the differential operator with respect to the outward normal η on $\partial\Omega$. By interchanging u and v and subtracting, we obtain *Green's second formula*

$$\int_{\Omega} (u\, \Delta v - v\, \Delta u) = \int_{\partial\Omega} (u\, \partial_\eta v - v\, \partial_\eta u)$$

Now we choose Ω so that it contains supp ϕ as well as the closed ball $\overline{B}(0,\varepsilon)$, for some $\varepsilon > 0$. Using Green's second formula on the open set $\Omega_\varepsilon = \Omega - \overline{B}(0,\varepsilon) = \{x \in \Omega : |x| > \varepsilon\}$, we obtain

$$\int_{\Omega_\varepsilon} \log |x| \Delta\phi(x)\, dx = \int_{\Omega_\varepsilon} \phi(x) \Delta \log |x|\, dx + \int_{\partial\Omega_\varepsilon} [\log |x|\, \partial_\eta \phi(x) - \phi(x)\, \partial_\eta \log |x|]\, d\sigma$$

where η is the outward normal on $\partial\Omega_\varepsilon$. See Figure 2.1.

Since ϕ and $\partial_\eta\phi$ vanish on the boundary $\partial\Omega$, we have

$$\int_{|x|\geq\varepsilon} \log |x|\, \Delta\phi(x)\, dx = \int_{|x|\geq\varepsilon} \phi(x)\, \Delta \log |x|\, dx + \int_{|x|=\varepsilon} [\log |x|\, \partial_\eta \phi(x) - \phi(x)\, \partial_\eta \log |x|]\, d\sigma. \quad (2.6)$$

With $|x| = (x_1^2 + x_2^2)^{1/2} = r$, we have $\partial_\eta = -\partial_r$ on the circle $|x| = \varepsilon$, and for all $x \neq 0$ we also have

$$\begin{aligned} \Delta \log |x| &= \partial_1\left(\frac{1}{|x|}\, \partial_1 |x|\right) + \partial_2\left(\frac{1}{|x|}\, \partial_2 |x|\right) \\ &= \partial_1\left(\frac{x_1}{|x|^2}\right) + \partial_2\left(\frac{x_2}{|x|^2}\right) \\ &= 0 \end{aligned}$$

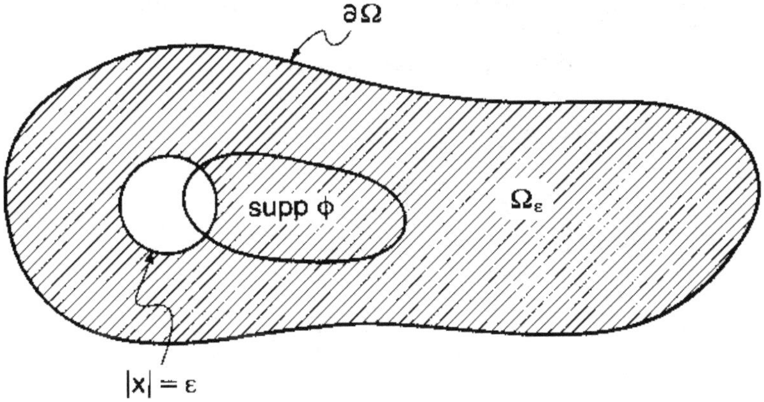

Figure 2.1

Thus the first integral on the right-hand side of (2.6) drops out, and we can write

$$\int_{|x|\geq\varepsilon} \log|x|\ \Delta\phi(x)\ dx = \int_{|x|=\varepsilon} \left[\frac{1}{\varepsilon}\phi(x) - \log\varepsilon\ \partial_r\phi(x)\right] d\sigma$$

Since ϕ is in $C_0^\infty(\mathbf{R}^2)$, its derivative ∂_r is bounded on $\mathbf{R}^2$ by some constant, say M; hence

$$\left|\log\varepsilon \int_{|x|=\varepsilon} \partial_r\phi(x)\ d\sigma\right| \leq 2\pi\varepsilon\ |\log\varepsilon|\ M \to 0 \qquad \text{as } \varepsilon \to 0$$

That leaves

$$\frac{1}{\varepsilon}\int_{|x|=\varepsilon} \phi(x)\ d\sigma = \frac{1}{\varepsilon}\int_{|x|=\varepsilon} [\phi(x) - \phi(0)]\ d\sigma + \frac{1}{\varepsilon}\phi(0)\int_{|x|=\varepsilon} d\sigma$$

which tends to $0 + 2\pi\phi(0)$ as $\varepsilon \to 0$, since the first integral on the right-hand side tends to 0 as $\varepsilon \to 0$, ϕ being continuous at $x = 0$.

Thus, in view of (2.5), $\langle\Delta \log|x|, \phi\rangle = 2\pi\phi(0)$ for all $\phi \in \mathcal{D}(\mathbf{R}^2)$ and we conclude that

$$\Delta \log|x| = 2\pi\delta$$

Example 2.15 Using the method of Example 2.14 we can also determine $\Delta(1/|x|)$ in $\mathbf{R}^3$.

In $\mathbf{R}^3$, $1/|x|$ is integrable in the neighborhood of 0, so we can write

$$\left\langle \Delta\frac{1}{|x|},\phi\right\rangle = \left\langle \frac{1}{|x|},\Delta\phi\right\rangle$$
$$= \lim_{\varepsilon\to 0}\int_{|x|\geq\varepsilon}\frac{1}{|x|}\,\Delta\phi(x)\,dx$$

$$\int_{|x|\geq\varepsilon}\frac{1}{|x|}\,\Delta\phi(x)\,dx = \int_{|x|\geq\varepsilon}\phi(x)\,\Delta\left(\frac{1}{|x|}\right)dx$$
$$+ \int_{|x|=\varepsilon}\left[\frac{1}{|x|}\,\partial_\eta\phi(x) - \phi(x)\,\partial_\eta\left(\frac{1}{|x|}\right)\right]d\sigma$$

Here, again, the first integral on the right-hand side vanishes because

$$\Delta\left(\frac{1}{|x|}\right) = (\partial_1^2 + \partial_2^2 + \partial_3^2)(x_1^2 + x_2^2 + x_3^2)^{-1/2}$$
$$= 0 \qquad \text{when } x \neq 0$$

Therefore, with $\partial_\eta = -\partial_r$,

$$\int_{|x|\geq\varepsilon}\frac{1}{|x|}\,\Delta\phi(x)\,dx = -\frac{1}{\varepsilon}\int_{|x|=\varepsilon}\partial_r\phi(x)\,d\sigma - \frac{1}{\varepsilon^2}\int_{|x|=\varepsilon}\phi(x)\,d\sigma$$

Since $\partial_r\phi$ is a bounded function in $\mathbf{R}^3$, there is a positive constant M such that $|\partial_r\phi(x)| \leq M$ for all $x \in \mathbf{R}^3$, and hence

$$\left|\frac{1}{\varepsilon}\int_{|x|=\varepsilon}\partial_r\phi(x)\,d\sigma\right| \leq \frac{M}{\varepsilon}\int_{|x|=\varepsilon}d\sigma$$
$$= 4\pi\varepsilon M \to 0 \qquad \text{as } \varepsilon \to 0.$$

We are left with

$$\frac{1}{\varepsilon^2}\int_{|x|=\varepsilon}\phi(x)\,dx = \frac{1}{\varepsilon^2}\int_{|x|=\varepsilon}[\phi(x) - \phi(0)]\,dx + \frac{1}{\varepsilon^2}\int_{|x|=\varepsilon}\phi(0)\,dx$$

The first integral on the right-hand side tends to 0 as $\varepsilon \to 0$, while the second is just $4\pi\phi(0)$.

Thus $\langle\Delta(1/|x|),\phi\rangle = -4\pi\phi(0)$ for every $\phi \in \mathcal{D}(\mathbf{R}^3)$, which means that

$$\Delta\frac{1}{|x|} = -4\pi\delta$$

2.4 CONVERGENCE OF DISTRIBUTIONS

Among the topologies that can be defined on the vector space $\mathcal{D}'(\Omega)$, the most important, from our point of view, is the one known as the *weak topology*. This is the locally convex topology defined by the family of seminorms

$$p_\phi(T) = |T(\phi)|$$

with $\phi \in \mathcal{D}(\Omega)$ and $T \in \mathcal{D}'(\Omega)$, and it leads to the following definition of (weak) convergence in $\mathcal{D}'(\Omega)$:

Definition The sequence (T_k) in $\mathcal{D}'(\Omega)$ *converges* to 0 if and only if, for every $\phi \in \mathcal{D}(\Omega)$, the sequence $(T_k(\phi))$ converges to 0 in $\mathbf{C}$.

This is really "pointwise" convergence on $\mathcal{D}(\Omega)$ and, as usual, we shall write

$$T_k \to T \qquad \text{in } \mathcal{D}'(\Omega)$$

if the sequence $(T_k - T)$ converges to 0 in the sense of the above definition.

The other type of convergence in $D'(\Omega)$ is where $T_k \to 0$ is equivalent to $T_k(\phi) \to 0$ uniformly on every bounded subset of $\mathcal{D}(\Omega)$. This type of convergence is called *strong*, or *uniform*, convergence and it clearly implies weak convergence. Convergence in $\mathcal{D}'(\Omega)$ will always be taken in the weak sense unless it is otherwise qualified.

Theorem 2.5 The space of distributions $\mathcal{D}'(\Omega)$ is (sequentially) complete.

Proof. If (T_k) is a Cauchy sequence in $\mathcal{D}'(\Omega)$ then it is bounded, which means that there is a neighborhood U of 0 in $\mathcal{D}(\Omega)$ and a positive number M such that $|T_k(\phi)| \leq M$ for all $\phi \in U$ and $k \in \mathbf{N}$. Furthermore $(T_k(\phi))$ is a Cauchy sequence in $\mathbf{C}$ for every $\phi \in \mathcal{D}(\Omega)$ and therefore its limit exists. Let T be defined by

$$T(\phi) = \lim T_k(\phi) \qquad \phi \in \mathcal{D}(\Omega)$$

T is clearly linear, and since

$$|T(\phi)| = \lim |T_k(\phi)| \leq M \qquad \phi \in U$$

it is bounded on U, and therefore continuous on $\mathcal{D}(\Omega)$. □

Corollary If $T_k \in \mathcal{D}'(\Omega)$ for every $k \in \mathbf{N}$ and $\lim T_k = T$, then $\lim \partial^\alpha T_k = \partial^\alpha T$ for every multi-index $\alpha \in \mathbf{N}_0^n$.

Proof. For any $\phi \in \mathcal{D}(\Omega)$ we have

$$\begin{aligned}\lim\,(\partial^\alpha T_k)(\phi) &= (-1)^{|\alpha|}\lim T_k(\partial^\alpha\phi)\\ &= (-1)^{|\alpha|}\,T(\partial^\alpha\phi)\\ &= \partial^\alpha T(\phi)\end{aligned}$$

□

For locally integrable functions, the following theorem relates convergence almost everywhere to convergence in $\mathcal{D}'(\Omega)$.

Theorem 2.6 If f_k is a sequence of functions in $L^1_{loc}(\Omega)$ which converges to f a.e. in Ω, and $|f_k| \leq g$ for some $g \in L^1_{loc}(\Omega)$, then $f_k \to f$ in $\mathcal{D}'(\Omega)$.

Proof. For every $\phi \in \mathcal{D}(\Omega)$, we have

$$T_{f_k}(\phi) = \langle f_k,\phi\rangle = \int_\Omega f_k\phi \to \int_\Omega f\phi \qquad \text{as } k \to \infty$$

by the Lebesgue dominated convergence theorem [4]. Since

$$\int_\Omega f\phi = T_f(\phi)$$

we conclude that $T_{f_k} \to T_f$. □

To show that convergence a.e. for a sequence of locally integrable functions does not imply its convergence in $\mathcal{D}'$, consider the sequence

$$f_k(x) = \begin{cases} k^2 & |x| < 1/k \\ 0 & |x| \geq 1/k \end{cases}$$

which converges to 0 a.e. If $\phi \in \mathcal{D}(\mathbf{R})$ is such that $\phi = 1$ in $(-1,1)$ then $\langle f_k,\phi\rangle = 2k$, which does not converge.

Moreover, distributional convergence does not imply pointwise convergence, as may be seen by considering the sequence $(\sin kx)$:

$$\begin{aligned}\langle \sin kx, \phi\rangle &= \int_{-\infty}^{\infty} \sin kx\,\phi(x)\,dx\\ &= \frac{1}{k}\int_{-\infty}^{\infty} \cos kx\,\phi'(x)\,dx\\ &\to 0 \qquad \text{as } k \to \infty\end{aligned}$$

for every $\phi \in \mathcal{D}(\mathbf{R})$.

Example 2.16 Let

$$T_n = n\delta - \left(\sum_1^n \frac{1}{k}\right)\delta' - \left(\sum_1^n \delta_{1/k}\right)$$

To show that T_n converges in $\mathcal{D}'(\mathbf{R})$ we must show that $\lim T_n(\phi)$ exists for every $\phi \in \mathcal{D}(\mathbf{R})$:

$$T_n(\phi) = n\phi(0) + \left(\sum_1^n \frac{1}{k}\right)\phi'(0) - \sum_1^n \phi\left(\frac{1}{k}\right).$$

By Taylor's formula, we can write $\phi(x) = \phi(0) + x\phi'(0) + x^2\psi(x)$, where ψ is a C^∞ function which is bounded by some constant, say M. Hence

$$T_n(\phi) = n\phi(0) + \left(\sum_1^n \frac{1}{k}\right)\phi'(0) - \sum_1^n \left[\phi(0) + \frac{1}{k}\phi'(0) + \frac{1}{k^2}\psi\left(\frac{1}{k}\right)\right]$$

$$= -\sum_1^n \frac{1}{k^2}\psi\left(\frac{1}{k}\right)$$

$$|T_n(\phi) - T_m(\phi)| \leq M \sum_m^n \frac{1}{k^2} \qquad \text{if } m < n$$

and $(T_n(\phi))$ is a Cauchy sequence in $\mathbf{C}$, so its limit exists.

Even when the sequence of functions f_k converges a.e. and in $\mathcal{D}'$, the two limits may not be equal. This is illustrated by the next example.

Example 2.17 Let

$$f_k(x) = \begin{cases} k & \text{if } |x| < 1/2k \\ 0 & \text{if } |x| \geq 1/2k \end{cases}$$

Then $\int f_k(x)\,dx = 1$ and $f_k \to 0$ a.e. on $\mathbf{R}$.

On the other hand, if ϕ is any function in $\mathcal{D}(\mathbf{R})$, then

$$\langle f_k, \phi\rangle = \phi(0) + k \int_{-1/2k}^{1/2k} [\phi(x) - \phi(0)]\,dx \to \phi(0) \qquad \text{as } k \to \infty$$

because ϕ is continuous at 0. Hence $\lim f_k = \delta$.

A sequence of functions, such as (f_k) in Example 2.17, which converges to δ in $\mathcal{D}'(\Omega)$ is called a *delta-convergent sequence*. The next theorem gives a prescription for constructing such a sequence.

Theorem 2.7 Let f be a nonnegative integrable function on $\mathbf{R}^n$ with $\int f(x)dx = 1$ and

$$f_\lambda(x) = \lambda^{-n} f\left(\frac{x}{\lambda}\right) = \lambda^{-n} f\left(\frac{x_1}{\lambda}, \ldots, \frac{x_n}{\lambda}\right) \qquad \lambda > 0.$$

Then $f_\lambda \to \delta$ in $\mathcal{D}'(\mathbf{R}^n)$ as $\lambda \to 0$.

Proof.

$$\begin{aligned}\int f_\lambda(x)\, dx &= \int f\left(\frac{x}{\lambda}\right) \lambda^{-n}\, dx \\ &= \int f(\xi)\, d\xi \\ &= 1\end{aligned}$$

Therefore

$$\begin{aligned}\lim_{\lambda\to 0} \langle f_\lambda, \phi\rangle &= \lim_{\lambda\to 0} \int f_\lambda(x)\phi(x)\, dx \\ &= \phi(0) + \lim_{\lambda\to 0} \int f_\lambda(x)[\phi(x) - \phi(0)]\, dx\end{aligned}$$

$$\begin{aligned}\left|\int f_\lambda(x)\,[\phi(x) - \phi(0)]\, dx\right| &\leq \int_{|x|\leq r} |f_\lambda(x)\,[\phi(x) - \phi(0)]|\, dx \\ &\quad + \int_{|x|\geq r} |f_\lambda(x)\,[\phi(x) - \phi(0)]|\, dx \\ &\leq \sup_{|x|\leq r} |\phi(x) - \phi(0)| \int_{|x|\leq r} f_\lambda(x)\, dx \\ &\quad + \sup_{|x|\geq r} |\phi(x) - \phi(0)| \int_{|x|\geq r} f_\lambda(x)\, dx \\ &\leq \sup_{|x|\leq r} |\phi(x) - \phi(0)| + M \int_{|\xi|\geq r/\lambda} f(\xi) d\xi\end{aligned}$$

where M is the maximum value of $|\phi(x) - \phi(0)|$ on $\mathbf{R}^n$.

Let $\varepsilon > 0$ be arbitrary. Because ϕ is continuous at 0 and f is integrable on $\mathbf{R}^n$, we can make the first term in the above expression less than $(1/2)\varepsilon$ by choosing r small enough; then we choose λ small enough so that the second term is less than $(1/2)\varepsilon$. □

In Theorem 2.7 we can obviously replace $\mathbf{R}^n$ by any open subset of $\mathbf{R}^n$ containing the origin.

Example 2.18 Using the equality

$$\int_{-\infty}^{\infty} \frac{dx}{1+x^2} = \pi$$

we define $f(x) = 1/(\pi(1+x^2))$ and use Theorem 2.7 to obtain the delta-convergent sequence in $\mathbf{R}$,

$$f_\lambda(x) = \frac{1}{\lambda} \frac{1}{\pi[1+(x/\lambda)^2]} = \frac{\lambda}{\pi\,(x^2+\lambda^2)}$$

by choosing $\lambda = 1/k$.

In the next example we use Theorem 2.7 to construct another delta-convergent sequence, which has special significance for the treatment of the heat equation (see Example 3.22).

Example 2.19 From the equality $\int_{-\infty}^{\infty} \exp(-x^2)\,dx = \pi^{1/2}$, we obtain

$$\begin{aligned}\int_{\mathbf{R}^n} \exp(-|x|^2)\,dx &= \int_{\mathbf{R}^n} \prod_{k=1}^{n} \exp(-x_k^2)\,dx_k \\ &= \prod_{k=1}^{n} \int_{-\infty}^{\infty} \exp(-x_k^2)\,dx_k \\ &= \pi^{n/2}\end{aligned}$$

Replacing the positive parameter λ in Theorem 2.7 by $\lambda^{1/2}$, we conclude that the function

$$f_\lambda(x) = (\pi\lambda)^{-n/2} \exp(-|x|^2/\lambda)$$

which is defined on $\mathbf{R}^n$ for all positive values of λ, converges to δ as $\lambda \to 0$.

2.5 MULTIPLICATION BY SMOOTH FUNCTIONS

We have no way of defining the product of two distributions as a natural extension of the product of two functions, but we can define the product of a C^∞ function with a distribution. For any $T \in \mathcal{D}'(\Omega)$ and $f \in C^\infty(\Omega)$ we define fT as the linear functional

$$(fT)(\phi) = T(f\phi) \qquad \phi \in \mathcal{D}(\Omega)$$

The right-hand side of this equation makes sense because the product $f\phi$ is in $\mathcal{D}(\Omega)$, and if the sequence ϕ_k converges to 0 in $\mathcal{D}(\Omega)$ then the sequence $f\phi_k$ also converges to 0 in $\mathcal{D}(\Omega)$. Therefore

$$(fT)(\phi_k) = T(f\phi_k) \to 0$$

and fT is a continuous linear functional on $\mathcal{D}(\Omega)$.

When T is a regular distribution defined by the locally integrable function g, we have

$$\begin{aligned}(fT_g)(\phi) &= \langle g, f\phi\rangle \\ &= \int gf\phi \\ &= \langle fg, \phi\rangle\end{aligned}$$

since fg is also locally integrable. Thus $fT_g = T_{fg}$ in this case.

Using the definition of the derivative of $T \in \mathcal{D}'(\Omega)$, it is a straightforward calculation to show that the ordinary rules of differentiating a product of two functions apply to fT when $f \in C^\infty(\Omega)$:

$$\begin{aligned}\partial_k(fT)(\phi) &= -fT(\partial_k\phi) \qquad \phi \in \mathcal{D}(\Omega) \\ &= -T(f\partial_k\phi) \\ &= -T(\partial_k(f\phi) - (\partial_k f)\phi) \\ &= f\partial_k T(\phi) + (\partial_k f)T(\phi)\end{aligned}$$

Therefore

$$\partial_k(fT) = (\partial_k f)T + f\partial_k T$$

More generally, we can use induction to show that Leibnitz' formula

$$\partial^\alpha(fT) = \sum_{\beta=0}^{\alpha} \frac{\alpha!}{\beta!(\alpha - \beta)!} (\partial^\beta f)(\partial^{(\alpha-\beta)}T) \qquad \alpha \in \mathbf{N}_0^n \tag{2.7}$$

remains valid for any $f \in C^\infty(\Omega)$ and $T \in \mathcal{D}'(\Omega)$. Note that the summation is over the *multi-indices* from $(0, \ldots, 0)$ to $(\alpha_1, \ldots, \alpha_n)$.

Example 2.20 The product $\sin x\ \delta$ is the distribution defined on $\mathcal{D}(\mathbf{R})$ by

$$\langle \sin x\ \delta, \phi\rangle = \langle \delta, \sin x\ \phi\rangle = \sin 0\ \phi(0) = 0$$

On the other hand, $\sin x\ \delta'$ is given by

$$\begin{aligned}\langle \sin x\ \delta', \phi\rangle &= \langle \delta', \sin x\ \phi\rangle \\ &= -\langle \delta, \cos x\ \phi + \sin x\ \phi'\rangle \\ &= -\phi(0)\end{aligned}$$

2.6 LOCAL PROPERTIES OF DISTRIBUTIONS

It does not make sense to assign a value to a distribution at a given point in Ω, but we can define what it means for a distribution to vanish on an open subset of Ω.

Definition For any $T \in \mathcal{D}'(\Omega)$ and any open subset G of Ω, we say that $T = 0$ on G if $T(\phi) = 0$ for every $\phi \in \mathcal{D}(G)$.

With this definition we can say that $T \in \mathcal{D}'(\Omega)$ is zero if $T = 0$ on Ω and that $T_1, T_2 \in \mathcal{D}'(\Omega)$ are equal if $T_1 - T_2 = 0$ on Ω. Referring to Example 2.20 we may therefore conclude that $\sin x\ \delta = 0$ and $\sin x\ \delta' = -\delta$ on $\mathbf{R}$. Earlier on, in example (2.3), we interpreted the equality $T = T_f$ on $\mathcal{D}(\mathbf{R} - \{0\})$ to mean that T is represented by f on $\mathbf{R} - \{0\}$, i.e. that $T = T_f$ on $\mathbf{R} - \{0\}$.

Example 2.21 Let $I = (a,b)$ be any interval in $\mathbf{R}$, including $\mathbf{R}$ itself. If $T \in \mathcal{D}'(I)$ is such that $T' = 0$, then we shall prove that T must be a constant.

Since $T'(\phi) = -T(\phi') = 0$ for all $\phi \in \mathcal{D}(I)$, we see that T vanishes at every test function which can be expressed as the derivative of some function in $\mathcal{D}(I)$. Let $\mathcal{D}_0(I)$ be the subspace of $\mathcal{D}(I)$ characterized by the condition that $\phi \in \mathcal{D}_0(I)$ if and only if there exists a $\psi \in \mathcal{D}(I)$ such that $\phi = \psi'$. It is then a simple matter to verify that $\phi \in \mathcal{D}_0(I)$ if and only if

$$\int_a^b \phi(x)\, dx = 0$$

This condition is clearly necessary; on the other hand, if it is satisfied then, by defining

$$\psi(x) = \int_a^x \phi(t)\, dt$$

we see that $\psi \in \mathcal{D}(I)$ and $\psi' = \phi$.

Now $T(\phi) = 0$, by hypothesis, for every $\phi \in \mathcal{D}_0(I)$. Let ϕ_0 be a fixed function in $\mathcal{D}(I)$ such that

$$\int_a^b \phi_0(x)\,dx = 1$$

Then, given any $\phi \in \mathcal{D}(I)$, the function

$$\phi - \left(\int_a^b \phi(x)\,dx\right)\phi_0$$

lies in $\mathcal{D}_0(I)$, and therefore

$$T\left(\phi - \phi_0 \int_a^b \phi(x)\,dx\right) = 0$$

or

$$T(\phi) = c \int_a^b \phi(x)\,dx$$

where c is the constant $T(\phi_0)$. This last equation implies that T is the constant function c.

Suppose $T \in \mathcal{D}'(I)$ satisfies $T' = c_1$ for some constant c_1, and define $S \in \mathcal{D}'(I)$ by $S = c_1x$. Then $(T - S)' = 0$, and therefore $T = c_1x + c_2$ for some constant c_2. If $T^{(m)} = 0$, we can use induction to conclude that T is a polynomial of degree $\leq m - 1$.

It would be natural now to define the support of a distribution $T \in \mathcal{D}'(\Omega)$ as the complement, in Ω, of the largest open set on which T vanishes. But first we have to show that such a largest open set exists. This will follow from Theorem 2.9, whose proof requires a few preliminary concepts.

Definition For any $f \in L^1_{\text{loc}}(\mathbf{R}^n)$ and $\phi \in C^\infty_K(\mathbf{R}^n)$, where K is a compact subset of $\mathbf{R}^n$, we define the *convolution* of f and ϕ as the function

$$\int f(x - y)\,\phi(y)\,dy = \int f(y)\,\phi(x - y)\,dy$$

which will be denoted by $(f * \phi)(x)$.

Note that $f * \phi$ is also defined if ϕ is merely continuous with compact support in $\mathbf{R}^n$, but it is not necessarily defined when supp ϕ is not compact, unless, of course, supp f is compact.

The C^∞ function

$$\alpha(x) = \begin{cases} \exp\left(-\dfrac{1}{1-|x|^2}\right) & \text{on } |x| < 1 \\ 0 & \text{on } |x| \geq 1 \end{cases}$$

has support in the closed unit ball $\bar{B}\ (0,1)$ and its integral over $\mathbf{R}^n$ is a finite positive number. The function

$$\beta(x) = \alpha(x)\left[\int \alpha(x)\, dx\right]^{-1}$$

is another C^∞ function with support in $\bar{B}\ (0,1)$, which satisfies $\int\beta(x)\, dx = 1$. Let

$$\beta_\lambda(x) = \frac{1}{\lambda^n}\beta\left(\frac{x}{\lambda}\right) \tag{2.8}$$

for any positive number λ. Then $\beta_\lambda \in \mathcal{D}\ (\mathbf{R}^n)$, supp $(\beta_\lambda) = \bar{B}\ (0,\lambda)$, and

$$\int \beta_\lambda(x)\, dx = \int \beta(x)\, dx = 1$$

Theorem (2.8)

(i) If $f \in L^1_{\text{loc}}\ (\mathbf{R}^n)$ then $f * \beta_\lambda \in C^\infty\ (\mathbf{R}^n)$.

(ii) If $f \in L^1\ (\mathbf{R}^n)$ with compact support K, then supp $(f * \beta_\lambda)$ is contained in a neighborhood of K defined by

$$K_\lambda = \bigcup_{x\in K} \bar{B}(x,\lambda) = K + \bar{B}(0,\lambda).$$

(iii) If $f \in C^0\ (\mathbf{R}^n)$ then, as $\lambda \to 0$, $f * \beta_\lambda \to f$ uniformly on every compact subset of $\mathbf{R}^n$.

Proof.

$$\text{(i)}\ (f * \beta_\lambda)(x) = \int f(y)\, \beta_\lambda(x-y)\, dy$$
$$= \int_{B(x,\lambda)} f(y)\, \beta_\lambda(x-y)\, dy$$

Since $B(x,\lambda)$ is bounded and β_λ is infinitely differentiable, $f * \beta_\lambda \in C^\infty(\mathbf{R}^n)$

(ii) When $x \not\in K_\lambda$ then

$$d(x,K) = \inf_{y \in K} |x - y| > \lambda$$

and so $\beta_\lambda(x - y) = 0$ for all $y \in K$. Consequently

$$(f * \beta_\lambda)(x) = \int_K f(y)\, \beta_\lambda(x - y)\, dy = 0$$

(iii) Since f is continuous on $\mathbf{R}^n$, it is uniformly continuous on any compact subset E of $\mathbf{R}^n$. Therefore, given $\varepsilon > 0$, there is a $\delta > 0$ such that

$$|f(x - y) - f(x)| < \varepsilon$$

for all $x \in E$ and all $y \in B(0,\delta)$. We then have

$$\begin{aligned} |(f * \beta_\lambda)(x) - f(x)| &= \left| \int [f(x - y) - f(x)]\, \beta_\lambda(y)\, dy \right| \\ &\leq \int_{B(0,\lambda)} |f(x - y) - f(x)|\, \beta_\lambda(y)\, dy \\ &< \varepsilon \end{aligned}$$

for all $x \in E$, provided $\lambda \leq \delta$. □

In this proof the only properties of β_λ that were used are: $\beta_\lambda \in C_0^\infty(\mathbf{R}^n)$, supp $\beta_\lambda \subset \overline{B}(0,\lambda)$, and $\int \beta_\lambda(x)\, dx = 1$. Hence β_λ may be replaced in the statement of Theorem 2.8 by any function with these properties. The theorem indicates that the convolution of f with β_λ smoothes out the discontinuities in f while preserving its general shape. For that reason the sequence of functions $f_k = f * \beta_{1/k}$ is called a *regularizing sequence*, or a *regularization*, of f. This property of the convolution product with a smooth function will become even more apparent when applied to distributions in the next chapter.

Corollary (1) If K is a compact subset of $\Omega \subset \mathbf{R}^n$, then there is a $\phi \in \mathcal{D}(\Omega)$ such that $0 \leq \phi \leq 1$ and $\phi = 1$ on K.

Proof. There is no loss of generality in taking Ω to be bounded. With $\delta = (1/3)d(K,\partial\Omega)$, let K_δ be the δ-neighborhood of K and I_{K_δ} be the characteristic function of K_δ. Then the C^∞ function

$$\phi(x) = (I_{K_\delta} * \beta_\delta)(x) = \int_{K_\delta} \beta_\delta(x - y)\, dy$$

is such that $\phi = 1$ on K, $0 \leq \phi \leq 1$ on $K_{2\delta}$, and $\phi = 0$ outside $K_{2\delta}$. □

Corollary (2) $\mathcal{D}(\mathbf{R}^n)$ is a dense subspace of $C_0^0(\mathbf{R}^n)$ with the identity map from $\mathcal{D}(\mathbf{R}^n)$ to $C_0^0(\mathbf{R}^n)$ continuous.

Proof. If the sequence ϕ_k converges in $\mathcal{D}(\mathbf{R}^n)$ to ϕ then there is a compact set $K \subset \mathbf{R}^n$ such that supp $\phi_k \subset K$ for all k, and ϕ_k converges uniformly to ϕ on K. But that implies $\phi_k \to \phi$ in $C_0^0(\mathbf{R}^n)$. Hence the identity map from $\mathcal{D}(\mathbf{R}^n)$ to C_0^0 is continuous.

On the other hand, if ϕ is any function in $C_0^0(\mathbf{R}^n)$ with supp $\phi = K$ then the sequence $\phi_k = \phi * \beta_{1/k}$ is supported in $K + \bar{B}(0,1)$ and, by Theorem 2.8, converges uniformly to ϕ on $K + \bar{B}(0,1)$. □

Theorem 2.9 If $\{G_\alpha : \alpha \in A\}$ is a collection of open subsets of Ω, and $T \in \mathcal{D}'(\Omega)$ is zero on every G_α, then T is zero on the union $\cup_{\alpha\in A} G_\alpha$.

Proof. Let $G = \cup G_\alpha$ and ϕ be in $\mathcal{D}(G)$ with supp $\phi = K$. Since $\{G_\alpha\}$ is an open covering of the compact set K, it contains a finite subcovering of K which may be denoted, after relabeling if necessary, by $G_1, \ldots, G_m$. For every $k \in \{1, \ldots, m\}$ we choose a compact set $K_k \subset G_k$ so that

$$K \subset \bigcup_{k=1}^{m} \overset{\circ}{K}_k$$

and we choose $\phi_k \in \mathcal{D}(G_k)$ so that $0 \leq \phi_k \leq 1$ and $\phi_k = 1$ on K_k. Now let

$$\psi_1 = \phi_1$$

$$\psi_k = \phi_k\,(1 - \phi_1) \cdots (1 - \phi_{k-1}) \qquad k = 2, \ldots, m$$

Then $\psi_k \in \mathcal{D}(G_k)$, $0 \leq \psi_k \leq 1$ for every $k \in \{1, \ldots, m\}$ and

$$\sum_{k=1}^{m} \psi_k = 1$$

on a neighborhood of K. Therefore

$$\phi = \sum_{k=1}^{m} \phi\psi_k$$

Since $\phi\psi_k \in \mathcal{D}(G_k)$ and $T = 0$ on G_k, we have

$$T(\phi) = \sum_{k=1}^{m} T(\phi\psi_k) = 0 \qquad \square$$

The set of functions $\{\psi_1, \ldots, \psi_m\}$ is called a C^∞ *partition of unity* subordinate to the open cover $\{G_1, \ldots, G_m\}$ of K.

Definition The *support* of $T \in \mathfrak{D}'(\Omega)$ is the complement in Ω of the largest open subset of Ω where $T = 0$.

The support of δ, for example, is $\{0\}$ because $\langle\delta,\phi\rangle = 0$ for every ϕ in $\mathfrak{D}(\Omega - \{0\})$. Note, however, that if T is a distribution and f is a C^∞ function which vanishes on supp T, it does not necessarily follow that $fT = 0$, for we have seen that $x\delta' = -\delta$. But when f vanishes on a *neighborhood* of supp T then we may conclude that $fT = 0$.

Theorem 2.10 Every distribution with compact support is of finite order.

Proof. If $T \in \mathfrak{D}'(\Omega)$ and supp T is compact, then there exists a $\psi \in \mathfrak{D}(\Omega)$ such that $\psi = 1$ on some open set containing supp T. For any $\phi \in \mathfrak{D}(\Omega)$ the support of $\phi - \psi\phi$ does not intersect supp T, which means

$$\text{supp } (\phi - \psi\phi) \subset \Omega - \text{supp } T$$

Therefore $T(\phi - \psi\phi) = 0$ or $T(\phi) = T(\psi\phi)$.

Suppose that supp $\psi = K$. By Theorem 2.4 there is a nonnegative integer m and a constant M_1 such that

$$T(\phi) \leq M_1|\phi|_m$$

for all $\phi \in \mathfrak{D}_K$. From the Leibnitz formula (2.7) for the derivative of $\psi\phi$, there is a constant M_2 such that

$$|\psi\phi|_m \leq M_2|\phi|_m$$

for all $\phi \in \mathfrak{D}(\Omega)$. For this choice of ψ and for every $\phi \in \mathfrak{D}(\Omega)$ we have

$$\begin{aligned}|T(\phi)| &= |T(\psi\phi)| \\ &\leq M_1|\psi\phi|_m \\ &\leq M_1M_2|\phi|_m\end{aligned} \qquad \square$$

A distribution of finite order, however, does not necessarily have compact support, since any locally integrable function defines a distribution of order 0. $\Sigma_0^\infty \delta_k^{(k)}$ is an example of a distribution of infinite order.

Example 2.22 The linear combination of derivatives of the Dirac measure on $\mathbf{R}^n$

$$T = \sum_{|\alpha| \leq m} c_\alpha \, \partial^\alpha \delta$$

has support $\{0\}$, and

$$T(\phi) = \sum_{|\alpha| \leq m} c_\alpha (-1)^{|\alpha|} \, \partial^\alpha \phi(0) \qquad \phi \in \mathcal{D}(\mathbf{R}^n)$$

Note that

$$\begin{aligned} |T(\phi)| &\leq \sum_{|\alpha| \leq m} |c_\alpha \partial^\alpha \phi(0)| \\ &\leq M_{mn} \max_{|\alpha| \leq m} |\partial^\alpha \phi(0)| \\ &\leq M_{mn} \, |\phi|_m \end{aligned}$$

where M_{mn} is a positive constant which depends on m and n. This implies that the order of T is m. In Chapter 3 we shall prove that, conversely, every distribution with support $\{0\}$ is a finite linear combination of derivatives of δ.

If $\mathbf{R}^n$ is replaced in Theorem 2.8 by any open set $\Omega \subset \mathbf{R}^n$, the conclusions of the theorem and its corollaries remain valid because the proofs depend only on the local properties of the functions involved.

In Exercises 1.11 and 1.12 we defined the space $L^p(I)$, where $I = (a,b)$ and $1 \leq p < \infty$, to be the completion of $C_0^0(I)$ in the norm

$$f \mapsto \|f\|_p = \left[\int_I |f(x)|^p \, dx \right]^{1/p}$$

More generally, for any open set $\Omega \subset \mathbf{R}^n$, we can also define $L^p(\Omega)$ to be the completion of $C_0^0(\Omega)$ in the norm $\|\cdot\|_p$ with I replaced by Ω. It is a standard result of real analysis that this definition is equivalent to the usual definition of $L^p(\Omega)$ as the linear space of measurable functions on Ω with finite norm $\|\cdot\|_p$ (see [4] for example). Since convergence in $C_0^0(\Omega)$ implies convergence in $L^p(\Omega)$, and in view of Corollary (2) to Theorem 2.8, we conclude that

Theorem (2.11) $\mathcal{D}(\Omega)$ is a dense subspace of $L^p(\Omega)$ for $1 \leq p < \infty$, with the identity map from $\mathcal{D}(\Omega)$ to $L^p(\Omega)$ continuous.

Theorem 2.11 does not prescribe how the approximating sequence in $\mathcal{D}$ of an L^p function is constructed. But, based on Theorem 2.8, we should expect the regularizing sequence to play such a role. This is shown to be the case in the next example. Recalling Definition 2.8 of β_λ, we shall use (γ_k) to denote the sequence $(\beta_{1/k})$.

Example 2.23 Let $u \in L^p(\mathbf{R}^n)$. First we shall show that $\|u * \gamma_k\|_p \leq \|u\|_p$, $1 \leq p < \infty$, and then conclude that $u * \gamma_k \to u$ in $L^p(\mathbf{R}^n)$ for any $u \in L^p(\mathbf{R}^n)$.

(i) Let $1 < p < \infty$. Then

$$\|u * \gamma_k\|_p^p = \int \left| \int \gamma_k(y)\, u(x-y)\, dy \right|^p dx$$

We can write $u\gamma_k = (u\gamma_k^{1/p})(\gamma_k^{1/q})$, where $1/p + 1/q = 1$, and use Hölder's inequality to obtain

$$\int \gamma_k(y)\, |u(x-y)|\, dy \leq \left[\int \gamma_k(y)\, |u(x-y)|^p\, dy \right]^{1/p} \left[\int \gamma_k(y)\, dy \right]^{1/q}$$

Since $\int \gamma_k(y)\, dy = 1$, we have

$$\begin{aligned} \|u * \gamma_k\|_p^p &\leq \int\int \gamma_k(y)\, |u(x-y)|^p\, dy\, dx \\ &= \int \gamma_k(y) \left[\int |u(x-y)|^p\, dx \right] dy \qquad \text{by Fubini's theorem} \\ &= \int \gamma_k(y)\, \|u\|_p^p\, dy \\ &= \|u\|_p^p \end{aligned}$$

If $p = 1$,

$$\begin{aligned} \|u * \gamma_k\|_1 &\leq \int\int \gamma_k(y)\, |u(x-y)|\, dy\, dx \\ &= \int \gamma_k(y) \left[\int |u(x-y)|\, dx \right] dy \\ &= \|u\|_1 \end{aligned}$$

Hence $\|u * \gamma_k\|_p \leq \|u\|_p$ for all $p \in [1, \infty)$.

(ii) Let $u \in L^p(\mathbf{R}^n)$ and $\varepsilon > 0$ be arbitrary. Since C_0^0 is dense in L^p we can choose $\phi \in C_0^0(\mathbf{R}^n)$ such that $\|u - \phi\|_p < \varepsilon$. It then follows, from (i), that

$$\|u * \gamma_k - \phi * \gamma_k\|_p = \|(u - \phi) * \gamma_k\|_p \leq \|u - \phi\|_p < \varepsilon$$

But since $\phi * \gamma_k$ and ϕ are supported in the compact set $K = \operatorname{supp} \phi +$

$\overline{B}(0,1)$, and since $\phi * \gamma_k \to \phi$ uniformly on K (by Theorem 2.8), we can write

$$\|\phi * \gamma_k - \phi\|_p = \left[\int_K |(\phi * \gamma_k)(x) - \phi(x)|^p \, dx\right]^{1/p}$$

$$\leq \sup_{x \in K} |(\phi * \gamma_k)(x) - \phi(x)| \left[\int_K dx\right]^{1/p}$$

$$< \varepsilon$$

provided k is large enough. Thus

$$\|u * \gamma_k - u\|_p \leq \|u * \gamma_k - \phi * \gamma_k\|_p + \|\phi * \gamma_k - \phi\|_p + \|\phi - u\|_p < 3\varepsilon$$

If $u \in L^p_{\text{loc}}(\mathbf{R}^n)$ and K is any compact set in $\mathbf{R}^n$, then the function $v = uI_K$, where I_K is the characteristic function of K, lies in $L^p(\mathbf{R}^n)$ and the sequence $v * \gamma_k$ converges to v in $L^p(\mathbf{R}^n)$. This means that $u * \gamma_k \to u$ in the L^p norm on every compact subset of $\mathbf{R}^n$. If convergence in $L^p_{\text{loc}}(\mathbf{R}^n)$ is understood in this sense, then we can say that $\mathcal{D}(\mathbf{R}^n)$ is also dense in $L^p_{\text{loc}}(\mathbf{R}^n)$.

Recall that every locally integrable function f defines a distribution T_f. If $f = g$ a.e. then clearly $T_f = T_g$. The next example shows that, conversely, if $T_f = T_g$ for two locally integrable functions f and g, then $f = g$ a.e.

Example 2.24 Given $f \in L^1_{\text{loc}}(\mathbf{R}^n)$ such that $T_f = 0$ in $\mathcal{D}'(\mathbf{R}^n)$, we shall prove that $f = 0$ a.e.

(i) First assume that $f \in L^1(\mathbf{R}^n)$. Then

$$(f * \gamma_k)(x) = \int f(y)\, \gamma_k(x - y)\, dy = 0$$

because $\gamma_k(x - y)$ lies in $\mathcal{D}(\mathbf{R}^n)$ for every fixed x and $T_f = 0$ on $\mathcal{D}(\mathbf{R}^n)$. Hence $f = \lim f * \gamma_k = 0$ in $L^1(\mathbf{R}^n)$, which means that $f = 0$ a.e.

(ii) Now let $f \in L^1_{\text{loc}}(\mathbf{R}^n)$ and suppose that K is any compact set in $\mathbf{R}^n$. Choose $\psi \in \mathcal{D}(\mathbf{R}^n)$ such that $0 \leq \psi \leq 1$ and $\psi = 1$ on K. This is always possible by Corollary (1) to Theorem 2.8. Thus $\psi f \in L^1(\mathbf{R}^n)$.

If $\phi \in \mathcal{D}(\mathbf{R}^n)$ then $T_{\psi f}(\phi) = T_f(\psi\phi) = 0$ by hypothesis. From part (i), therefore, we conclude that $\psi f = 0$ a.e. in $\mathbf{R}^n$, which implies that $f = 0$ a.e. on K. K being arbitrary, this means that $f = 0$ a.e.

Note how this proof depends essentially on the fact that $\mathcal{D}(\mathbf{R}^n)$ is dense in $L^1_{\text{loc}}(\mathbf{R}^n)$.

2.7 DISTRIBUTIONS OF FINITE ORDER

Recall that $\mathcal{D}^m(\Omega)$, $m \in \mathbf{N}_0$, is the linear space $C_0^m(\Omega)$ equipped with the inductive limit topology of $\{C_K^m(\Omega) : K \subset \Omega\}$. This is the locally convex topology in which a set is open if and only if its intersection with $C_K^m(\Omega)$ is open, for every compact $K \subset \Omega$. Here, of course $C_K^m(\Omega)$ carries its natural locally convex topology defined by the seminorms in equation (1.4). This topology on $\mathcal{D}^m(\Omega)$ is weaker than the topology of $\mathcal{D}(\Omega)$, with the result that the inclusion $\mathcal{D}(\Omega) \subset \mathcal{D}^m(\Omega)$ is in fact a continuous injection. Consequently, the dual space $\mathcal{D}^{m\prime}(\Omega)$ is a subspace of $\mathcal{D}'(\Omega)$ which can be characterized by the following theorem:

Theorem (2.12) $\mathcal{D}^{m\prime}(\Omega)$ consists of all the distributions in $\mathcal{D}'(\Omega)$ of order $\leq m$.

Proof. If $T \in \mathcal{D}^{m\prime}(\Omega)$ then, by definition, there is a constant M such that $|T(\phi)| \leq M|\phi|_m$ for all $\phi \in D^m(\Omega)$. The restriction of T to $\mathcal{D}(\Omega)$ is therefore a distribution of order $\leq m$.

Conversely, if $T \in \mathcal{D}'(\Omega)$ is of order m, then there is a constant M such that $|T(\phi)| \leq M|\phi|_m$ for all $\phi \in \mathcal{D}(\Omega)$. Since $\mathcal{D}(\Omega) \subset \mathcal{D}^m(\Omega) \subset \mathcal{D}^0(\Omega)$ and $\mathcal{D}(\Omega)$ is dense in $\mathcal{D}^0(\Omega) = C_0^0(\Omega)$, by Corollary (2) to Theorem 2.8, $\mathcal{D}(\Omega)$ is dense in $\mathcal{D}^m(\Omega)$, and the continuous linear functional T may be extended by continuity to $\mathcal{D}^m(\Omega)$ with the inequality $|T(\phi)| \leq M|\phi|_m$ still valid. Thus $T \in \mathcal{D}^{m\prime}(\Omega)$. □

Let $\mathcal{D}_F(\Omega)$ be the set $\cap_{m=0}^{\infty} C_0^m(\Omega) = C_0^\infty(\Omega)$ equipped with the weakest topology in which the identity map $i_m : \mathcal{D}_F(\Omega) \rightarrow \mathcal{D}^m(\Omega)$ is continuous for every $m \in \mathbf{N}_0$. This is a locally convex topology which is induced by the topologies of $\mathcal{D}^m(\Omega)$ under the inverse maps i_m^{-1}. If $\mathcal{U}_m$ is a base of 0-neighborhoods in $\mathcal{D}^m(\Omega)$, the finite intersections of the sets $i_m^{-1}(U_m)$, where $U_m \in \mathcal{U}_m$ and $m \in \mathbf{N}_0$, form a base of 0-neighborhoods for the topology of $\mathcal{D}_F$. This topology on $\mathcal{D}_F$ is called the *projective limit* of the topologies of $\{\mathcal{D}^m(\Omega)\}$. It is in fact a dual topology to the inductive limit which was defined in Section 2.1, in a sense which will soon be clarified.

Thus, although $\mathcal{D}_F(\Omega)$ and $\mathcal{D}(\Omega)$ represent the same *set*, namely $C_0^\infty(\Omega)$, they are different topological spaces, the topology of $\mathcal{D}(\Omega)$ being the stronger. To see this, take any sequence ϕ_k in $\mathcal{D}(\Omega)$ which converges to ϕ. By Theorem 2.3 there is a compact set $K \subset \Omega$ which contains supp ϕ_k, for all k, and $|\phi_k - \phi|_m \rightarrow 0$ for all m. But this implies that $\phi_k \rightarrow \phi$ in $\mathcal{D}^m(\Omega)$ for every m, and hence $\phi_k \rightarrow \phi$ in $\mathcal{D}_F(\Omega)$. Consequently the identity map from $\mathcal{D}(\Omega)$ to $\mathcal{D}_F(\Omega)$ is continuous and the corresponding dual spaces $\mathcal{D}_F'(\Omega)$ and $\mathcal{D}'(\Omega)$

are related by the (proper) inclusion $\mathfrak{D}'_F(\Omega) \subset \mathfrak{D}'(\Omega)$. This is brought out by the next theorem, which also characterizes the elements of $\mathfrak{D}'_F(\Omega)$.

Theorem 2.13 $\mathfrak{D}'_F(\Omega)$ consists of all the distributions in $\mathfrak{D}'(\Omega)$ of finite order. In other words,

$$\mathfrak{D}'_F(\Omega) = \bigcup_{m=0}^{\infty} \mathfrak{D}^{m\prime}(\Omega).$$

Proof. If $T \in \mathfrak{D}'(\Omega)$ is of finite order, say m, then $T \in \mathfrak{D}^{m\prime}(\Omega)$ by Theorem 2.12. Its restriction to $C_0^\infty(\Omega)$ is therefore continuous in the topology of $\mathfrak{D}_F(\Omega)$, and hence $T \in \mathfrak{D}'_F(\Omega)$.

Now let $T \in \mathfrak{D}'_F(\Omega)$. Then there is a neighborhood U of $0 \in \mathfrak{D}_F(\Omega)$ such that $|T(\phi)| \leq M$ for all $\phi \in U$. But U contains a neighborhood of the form $U_1 \cap \cdots \cap U_k \cap C_0^\infty(\Omega)$ where U_i is a neighborhood of $0 \in \mathfrak{D}^{m_i}(\Omega)$, i.e. of the form $\{\phi \in C_0^\infty(\Omega) : |\phi|_{m_i} \leqslant \varepsilon_i\}$. If $\varepsilon = \min\{\varepsilon_1, \ldots, \varepsilon_k\}$ and $m = \max\{m_1, \ldots, m_k\}$, then

$$\{\phi \in C_0^\infty(\Omega) : |\phi|_m \leqslant \varepsilon\} \subset \{\phi \in C_0^\infty(\Omega) : |\phi|_{m_i} \leqslant \varepsilon_i\} \subset U$$

Thus the linear functional T satisfies

$$|T(\phi)| \leqslant M \qquad \text{for all } \phi \in C_0^\infty(\Omega) \text{ such that } |\phi|_m \leqslant \varepsilon$$

which means that T is a continuous linear functional on $C_0^\infty(\Omega)$ in the topology induced by $\mathfrak{D}^m(\Omega)$. Therefore T is a distribution of order m. □

With $\mathfrak{D}'_F(\Omega) = \bigcup \mathfrak{D}^{m\prime}(\Omega)$ we can also define a topology on $\mathfrak{D}'_F(\Omega)$ through the inductive limit of the topologies of $\{\mathfrak{D}^{m\prime}(\Omega)\}$. It turns out that this topology coincides with the one that we have already defined on $\mathfrak{D}'_F(\Omega)$ as the dual of $\mathfrak{D}_F(\Omega)$ (see [2]). Since the topology of $\mathfrak{D}_F(\Omega)$ is the projective limit of the topologies of $\{\mathfrak{D}^m(\Omega)\}$, we see that these two methods of defining a topology are naturally suited to dual spaces, in this case $\mathfrak{D}_F(\Omega)$ and $\mathfrak{D}'_F(\Omega)$.

With this background we can now say a little more about the connection between distributions and measures on open subsets of $\mathbf{R}^n$. A *Radon measure* on an open set $\Omega \subset \mathbf{R}^n$ is an element of $\mathfrak{D}^{0\prime}(\Omega)$, i.e., a continuous linear functional on $\mathfrak{D}^0(\Omega) = C_0^0(\Omega)$, or a distribution of order 0. As a continuous linear functional on $C_0^0(\Omega)$ it is also represented, according to the Riesz representation theorem, by a regular Borel measure on Ω, as we have already indicated in Section 2.2.

Definition A real linear functional T on a real linear space of functions F is said to be *positive* if $T(f) \geq 0$ whenever $f \in F, f \geq 0$.

If T is a positive linear functional on $C_0^0(\Omega)$, we can show that T is continuous on $C_0^0(\Omega)$ and hence defines a (positive) Radon measure on Ω. By the corollary to Theorem 2.3 it suffices to prove that, if $\phi_k \in C_0^0(\Omega)$, with supp ϕ_k contained in some fixed compact set $K \subset \Omega$ and

$$|\phi_k|_0 = \sup_{x \in K} |\phi_k(x)| \to 0 \qquad \text{as } k \to \infty,$$

then $T(\phi k) \to 0$. Choose $\psi \in C_0^0(\Omega)$ such that $0 \leq \psi \leq 1$ and $\psi = 1$ on K. Then $|\phi_k| \leq |\phi_k|_0 \psi$, and therefore

$$-|\phi_k|_0 \psi \leq \phi_k \leq |\phi_k|_0 \psi$$

Since T is a positive linear functional, we conclude that

$$-|\phi_k|_0 T(\psi) \leq T(\phi_k) \leq |\phi_k|_0 T(\psi)$$

and hence $\lim T(\phi_k) = 0$.

Using the definition above we can therefore say that $T \in \mathcal{D}'(\Omega)$ is positive, and write $T \geq 0$, if $T(\phi) \geq 0$ for all $\phi \geq 0$ in $\mathcal{D}(\Omega)$. Example 2.6 illustrates that not every distribution is a (Radon) measure. But in the next example we shall show that every positive distribution is a (positive) Radon measure.

Example 2.25 Let T be a positive distribution on Ω. To show that T is a Radon measure on Ω we first need to extend T from $\mathcal{D}(\Omega)$ to $\mathcal{D}^0(\Omega)$, and then prove that it is continuous as a linear functional on $\mathcal{D}^0(\Omega)$.

Let $\phi \in \mathcal{D}^0(\Omega)$ be arbitrary. By Corollary (2) to Theorem 2.8, there is a sequence $\phi_k \in \mathcal{D}(\Omega)$ such that $\phi_k \to \phi$ in $\mathcal{D}^0(\Omega)$. This means, according to the corollary to Theorem 2.3, that supp ϕ_k is contained in some compact set $K \subset \Omega$ and that $|\phi_k - \phi|_0 \to 0$ on K.

Choose $\psi \in \mathcal{D}(\Omega)$ such that $0 \leq \psi \leq 1$ in Ω and $\psi = 1$ on K. Now

$$|\phi_j(x) - \phi_k(x)| \leq |\phi_j - \phi_k|_0 \psi(x)$$

Since $T \geq 0$ and $|\phi_j - \phi_k|_0 \to 0$ as $j,k \to \infty$ we obtain

$$|T(\phi_j) - T(\phi_k)| \leq |\phi_j - \phi_k|_0 T(\psi) \to 0$$

Hence $\lim T(\phi_k)$ exists and we denote it by $T(\phi)$.

If $\psi_k \in \mathcal{D}(\Omega)$ is another sequence which tends to ϕ in $\mathcal{D}^0(\Omega)$, the above argument implies that $T(\phi_k) - T(\psi_k) \to 0$ as $k \to \infty$. Therefore the limit $T(\phi)$ does not depend on the particular choice of the sequence ϕ_k, and we have shown that T has an extension to $\mathcal{D}^0(\Omega)$, which is clearly linear.

To show that T is continuous on $\mathcal{D}^0(\Omega)$ it suffices to show that T is positive on $\mathcal{D}^0(\Omega)$. Let ϕ be any function in $C_0^0(\Omega)$ and $\phi \geqslant 0$. Then $\phi * \gamma_k \in \mathcal{D}(\Omega)$, if k is large enough, and $\phi * \gamma_k \geqslant 0$. Hence $T(\phi * \gamma_k) \geqslant 0$. As $k \to \infty$, $\phi * \gamma_k \to \phi$ in $\mathcal{D}^0(\Omega)$ and $T(\phi) = \lim T(\phi * \gamma_k) \geqslant 0$.

2.8 DISTRIBUTIONS DEFINED BY POWERS OF x

In this section we look at some examples of distributions on $\mathbf{R}$ which are defined by powers of x, and in Section 4.7 we shall extend some of our results to $\mathbf{R}^n$ by considering powers of $|x|$. Such functions will, in general, be singular at $x = 0$. If the singularity is such that the function is locally integrable, then it defines a distribution by equation (2.2); otherwise the resulting integral has to be regularized so as to make it converge. Example 2.12 indicates how this can be done for x^{-1} and x^{-2}. Here we take a slightly different approach.

First we shall define what it means for a distribution which depends on a complex parameter λ to be analytic on a subset Λ of the complex plane. Let

$$\lambda \mapsto T_\lambda$$

be a mapping from $\mathbf{C}$ to $\mathcal{D}'(\Omega)$. We shall say that T_λ *is analytic in* Λ *if the function*

$$\lambda \mapsto \langle T_\lambda, \phi\rangle$$

is analytic in Λ for every $\phi \in \mathcal{D}(\Omega)$. That this definition extends the usual meaning of the analytic dependence of a function on a complex variable λ is clear from the equality

$$\lim_{\lambda \to \lambda_0} \frac{\langle T_\lambda, \phi\rangle - \langle T_{\lambda_0}, \phi\rangle}{\lambda - \lambda_0} = \left\langle \lim_{\lambda \to \lambda_0} \frac{T_\lambda - T_{\lambda_0}}{\lambda - \lambda_0}, \phi \right\rangle \qquad \phi \in \mathcal{D}(\Omega) \tag{2.9}$$

Thus, when T_λ is a function of λ which is differentiable at λ_0, $\langle T_\lambda, \phi\rangle$ is differentiable at λ_0. When the limit in equation (2.9) exists, it defines a distribution which is denoted by $(\partial_\lambda T)_{\lambda_0}$.

Now $|x|^\lambda = \exp(\lambda \log |x|)$ defines a regular distribution which is analytic in λ on $\operatorname{Re} \lambda > -1$, because the function $|x|^\lambda$ is locally integrable when $\operatorname{Re} \lambda > -1$. We shall exploit the above definition of analyticity in the distributional sense to extend $|x|^\lambda$ as a distribution beyond Re $\lambda > -1$ by continuing the function $\langle |x|^\lambda, \phi \rangle$ analytically for every $\phi \in \mathcal{D}(\mathbf{R})$ to a larger connected subset of the complex λ-plane.

Example 2.26 Let

$$x_+^\lambda = \begin{cases} x^\lambda & x > 0 \\ 0 & x \leq 0 \end{cases}$$

This is a locally integrable function for Re $\lambda > -1$ which determines the distribution

$$\langle x_+^\lambda, \phi \rangle = \int_0^\infty x^\lambda \phi(x)\, dx \qquad \phi \in \mathcal{D}(\mathbf{R}) \tag{2.10}$$

Since the right-hand side of equation (2.10) is analytic in Re $\lambda > -1$ for every $\phi \in \mathcal{D}(\mathbf{R})$, the distribution x_+^λ is also analytic in Re $\lambda > -1$. When Re $\lambda > -1$ and $\phi \in \mathcal{D}(\mathbf{R})$, we can write

$$\begin{aligned} \int_0^\infty x^\lambda \phi(x)\, dx &= \int_0^\infty x^\lambda [\phi(x) - \phi(0) H(1 - x)] dx + \phi(0) \int_0^1 x^\lambda\, dx \\ &= \int_0^1 x^\lambda [\phi(x) - \phi(0)] dx + \int_1^\infty x^\lambda \phi(x) dx + (\lambda + 1)^{-1} \phi(0) \end{aligned} \tag{2.11}$$

Now the first integral on the right-hand side of (2.11) is convergent provided Re $\lambda > -2$, for then the function

$$x^\lambda [\phi(x) - \phi(0)] = x^{1+\lambda} \frac{\phi(x) - \phi(0)}{x}$$

is integrable on $[0,1]$, ϕ being differentiable at 0. The second term in equation (2.11) is finite for all $\lambda \in \mathbf{C}$, and the third term for all $\lambda \neq -1$. Therefore x^λ can be continued analytically to

$$\Lambda = \{\lambda \in \mathbf{C} : \text{Re } \lambda > -2,\ \lambda \neq -1\}$$

Note that the subtraction of $\phi(0) H(1 - x)$ from $\phi(x)$ is designed to reduce the order of the singularity of x^λ at $x = 0$ while still preserving compact support for the integrand. This process can be repeated with higher order terms from the Taylor expansion of ϕ at $x = 0$. In the mth step,

$$\langle x_+^\lambda, \phi \rangle = \int_0^\infty x^\lambda \left[\phi(x) - \left\{ \phi(0) + \cdots + \frac{x^{m-1}}{(m-1)!} \right.\right.$$

$$\left.\left. \times \phi^{(m-1)}(0) \right\} H(1-x) \right] dx + \sum_{k=1}^{m} \phi^{(k-1)}(0) \int_0^1 \frac{x^{\lambda+k-1}}{(k-1)!}\, dx$$

$$= \int_0^1 x^\lambda \left[\phi(x) - \phi(0) - x\phi'(0) - \cdots - \frac{x^{m-1}}{(m-1)!} \right.$$

$$\left. \times \phi^{(m-1)}(0) \right] dx$$

$$+ \int_1^\infty x^\lambda \phi(x)\, dx + \sum_{k=1}^{m} \frac{1}{(\lambda+k)(k-1)!} \phi^{(k-1)}(0) \qquad (2.12)$$

where the first integral on the right-hand side converges for $\operatorname{Re} \lambda > -m-1$ because the expression

$$\phi(x) - \sum_{k=1}^{m} x^{k-1} \phi^{(k-1)}(0)/(k-1)!$$

is of order x^m in the neighborhood of 0, and when it is multiplied by x^λ, with $\operatorname{Re} \lambda > -m-1$, the resulting function is integrable in the neighborhood of $x = 0$. Since the third term on the right-hand side of (2.12) has simple poles at $\lambda = -1, -2, \ldots, -m$, the distribution x_+^λ may be continued analytically into $\operatorname{Re} \lambda > -m-1$, $\lambda \neq -1, -2, \ldots, -m$. But m is arbitrary, so x_+^λ is defined for all $\lambda \in \mathbf{C} - \mathbf{Z}^-$, where $\mathbf{Z}^-$ is the set of negative integers.

Example 2.27 The function

$$x_-^\lambda = \begin{cases} (-x)^\lambda & x < 0 \\ 0 & x \geqslant 0 \end{cases}$$

is also in $\mathcal{D}'(\mathbf{R})$ for Re $\lambda > -1$, where it is analytic, and may be continued analytically into Re $\lambda > -m-1$, $\lambda \neq -1, \ldots, -m$ through the equation

$$\begin{aligned}
\langle x_-^\lambda, \phi \rangle &= \int_{-\infty}^{0} (-x)^\lambda \, \phi(x) \, dx \\
&= \int_0^\infty x^\lambda \, \phi(-x) \, dx \\
&= \int_0^1 x^\lambda \left[\phi(-x) - \sum_{k=1}^{m} \frac{(-x)^{k-1}}{(k-1)!} \phi^{(k-1)}(0) \right] dx \\
&\quad + \int_1^\infty x^\lambda \, \phi(-x) dx + \sum_{k=1}^{m} \frac{(-1)^{k-1}}{(\lambda + k)(k-1)!} \phi^{(k-1)}(0)
\end{aligned} \tag{2.13}$$

Hence the distribution x_-^λ is also defined for all $\lambda \in \mathbf{C} - \mathbf{Z}^-$.

Example 2.28 Let

$$|x|^\lambda = \begin{cases} x^\lambda & x > 0 \\ (-x)^\lambda & x < 0 \\ 0 & x = 0 \end{cases}$$

$$\operatorname{sgn} x = \begin{cases} 1 & x > 0 \\ -1 & x < 0 \\ 0 & x = 0 \end{cases}$$

Then

$$|x|^\lambda \operatorname{sgn} x = \begin{cases} x^\lambda & x > 0 \\ -(-x)^\lambda & x < 0 \\ 0 & x = 0 \end{cases}$$

It then follows that

$$|x|^\lambda = x_+^\lambda + x_-^\lambda$$

$$|x|^\lambda \operatorname{sgn} x = x_+^\lambda - x_-^\lambda$$

and the poles of $|x|^\lambda$ are located at $\lambda = -1, -3, -5 \ldots$, while $|x|^\lambda \operatorname{sgn} x$ has its poles at $\lambda = -2, -4, -6, \ldots$ because of the cancellations that result from adding and subtracting the sums in (2.12) and (2.13). Thus $|x|^{-2m} = x^{-2m}$ is defined at $m = 1,2,3, \ldots$ and $|x|^{-2m-1} \operatorname{sgn} x = x^{-2m-1}$ is defined at $m = 0,1,2, \ldots$. Consequently x^{-n} is defined in $\mathcal{D}'(\mathbf{R})$ for all positive integers n.

In the strip $-m-1 < \operatorname{Re} \lambda < -m$, we obtain from (2.12) and (2.13)

$$\langle x_+^\lambda, \phi\rangle = \int_0^\infty x^\lambda \left[\phi(x) - \phi(0) - x\phi'(0) - \cdots - \frac{x^{m-1}}{(m-1)!}\phi^{(m-1)}(0)\right] dx$$

$$\langle x_-^\lambda, \phi\rangle = \int_0^\infty x^\lambda \left[\phi(-x) - \phi(0) + x\phi'(0) - \cdots + (-1)^{m-1}\frac{x^{m-1}}{(m-1)!}\phi^{(m-1)}(0)\right] dx$$

Therefore

$$\langle |x|^\lambda, \phi\rangle = \int_0^\infty x^\lambda \left[\phi(x) + \phi(-x) - 2\left\{\phi(0) + \frac{x^2}{2!}\phi''(0) + \cdots + \frac{x^{2m-2}}{(2m-2)!}\phi^{(2m-2)}(0)\right\}\right] dx \tag{2.14}$$

in the strip $-2m-1 < \operatorname{Re} \lambda < -2m + 1$, and

$$\langle |x|^\lambda \operatorname{sgn} x, \phi\rangle = \int_0^\infty x^\lambda \left[\phi(x) - \phi(-x) - 2\left\{x\phi'(0) + \cdots + \frac{x^{2m-1}}{(2m-1)!}\phi^{(2m-1)}(0)\right\}\right] dx \tag{2.15}$$

in $-2m - 2 < \operatorname{Re} \lambda < -2m$.

Thus we obtain explicit expressions for x^{-2m} and x^{-2m-1} by direct substitution into the above formulas. Setting $\lambda = -2m$ in equation (2.14) and $\lambda = -2m - 1$ in (2.15), we arrive at

$$\langle x^{-2m},\phi\rangle = \int_0^\infty x^{-2m}\Big[\phi(x) + \phi(-x) - 2\Big\{\phi(0) + \frac{x^2}{2!}\phi''(0) + \cdots + \frac{x^{2m-2}}{(2m-2)!}\phi^{(2m-2)}(0)\Big\}\Big]\,dx$$

$$\langle x^{-2m-1},\phi\rangle = \int_0^\infty x^{-2m-1}\Big[\phi(x) - \phi(-x) - 2\Big\{x\phi'(0) + \cdots + \frac{x^{2m-1}}{(2m-1)!}\phi^{(2m-1)}(0)\Big\}\Big]\,dx$$

or

$$\langle x^{-n},\phi\rangle = \int_{-\infty}^\infty x^{-n}\Big[\phi(x) - \Big\{\phi(0) + x\phi'(0) + \frac{x^2}{2!}\phi''(0) + \cdots + \frac{x^{n-1}}{(n-1)!}\phi^{(n-1)}(0)\Big\}\Big]\,dx$$

for all positive integers n.

In particular, when $n = 1$ and $n = 2$, we retrieve some earlier results:

$$\begin{aligned}\langle x^{-1},\phi\rangle &= \int_{-\infty}^\infty x^{-1}[\phi(x) - \phi(0)]\,dx \\ &= \lim_{\varepsilon\to 0}\int_{|x|\geq\varepsilon} x^{-1}[\phi(x) - \phi(0)]\,dx \\ &= pv\int_{-\infty}^\infty x^{-1}\phi(x)\,dx \\ &= \Big\langle \frac{d}{dx}\log|x|,\phi\Big\rangle\end{aligned}$$

$$\begin{aligned}\langle x^{-2},\phi\rangle &= \int_{-\infty}^\infty x^{-2}[\phi(x) - \phi(0) - x\phi'(0)]\,dx \\ &= \lim_{\varepsilon\to 0}\int_{|x|\geq\varepsilon} x^{-2}[\phi(x) - \phi(0) - x\phi'(0)]\,dx\end{aligned}$$

$$= \lim_{\varepsilon\to 0} \int_{|x|\geq\varepsilon} x^{-2}[\phi(x) - \phi(0)]\, dx$$

$$= \lim_{\varepsilon\to 0} x^{-1}[\phi(x) - \phi(0)] \Big|_{-\varepsilon}^{\varepsilon} + \lim_{\varepsilon\to 0} \int_{|x|\geq\varepsilon} x^{-1}\, \phi'(x)\, dx$$

$$= 0 + \langle x^{-1},\phi'\rangle$$

$$= \left\langle -\frac{d}{dx}\, x^{-1},\phi\right\rangle$$

Therefore $x^{-1} = (d/dx) \log |x|$ and $x^{-2} = -(d/dx)\, x^{-1}$ in $\mathcal{D}'(\mathbf{R})$.

In general, we have

$$\frac{d}{dx}\, x_+^{\lambda} = \lambda x_+^{\lambda-1} \qquad \lambda \neq 0,-1,-2,\ldots \tag{2.16}$$

$$\frac{d}{dx}\, x_-^{\lambda} = \lambda x_-^{\lambda-1} \qquad \lambda \neq 0,-1,-2,\ldots \tag{2.17}$$

$$\frac{d}{dx}\, |x|^{\lambda} = \lambda |x|^{\lambda-1} \operatorname{sgn} x \qquad \lambda \neq -1,-3,-5,\ldots \tag{2.18}$$

$$\frac{d}{dx}\, |x|^{\lambda} \operatorname{sgn} x = \lambda |x|^{\lambda-1} \qquad \lambda \neq 0,-2,-4,\ldots \tag{2.19}$$

Equation (2.16) is clearly valid when $\operatorname{Re} \lambda > 0$, where

$$\left\langle \frac{d}{dx}\, x_+^{\lambda},\phi\right\rangle = -\langle x_+^{\lambda},\phi'\rangle = \langle \lambda x_+^{\lambda-1},\phi\rangle$$

Now both sides of this equality can be continued analytically into $\mathbf{C} - \{0,-1,-2,\ldots\}$. Similarly (2.17) is valid in $\mathbf{C} - \{0,-1,-2,\ldots\}$, and the other formulas easily follow. Equations (2.18) and (2.19) imply that

$$\frac{d}{dx}\, x^{-n} = -nx^{-n-1} \qquad n \in \mathbf{N} \tag{2.20}$$

When $\lambda = 0$ we have $x_+^{\lambda} = H(x)$ and $x_-^{\lambda} = H(-x)$, whose derivatives are δ and $-\delta$ respectively. Hence $x^0 = H(x) + H(-x)$ and $(d/dx)\, x^0 = \delta - \delta$, and equation (2.20) is also valid when $n = 0$.

The formulas (2.16) and (2.17) for the derivatives $x_\pm^{\lambda}$ are not valid when $\lambda = -n$ and $n \in \mathbf{N}_0$, since the formulas (2.12) and (2.13) for $x_\pm^{\lambda}$ have poles at these values of λ. In order to define the distribution x_+^{-n}, for example, we have to resort to a more suitable regularization of the integral $\int_0^\infty x^{-n}\phi(x)\, dx$

than that provided by (2.12). In effect, we need to get rid of the divergent term

$$\phi^{(m-1)}(0)\,\frac{1}{(\lambda + m)(m - 1)!}$$

on the right hand side of that equation. One way to achieve that is to modify the definition (2.12) to

$$\langle x_+^{-n},\phi\rangle = \int_0^\infty x^{-n}\Big[\phi(x) - \phi(0) - \cdots - \frac{x^{n-1}}{(n-1)!}\phi^{(n-1)}(0)H(1 - x)\Big]\,dx \qquad (2.21)$$

where the integral is finite because the integrand remains bounded as $x \to 0^+$ and is of order x^{-2} as $x \to \infty$. Thus

$$\Big\langle \frac{d}{dx}x_+^{-n},\phi\Big\rangle = -\langle x_+^{-n},\phi'\rangle$$

$$= -\int_0^\infty x^{-n}\Big[\phi'(x) - \phi'(0) - \cdots - \frac{x^{n-2}}{(n-2)!}\phi^{(n-1)}(0) - \frac{x^{n-1}}{(n-1)!}\phi^{(n)}(0)H(1 - x)\Big]\,dx$$

$$= -\int_0^1 x^{-n}\Big[\phi'(x) - \phi'(0) - \cdots - \frac{x^{n-1}}{(n-1)!} \times \phi^{(n)}(0)\Big]\,dx - \int_1^\infty x^{-n}\Big[\phi'(x) - \phi'(0) - \cdots - \frac{x^{n-2}}{(n-2)!}\phi^{(n-1)}(0)\Big]\,dx$$

Integrating by parts,

$$\Big\langle \frac{d}{dx}x_+^{-n},\phi\Big\rangle = \int_0^\infty (-n)x^{-n-1}\Big[\phi(x) - \phi(0) - \cdots - \frac{x^n}{n!}\phi^{(n)}(0)H(1 - x)\Big]dx + \frac{\phi^{(n)}(0)}{n!}$$

Hence

$$\frac{d}{dx} x_+^{-n} = -nx_+^{-n-1} + \frac{(-1)^n}{n!} \delta^{(n)} \tag{2.22}$$

Similarly,

$$\frac{d}{dx} x_-^{-n} = -nx_-^{-n-1} - \frac{(-1)^n}{n!} \delta^{(n)} \tag{2.23}$$

Given a distribution $T \in \mathscr{D}'(\mathbf{R})$, any distribution S which satisfies $S'(\phi) = T(\phi)$ for every $\phi \in \mathscr{D}(\mathbf{R})$ is called a *primitive* of T. Using formulas (2.16) to (2.20), the primitive distributions of x_+^λ, x_-^λ, and x^{-n} may be determined, up to an additive constant, by the familiar rule for integrating a power of x. In fact, the process of extending the definitions of x_+^λ and x_-^λ that we have outlined above is equivalent to regularizing the corresponding divergent integrals

$$\int_0^\infty x^\lambda \, \phi(x)dx \qquad \int_0^\infty x^\lambda \, \phi(-x)dx$$

in the sense of Section 2.3. That is, we have defined the distribution x^λ for values of λ where the integral $\int_0^\infty x^\lambda \phi(x)dx$ is divergent by a process of successive integrations of x^λ to the point when it becomes locally integrable. This process, of course, involves differentiating ϕ at every step and choosing suitable integration constants along the way, the choice being dictated by convergence criteria. To the extent that this still leaves some arbitrariness in our choice, we can have more than one definition of the distributions $x_\pm^\lambda$. Naturally, we choose the definition which best preserves the properties of the corresponding *functions* $x_\pm^\lambda$. For example the distribution x_+^{-n} could have been defined by

$$\langle x_+^{-n}, \phi \rangle = \int_0^\infty x^{-n} \left[\phi(x) - \left\{ \phi(0) + \cdots + \frac{x^{n-1}}{(n-1)!} \phi^{(n-1)}(0) \right\} H(1-x) \right] dx$$

instead of the expression (2.21), but then the formula (2.22) for its derivative would have to be modified accordingly, in fact by adding the sum $\Sigma_0^{n-1}(-1)^k \delta^{(k)}/k!$ to the right-hand side.

Finally, a word of warning regarding the notation we have used is in order. The extension of the function x_+^λ as a distribution outside Re $\lambda > -1$ should not be confused with the *function* x_+^λ which is well defined on $\mathbf{R} - \{0\}$ for all values of λ. In other words, the distribution x_+^λ and the function x_+^λ are quite different when Re $\lambda < -1$. The first is obtained from the second by discarding the divergent part of the integral $\int x_+^\lambda \phi(x)\,dx$.. For values of λ in $\mathbf{C} - \mathbf{Z}^-$ this was achieved by analytic continuation. $|x|^\lambda$ and $|x|^\lambda$ sgn x were then defined on the negative odd and even integers, respectively, by cancelling out the poles at those points, while x^λ was defined for all values of λ in $\mathbf{Z}^-$ by cancelling out the singularities of $|x|^{-2m}$ and $|x|^{-2m-1}$, a process which would not be justified in integration theory. On the other hand, the definition of x_+^{-n}, as given by equation (2.21), involved simply discarding the pole at $\lambda = -n$. In all this it should come as no surprise that the more we have to change the integral $\int x^\lambda \phi(x)\,dx$ to arrive at a definition of $\langle x^\lambda,\phi\rangle$, the more the resulting distribution will deviate from the function x^λ. Thus the differentiation formulas (2.16) to (2.20) conform to the familiar rule for differentiating a power of x, but in (2.22) this rule was lost.

Some books use the notation $[x_+^\lambda]$ or pf x_+^λ, the "pseudo-function" x_+^λ, to designate the distribution x_+^λ. When $\lambda = -1$ the distribution x^{-1} is often denoted by *pv* $1/x$ because $\langle x^{-1},\phi\rangle$ turns out to be the Cauchy principal value of the divergent integral $\int x^{-1}\phi(x)\,dx$, as we found in Example 2.12. In our convention, T_f was used to denote the distribution determined by the function f. But in this section we did not distinguish between the powers of x as functions and as distributions because here we are only interested in them as distributions, and it would greatly complicate the notation if we had to make the distinction.

EXERCISES

2.1 Prove that $C_0^\infty(\Omega)$ is dense in $C_0^m(\Omega)$ for every $m \in \mathbf{N}_0$.

2.2 Prove that $C_0^\infty(\Omega)$ is a closed subspace of $C^\infty(\Omega)$.

2.3 Since the mapping $\phi \mapsto |\phi|_m = \sup\{|\partial^\alpha\phi(x)| : x \in \Omega, |\alpha| \leq m\}$, $m \in \mathbf{N}_0$, defines a norm on $\mathcal{D}(\Omega)$, why is $\mathcal{D}(\Omega)$ not a normed linear space?

2.4 Prove that the partial differential operator ∂_k, $k \in \{1, \ldots, n\}$, is a continuous linear mapping from $C_0^\infty(\Omega)$ to $C_0^\infty(\Omega)$.

2.5 Prove Leibnitz' formula (2.7).

2.6 Since the Heaviside function on $\mathbf{R}$ is differentiable on $\mathbf{R} - \{0\}$, where its derivative vanishes, it follows that, for every $\phi \in C_0^\infty(\mathbf{R})$, $H'\phi = 0$ almost everywhere. Consequently $\int H'\phi = 0$ for all $\phi \in C_0^\infty(\mathbf{R})$. How is this consistent with the result of example (2.8) where $\langle H',\phi\rangle = \langle\delta,\phi\rangle = \phi(0)$?

2.7 Indicate which of the following functions define distributions in $\mathcal{D}'(\mathbf{R}^n)$ by the formula (2.2):

(a) $(\sin x)/x$ on $\mathbf{R}$

(b) $(x + iy)^{-1}$ on $\mathbf{R}^2$

(c) $(x^2 + y^2)^{-1}$ on $\mathbf{R}^2$.

2.8 Differentiate each of the following distributions on $\mathbf{R}$:

(a) $\sin^2 x\, H(x)$

(b) $\cos x\, \delta$

(c) $(\sin x)/|x|$

(d) $\log x_+ = \begin{cases} \log x & \text{for } x > 0 \\ 0 & \text{for } x \leq 0 \end{cases}$

2.9 Why does the product $H\delta$ not make sense on $\mathbf{R}$? Since $H^2 = H$, what is wrong with writing $(H^2)' = 2HH' = 2H\delta$?

2.10 With $x \in \mathbf{R}$ and c constant, show that

(a) $\cos cx\, \delta' = \delta'$

(b) $$x^n\delta^{(m)} = \begin{cases} \dfrac{(-1)^n m!}{(m-n)!}\,\delta^{(m-n)} & \text{if } n \leq m \\ 0 & \text{if } n > m \end{cases}$$

(c) $$e^{cx}\,\delta_\xi^{(m)} = e^{c\xi}\sum_{k=0}^{m}\binom{m}{k}(-c)^{m-k}\,\delta_\xi^{(k)}.$$

2.11 Verify that each of the following sequences converges to δ as $n \to \infty$:

(a) $f_n(x) = \frac{1}{2}\, n\, e^{-n|x|}$

(b) $$g_n(x) = \frac{1}{2\pi}\int_{-n}^{n} e^{ix\xi}\, d\xi = \frac{\sin nx}{\pi x} \qquad x \in \mathbf{R} - \{0\}$$

(c) $$h_n(x) = \frac{1}{2n|x|^{(n-1)/n}} \qquad x \in \mathbf{R} - \{0\}.$$

2.12 Regularize the divergent integral $\int_0^\infty x^{-3/2}\,\phi(x)\,dx$, $\phi \in \mathcal{D}(\mathbf{R})$, and thereby show that the equality $(d/dx)(x_+^{-1/2}) = (-1/2)x_+^{-3/2}$ holds in $\mathcal{D}'(\mathbf{R})$.

2.13 Let

$$(x \pm i0)^\lambda = \lim_{y \to \pm 0} (x + iy)^\lambda$$
$$= \lim_{y \to \pm 0} (x^2 + y^2)^{\lambda/2} \exp\,[i\lambda \arg\,(x + iy)],$$

where $-\pi < \arg(x + iy) < \pi$ and λ is a complex number. Show that, for Re $\lambda > -1$, $(x \pm i0)^\lambda$ are distributions which are related to $x_\pm^\lambda$ by

$$(x \pm i0)^\lambda = x_+^\lambda + e^{\pm i\lambda\pi} x_\pm^\lambda.$$

Use the analytic continuations of x_+^λ and x_-^λ to extend $(x \pm i0)^\lambda$ into $\lambda \neq -1, -2, \ldots$, and then show that the singularities at these points are removable. Thus the distributions $(x + i0)^\lambda$ and $(x - i0)^\lambda$ are entire functions of λ.

2.14 Determine which of the following functionals on $\mathcal{D}(\mathbf{R})$ are distributions:

(a) $\phi \to \phi(0)]^2$

(b) $\phi \to \sum_{k=0}^{\infty} \phi(k)$

(c) $\phi \to \sum_{k=0}^{\infty} \phi^{(k)}(0)$

(d) $\phi \to \sum_{k=0}^{\infty} \phi^{(k)}(k)$

(e) $\phi \to \int_{-\infty}^{\infty} |\phi(x)|\, dx$

2.15 Determine the order of each distribution in exercise (2.14).

2.16 Show that the order of $pv\,(1/x)$ in the interval $(-a,a)$, with $a > 0$, is 1, and that its order in any open interval not containing 0 is 0.

2.17 Show that the distribution $T \in \mathcal{D}'(-1,1)$ defined by $T(\phi) = \int_{-1}^{1} |x|\phi'(x)\, dx$ is of order 0. Determine the measure which represents T.

2.18 If $T \in \mathcal{D}'(\mathbf{R})$ is of order m show that T' is of order $m + 1$.

2.19 Prove the corollary to Theorem 2.3.

2.20 Prove that the topology of $\mathcal{D}'_F(\Omega)$ as the dual space of $\mathcal{D}_F(\Omega)$ coincides with the inductive limit of the topologies of $\mathcal{D}^{m\prime}(\Omega)$.

2.21 Verify formula (2.23).

3

Distributions with Compact Support and Convolutions

3.1 THE DUAL SPACE OF $C^\infty(\Omega)$

We shall use $\mathscr{E}(\Omega)$ to denote the Fréchet space $C^\infty(\Omega)$ topologized by the system of seminorms

$$p_{m,K}(\phi) = \sup \{|\partial^\alpha \phi(x)| : x \in K, |\alpha| \leq m\}$$

where $m \in \mathbf{N}_0$ and K runs through the compact subsets of Ω. We have already seen in Chapter 1 that $\mathscr{D}_K$ is a closed subspace of $\mathscr{E}(\Omega)$ for every compact set $K \subset \Omega$, and that the topology defined on $\mathscr{D}_K$ is the subspace topology inherited from $\mathscr{E}(\Omega)$. Therefore the identity map from $\mathscr{D}_K$ to $\mathscr{E}(\Omega)$ is continuous, so that every continuous linear functional on $\mathscr{E}(\Omega)$ is also a continuous linear functional on $\mathscr{D}_K$. Since this is true for every $K \subset \Omega$, this means, by Theorem 2.1, that every continuous linear function on $\mathscr{E}(\Omega)$ is a continuous linear functional on $\mathscr{D}(\Omega)$. Thus every element in $\mathscr{E}'(\Omega)$, the dual space of $\mathscr{E}(\Omega)$, is a distribution, and the following theorem characterizes such distributions.

Theorem 3.1 For any open set Ω in $\mathbf{R}^n$, $\mathscr{E}'(\Omega)$ is the subspace of $\mathscr{D}'(\Omega)$ consisting of distributions with compact support.

Proof. We have already seen that every element of $\mathscr{E}'(\Omega)$ defines a distribution in $\mathscr{D}'(\Omega)$, and we shall now show that different elements in $\mathscr{E}'(\Omega)$

define different distributions by showing that $\mathfrak{D}(\Omega)$ is a dense subspace of $\mathscr{E}(\Omega)$.

Let (K_i) be an increasing sequence of compact subsets of Ω whose union is Ω, and (ϕ_i) a corresponding sequence in $\mathfrak{D}(\Omega)$ such that $\phi_i = 1$ on a neighborhood of K_i. For any $\psi \in \mathscr{E}(\Omega)$ the function $\psi_i = \phi_i\psi$ is in $\mathfrak{D}(\Omega)$ and $\psi_i \to \psi$ in $\mathscr{E}(\Omega)$. Thus if $T = 0$ in $\mathfrak{D}'(\Omega)$ then $T(\phi) = 0$ for all $\phi \in \mathfrak{D}(\Omega)$. For any $\psi \in \mathscr{E}(\Omega)$, $T(\psi) = \lim T(\psi_i) = 0$, hence $T = 0$ in $\mathscr{E}'(\Omega)$.

Let $T \in \mathscr{E}'(\Omega)$. Then there is a bounded neighborhood of 0 in $\mathscr{E}(\Omega)$ which is mapped by T into the unit disc in $\mathbf{C}$. By Theorem 1.8, that means there is an integer $m \in \mathbf{N}_0$, a compact set $K \subset \Omega$ and a positive number r such that the neighborhood of 0 in $\mathscr{E}(\Omega)$ defined by

$$U = \{\phi \in \mathscr{E}(\Omega) : p_{m,K}(\phi) < r\}$$

satisfies $|T(\phi)| \leq 1$ for every $\phi \in U$. If $\phi \in \mathscr{E}(\Omega)$ and $p_{m,K}(\phi) = 0$, then $\lambda\phi \in U$ for every $\lambda > 0$ and $|T(\lambda\phi)| = \lambda|T(\phi)| \leq 1$. Therefore $|T(\phi)| \leq 1/\lambda$ for every $\lambda > 0$, which means that $T(\phi) = 0$. Since $p_{m,K}(\phi) = 0$ for every $\phi \in \mathfrak{D}(\Omega - K)$ we conclude that $T = 0$ on $\Omega - K$, that is, supp $T \subset K$. □

Example 3.1 The sequence $T_n = \sum_1^n a^k\delta_k$, with $a > 0$, converges in $\mathfrak{D}'(\mathbf{R})$ but not in $\mathscr{E}'(\mathbf{R})$. To see that, let $\phi \in \mathfrak{D}(\mathbf{R})$. Then there exists an integer m such that $\phi = 0$ outside $[-m,m]$, and

$$\langle T_n,\phi\rangle = \sum_1^m a^k\phi(k) \qquad \text{if } n \geq m$$

Consequently $\lim_{n\to\infty} \langle T_n,\phi\rangle = \sum_1^m a^k\phi(k)$ exists in $\mathfrak{D}'(\mathbf{R})$.

The sequence (T_n) also lies in $\mathscr{E}'(\mathbf{R})$ but it does not converge there, for if we choose the test function $\phi(x) = a^{-x} \in \mathscr{E}(\mathbf{R})$ then $\langle T_n,\phi\rangle = \sum_1^n a^k a^{-k} = n \to \infty$. Thus the infinite sum $\sum_1^\infty a^k\delta_k$ lies in $\mathfrak{D}'(\mathbf{R})$ but not in $\mathscr{E}'(\mathbf{R})$.

Theorem 3.2 Every distribution whose support is $\{0\}$ may be represented by a unique finite linear combination of the derivatives of the Dirac measure δ.

Proof. Suppose $T \in \mathfrak{D}'(\Omega)$, $0 \in \Omega$ and supp $T = \{0\}$. From Theorem 2.10 T has finite order, say m, and from Theorem 3.1 T lies in $\mathscr{E}'(\Omega)$. For any $\phi \in \mathscr{E}(\Omega)$, Taylor's formula gives

$$\phi(x) = \sum_{|\alpha| \leq m} \frac{1}{\alpha!} \partial^\alpha \phi(0)\, x^\alpha + R_m(x) \tag{3.1}$$

where $R_m \in \mathscr{E}(\Omega)$ and $\partial^\alpha R_m(0) = 0$ for all $|\alpha| \leq m$. Since $\partial^\alpha R_m$ is continuous at 0 for every α, the derivatives $|\partial^\alpha R_m(x)|$ with $|\alpha| \leq m$ can be made arbitrarily small by taking $|x|$ small enough. Thus for every $\varepsilon > 0$ there is an $r > 0$ such that

$$|\partial^\alpha R_m(x)| < \varepsilon \tag{3.2}$$

when $x \in B(0,r) = \{x : |x| < r\}$, and $|\alpha| \leq m$.

Using Corollary 1 to Theorem 2.8, we can choose $\psi_r \in \mathscr{D}(\Omega)$ such that $\operatorname{supp} \psi_r \subset B(0,r)$ and $\psi_r = 1$ on $\bar{B}(0,\frac{1}{2}r)$. The function $\phi_r = \psi_r R_m$ lies in $\mathscr{D}(\Omega)$ and, by Leibnitz' formula (2.7), its derivative $\partial^\alpha \phi_r$ is a finite linear combination of products of the form $\partial^{\alpha-\beta}\psi_r \partial^\beta R_m$, as β runs through the multi-indices with $|\beta| \leq |\alpha| \leq m$. Since $\partial^{\alpha-\beta}\psi_r$ is bounded for all $|\alpha| \leq m$, there is a constant M_1 (which depends on r and m) such that

$$\begin{aligned} |\partial^\alpha \phi_r(x)| &\leq M_1 |\partial^\beta R_m(x)|, \qquad |\beta| \leq |\alpha| \leq m \\ &\leq \varepsilon M_1 \end{aligned}$$

for all $x \in \Omega$ by (3.2).

With $R_m = \phi_r$ on a neighborhood of supp T, we can write

$$|T(R_m)| = |T(\phi_r)| \leq M_2\, |\phi_r|_m$$

for some constant M_2, since T is of order m. Thus

$$|T(R_m)| \leq \varepsilon M_1 M_2$$

and, since $\varepsilon > 0$ was arbitrary, we conclude that $T(R_m) = 0$.

Using (3.1), we can therefore write

$$\begin{aligned} T(\phi) &= \sum_{|\alpha| \leq m} \frac{1}{\alpha!} T(x^\alpha)\, \partial^\alpha \phi(0) \\ &= \sum_{|\alpha| \leq m} c_\alpha \partial^\alpha \delta(\phi) \end{aligned}$$

where $c_\alpha = (-1)^{|\alpha|}\, T(x^\alpha)/\alpha!$. The uniqueness of this representation follows from the fact that, if

$$\sum_{|\alpha| \leq m} c_\alpha \partial^\alpha\, \delta(\phi) = \sum_{|\alpha| \leq m} (-1)^{|\alpha|}\, c_\alpha \partial^\alpha \phi(0) = 0$$

for every $\phi \in \mathscr{E}(\Omega)$, then $c_\alpha = 0$ for all $|\alpha| \leq m$, for we can choose $\phi(x) = x^\beta$, $|\beta| \leq m$, to obtain $0 = \sum_{|\alpha|\leq m} (-1)^{|\alpha|} c_\alpha \partial^\alpha \phi(0) = (-1)^{|\beta|} c_\beta \beta!$. □

Example 3.2 Let m be a positive integer and T be a distribution on $\mathbf{R}$. We can show that $x^m T = 0$ if and only if T is a linear combination of $\delta, \delta', \ldots, \delta^{(m-1)}$ with constant coefficients.

Since $x^m T(\phi) = T(x^m \phi)$, $\phi \in \mathscr{D}(\mathbf{R})$, a linear combination of δ and its derivatives up to order $m - 1$ will annihilate $x^m \phi$, so the "if" part is clear.

Now let $T(x^m\phi) = 0$ for every $\phi \in \mathscr{D}(\mathbf{R})$. We wish to prove that supp $T = \{0\}$. To that end, let Ω be an arbitrary open subset of $\mathbf{R} - \{0\}$ and ψ be in $\mathscr{D}(\Omega)$. Then the function defined by

$$\phi(x) = \begin{cases} x^{-m}\psi & \text{on supp } \psi \\ 0 & \text{on } \mathbf{R} - \text{supp } \psi \end{cases}$$

clearly lies in $\mathscr{D}(\mathbf{R})$. But

$$T(\psi) = T(x^m\phi) = 0$$

implies that T vanishes on every open subset of $\mathbf{R} - \{0\}$, and hence on $\mathbf{R} - \{0\}$ itself. Therefore supp $T = \{0\}$ and, by Theorem 3.2, T may be represented by a finite sum of the form

$$T = \sum_{k=0}^{l} c_k \delta^{(k)}$$

It remains to show that $c_k = 0$ for $k \geq m$. Note that $\langle \delta^{(k)}, x^j\phi \rangle = 0$ when $k < j$ and $\langle \delta^{(k)}, x^k\phi \rangle = (-1)^k k! \phi(0)$. Hence if $c_l \neq 0$ for $l \geq m$ in the above representation of T, then for $\phi \in \mathscr{D}(\mathbf{R})$ such that $\phi(0) \neq 0$ we have

$$\begin{aligned} 0 &= \langle x^m T, x^{l-m}\phi \rangle \\ &= \langle T, x^l\phi \rangle \\ &= \sum_{k=0}^{l} \langle c_k \delta^{(k)}, x^l \phi \rangle \\ &= c_l (-1)^l l! \, \phi(0) \end{aligned}$$

a contradiction. Thus $c_k = 0$ for $k \geq m$.

3.2 TENSOR PRODUCT

Let Ω_1 be an open set in $\mathbf{R}^{n_1}$ and Ω_2 be an open set in $\mathbf{R}^{n_2}$. Then the product

$$\Omega_1 \times \Omega_2 = \{(x,y) : x \in \Omega_1, y \in \Omega_2\}$$

is an open set in the Euclidean space $\mathbf{R}^{n_1+n_2} = \mathbf{R}^{n_1} \times \mathbf{R}^{n_2}$. If f is a function on Ω_1 and g a function on Ω_2 then we define the *direct*, or *tensor*, product $f \otimes g$ on $\Omega_1 \times \Omega_2$ by

$$(f \otimes g)(x,y) = f(x)g(y).$$

Clearly $(f \otimes g)\,(x,y) = (g \otimes f)\,(y,x)$ for every pair $(x,y) \in \Omega_1 \times \Omega_2$.

We shall use $C_0^\infty(\Omega_1) \times C_0^\infty(\Omega_2)$ to denote the linear space of functions $\phi(x,y)$ that can be represented as finite sums of products of the form $\phi_1(x)\phi_2(y)$ with $\phi_i \in C_0^\infty(\Omega_i)$, $i = 1,2$. It is clearly a subspace of the linear space $C_0^\infty(\Omega_1 \times \Omega_2)$, and moreover it is a dense subspace. The proof of this last assertion may be found in reference [6].

T_i will denote a distribution in Ω_i. For a fixed $y \in \Omega_2$ the function $\phi(\cdot,y)$ belongs to $C_0^\infty(\Omega_1)$ and T_1 maps $\phi(\cdot,y)$ to the number $T_1(\phi(\cdot,y))$, which we shall denote by $T_1(\phi)(y)$. Thus $T_1(\phi)$ is a function on Ω_2. Similarly, $T_2(\phi)$ is a function on Ω_1. The next theorem shows that $T_1(\phi)$ and $T_2(\phi)$ preserve all the smoothness properties of the test function space $\mathcal{D}$.

Theorem 3.3 If $\phi(x,y) \in \mathcal{D}(\Omega_1 \times \Omega_2)$ and $T_1 \in \mathcal{D}'(\Omega_1)$ then $T_1(\phi) \in \mathcal{D}(\Omega_2)$ and $\partial_y^\beta T_1(\phi) = T_1(\partial_y^\beta \phi)$ for all $\beta \in \mathbf{N}_0^{n_2}$.

Proof. For any point $y \in \Omega_2$, let h be any nonzero real number such that $B(y,2|h|) \subset \Omega_2$, and let $h_k = (0, \ldots, h, \ldots, 0)$ be the point in $\mathbf{R}^{n_2}$ with all coordinates 0 except the kth. Let $\phi \in C_0^\infty(\Omega_1 \times \Omega_2)$. Since the function ϕ is differentiable with respect to y, we can write

$$\phi(x,y + h_k) = \phi(x,y) + \partial_{y_k} \phi(x,y)h + R(x,y,h)$$

where $(1/h)\,|R(x,y,h)| \to 0$ as $h \to 0$. Using the linearity and continuity of T_1 we see that $T_1(\phi(x,y))$ has a kth partial derivative, as a function of y, and that

$$\partial_{y_k} T_1(\phi(\cdot,y)) = T_1(\partial_{y_k} \phi(\cdot,y))$$

and the formula $\partial_y^\beta T_1(\phi) = T_1(\partial_y^\beta \phi)$ follows by induction.

The assumption $\phi \in C_0^\infty(\Omega_1 \times \Omega_2)$ also implies that, for every x in a compact subset of Ω_1, the function $\partial_y^\beta \phi$ is continuous on Ω_2. Hence, by the continuity of T_1, so is $T_1(\partial_y^\beta \phi)$. Since $\phi(x,y)$ has compact support in $\Omega_1 \times \Omega_2$, it is obvious that the function $T_1(\phi(\cdot,y))$ has compact support in Ω_2. □

Corollary 1 If $\phi(x,y) \in \mathscr{E}(\Omega_1 \times \Omega_2)$ and $T_1 \in \mathscr{E}'(\Omega_1)$ then $T_1(\phi) \in \mathscr{E}(\Omega_2)$ and $\partial_y^\beta T_1(\phi) = T_1(\partial_y^\beta \phi)$ for all $\beta \in \mathbf{N}_0^{n_2}$.

This result may be proved by replacing ϕ by $\psi\phi$, where $\psi \in C_0^\infty(\Omega)$ equals 1 on a neighborhood of supp T_1, and using Theorem 3.3.

Corollary 2 If $\phi(x,y) \in C^\infty(\Omega_1 \times \Omega_2)$ has compact support as a function of x and y separately, then $T_1(\phi)(y) \in C^\infty(\Omega_2)$ for every $T_1 \in \mathscr{D}'(\Omega_1)$ and $T_2(\phi)(x) \in C^\infty(\Omega_1)$ for every $T_2 \in \mathscr{D}'(\Omega_2)$.

Example 3.3 (i) Let ϕ, $\psi \in \mathscr{D}(\mathbf{R})$. Their tensor product $(\phi \otimes \psi)(x,y) = \phi(x)\psi(y)$ is in $\mathscr{D}(\mathbf{R}^2)$, and, for any $T \in \mathscr{D}'(\mathbf{R})$,

$$T(\phi \otimes \psi) = T(\phi)\psi(y)$$

is a function in $\mathscr{D}(\mathbf{R})$.

The function $\phi(x + y)$, on the other hand, which lies in $C^\infty(\mathbf{R}^2)$, does not have compact support since $\phi \neq 0$ on the line $x + y = c$ in $\mathbf{R}^2$ whenever $\phi(c) \neq 0$. But, as a function of x and y separately, $\phi(x + y)$ clearly has compact support. Here we obtain

$$\begin{aligned}\langle 1_x, \phi(x,y)\rangle &= \int \phi(x + y)\, dx \\ &= \int \phi(\xi)\, d\xi \\ &= \text{constant}\end{aligned}$$

which is a $C^\infty(\mathbf{R})$ function, in agreement with Corollary 2. See Figure 3.1.

We also have

$$\langle \delta_x, \phi(x + y)\rangle = \phi(y)$$

which lies in $C_0^\infty(\mathbf{R})$. This would seem to suggest that if T_i in Corollary 2 is taken in $\mathscr{E}'(\Omega_i)$ then $T_i(\phi)$ will have compact support.

When $f \in L^1_{\text{loc}}(\Omega_1)$ and $g \in L^1_{\text{loc}}(\Omega_2)$ then $f \otimes g$ is clearly in $L^1_{\text{loc}}(\Omega_1 \times \Omega_2)$. If $\phi_i \in \mathscr{D}(\Omega_i)$ then $\phi_1 \otimes \phi_2 \in \mathscr{D}(\Omega_1 \times \Omega_2)$ and we have

$$\langle f \otimes g, \phi_1 \otimes \phi_2\rangle = \int_{\Omega_1 \times \Omega_2} f(x)g(y)\phi_1(x)\phi_2(y)\, dx\, dy$$

$$= \int_{\Omega_1} f(x)\phi_1(x)\,dx \int_{\Omega_2} g(y)\phi_2(y)\,dy$$

$$= \langle f,\phi_1\rangle\langle g,\phi_2\rangle$$

The next theorem generalizes this result.

Theorem 3.4 If $T_i \in \mathcal{D}'(\Omega_i)$, $i = 1,2$, then there is a unique distribution $T_1 \otimes T_2 \in \mathcal{D}'(\Omega_1 \times \Omega_2)$ defined by

$$(T_1 \otimes T_2)(\phi_1 \otimes \phi_2) = T_1(\phi_1)T_2(\phi_2)$$

for all tensor products $\phi_1 \otimes \phi_2$, where $\phi_i \in \mathcal{D}(\Omega_i)$, and such that

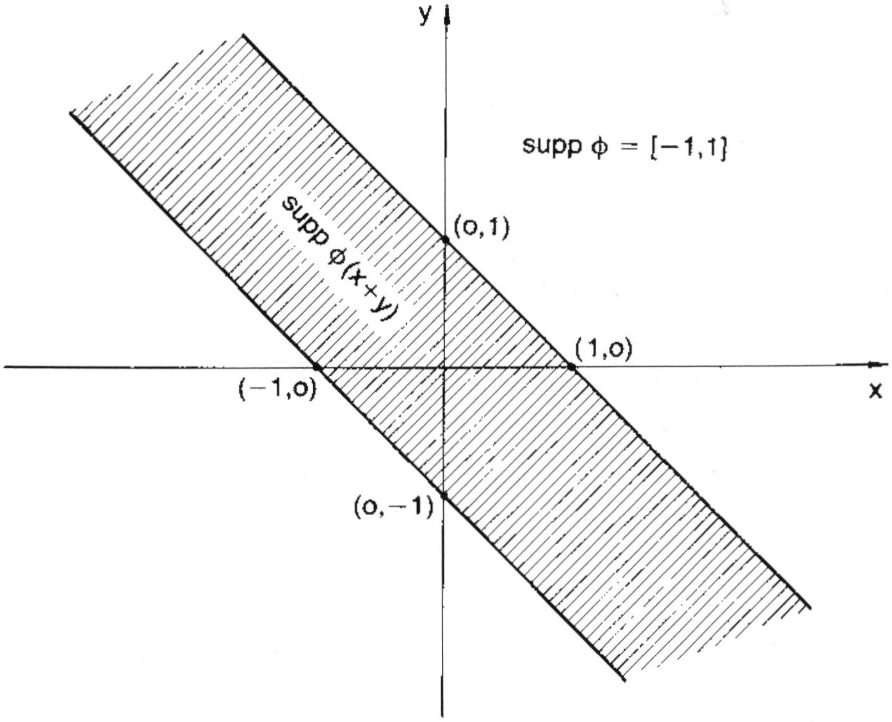

Figure 3.1

$$(T_1 \otimes T_2)(\phi) = T_1(T_2(\phi)) = T_2(T_1(\phi))$$

for all $\phi \in \mathfrak{D}(\Omega_1 \times \Omega_2)$.

Proof. The uniqueness of $T_1 \otimes T_2$ follows from the fact that $\mathfrak{D}(\Omega_1) \times \mathfrak{D}(\Omega_2)$ is dense in $\mathfrak{D}(\Omega_1 \times \Omega_2)$. To show that $T_1 \otimes T_2$ is a distribution of $\Omega_1 \times \Omega_2$, let K_i be a compact subset of Ω_i. By Theorem 2.4 there is a nonnegative integer m_i and a nonnegative constant M_i, $i = 1,2$, such that

$$|T_i(\phi_i)| \leq M_i|\phi_i|_{m_i}$$

for every $\phi_i \in \mathfrak{D}_{K_i}$. If $\phi \in \mathfrak{D}_K$, where $K = K_1 \times K_2$, Theorem 3.3 implies that $T_2(\phi)$ is in $\mathfrak{D}_{K_1}$ and therefore $T_1(T_2(\phi))$ is well defined and satisfies

$$|T_1(T_2(\phi))| \leq M_1|T_2(\phi)|_{m_1} \tag{3.3}$$

Since $\partial_x^\alpha T_2(\phi(x,\cdot)) = T_2(\partial_x^\alpha \phi(x,\cdot))$, we have

$$\begin{aligned}|T_2(\phi(x,\cdot))|_{m_1} &= \sup_{x\in K_1}\{|\partial_x^\alpha T_2(\phi(x,\cdot))| : |\alpha| \leq m_1\}\\ &= \sup_{x\in K_1}\{|T_2(\partial_x^\alpha \phi(x,\cdot))| : |\alpha| \leq m_1\}\\ &\leq M_2 \sup_{\substack{|\alpha|\leq m_1\\ x\in K_1}} |\partial_x^\alpha \phi(x,\cdot)|_{m_2}\\ &\leq M_2 \sup_{x\in K_1, y\in K_2} \{|\partial_x^\alpha \partial_y^\beta \phi(x,y)| : |\alpha| \leq m_1, |\beta| \leq m_2\}\end{aligned}$$

Using (3.3), we obtain

$$|T_1(T_2(\phi))| \leq M_1 M_2 \sup_{\substack{|\gamma|\leq m\\ x\in K}} |\partial^\gamma \phi| = M_1 M_2 |\phi|_m \tag{3.4}$$

where $\gamma = \alpha + \beta$ and $m = m_1 + m_2$. The inequality (3.4), which is true for every $\phi \in \mathfrak{D}_K$ and every $K = K_1 \times K_2 \subset \Omega_1 \times \Omega_2$, implies, by Theorem 2.4, that the linear functional defined on $\mathfrak{D}(\Omega_1 \times \Omega_2)$ by

$$\phi \mapsto T_1(T_2(\phi))$$

is a distribution in $\Omega_1 \times \Omega_2$. By interchanging the roles of T_1 and T_2 in the above argument we conclude that the linear functional defined on $\mathfrak{D}(\Omega_1 \times \Omega_2)$ by

$$\phi \mapsto T_2(T_1(\phi))$$

also lies in $\mathfrak{D}'(\Omega_1 \times \Omega_2)$.

Since

$$\begin{aligned} T_1(T_2(\phi_1 \otimes \phi_2)) &= T_1(\phi_1)T_2(\phi_2) \\ &= T_2(T_1(\phi_1 \otimes \phi_2)) \end{aligned}$$

for all $\phi_i \in \mathfrak{D}(\Omega_i)$, we must have

$$T_1(T7_2(\phi)) = T_2(T_1(\phi))$$

for all $\phi \in \mathfrak{D}(\Omega_1 \times \Omega_2)$, by uniqueness. □

With $T_i \in \mathfrak{D}'(\Omega_i)$ the distribution $T_1 \otimes T_2 = T_2 \otimes T_1$ is called the *direct*, or *tensor*, product of T_1 and T_2. Strictly speaking, $T_1 \otimes T_2$ and $T_2 \otimes T_1$ act on two different spaces and their equality should be understood as the equality of their images. If supp $T_i = K_i$ and $\phi \in \mathfrak{D}(\Omega_1 \times \Omega_2)$ with supp ϕ in $(\Omega_1 - K_1) \times \Omega_2$ or $\Omega_1 \times (\Omega_2 - K_2)$ then clearly $(T_1 \otimes T_2)(\phi) = 0$. Therefore $(T_1 \otimes T_2)(\phi) = 0$ if supp ϕ is in $(\Omega_1 \times \Omega_2) - (K_1 \times K_2)$. Thus

$$\operatorname{supp}(T_1 \otimes T_2) \subset K_1 \times K_2$$

On the other hand, the equality $(T_1 \otimes T_2)(\phi_1 \otimes \phi_2) = T_1(\phi_1)T_2(\phi_2)$ for any pair of functions $\phi_i \in \mathfrak{D}(\Omega_i)$ implies that

$$K_1 \times K_2 \subset \operatorname{supp}(T_1 \otimes T_2)$$

and we therefore conclude that

$$\operatorname{supp}(T_1 \otimes T_2) = (\operatorname{supp} T_1) \times (\operatorname{supp} T_2)$$

Example 3.4 Given $\xi \in \Omega_1$ and $\eta \in \Omega_2$, we have

$$\begin{aligned} \operatorname{supp} \delta_\xi &= \{\xi\} \\ \operatorname{supp} (\delta_\xi \otimes \delta_\eta) &= \{(\xi,\eta)\} \end{aligned}$$

which implies that $\delta_\xi \otimes \delta_\eta = \delta_{(\xi,\eta)}$.

3.3 CONVOLUTION

In Section 2.6 we defined the convolution of a C_0^∞ function with a locally integrable function. We now wish to extend this definition to the convolution of two distributions T_1 and T_2 on $\mathbf{R}^n$ by the formula

$$\begin{aligned} (T_1 * T_2)(\phi) &= (T_1 \otimes T_2)(\phi(x + y)) \qquad (3.5) \\ &= T_1(T_2(\phi(x + y)) \qquad \phi \in \mathfrak{D}(\mathbf{R}^n) \end{aligned}$$

Here $\phi(x + y) = \psi(x,y)$ is a C^∞ function in $\mathbf{R}^{2n}$ whenever ϕ is in $\mathcal{D}(\mathbf{R}^n)$. But since the boundedness of supp ϕ does not guarantee the boundedness of the set $\{(x,y) \in \mathbf{R}^{2n} : x + y \in \text{supp } \phi\}$, $\phi(x + y)$ as a function of (x,y) does not have compact support in $\mathbf{R}^{2n}$. Therefore the right-hand side of (3.5) is not necessarily bounded unless supp $(T_1 \otimes T_2) = (\text{supp } T_1) \times (\text{supp } T_2)$ intersects supp $(\phi(x + y))$ in a bounded set. Let K be the support of ϕ; then

$$\text{supp } (\phi(x + y)) = \{(x,y) \in \mathbf{R}^{2n} : x + y \in K\}$$

If either T_1 or T_2 has compact support then the intersection of (supp T_1) $\times$ (supp T_2) with supp $(\phi(x + y))$ is also compact, because if either x or y is bounded and $x + y$ is bounded, then both x and y are bounded. In that case the right-hand side of (3.5) is well defined, and we have $T_1 * T_2 = T_2 * T_1$ since $T_1 \otimes T_2 = T_2 \otimes T_1$.

Equation (3.5) therefore defines the convolution of two distributions $T_1, T_2 \in \mathcal{D}(\mathbf{R}^n)$ provided at least one of them has compact support.

Let T_i be defined by $f_i \in L^1_{\text{loc}}(\mathbf{R}^n)$, $i = 1,2$, and either f_1 or f_2 has compact support. Then

$$\begin{aligned}(T_1 * T_2)(\phi) &= T_1(T_2(\phi(x + y))) \\ &= \int f_1(x) \int f_2(y)\, \phi(x + y)\, dy\, dx \\ &= \int\int f_1(x - y)\, f_2(y)\, \phi(x)\, dy\, dx \\ &= \langle f_1 * f_2, \phi\rangle\end{aligned}$$

where

$$(f_1 * f_2)(x) = \int f_1(x - y)\, f_2(y)\, dy = \int f_1(y)\, f_2(x - y)\, dy$$

is a locally integrable function which represents the distribution $T_1 * T_2$ and extends the definition of $f_1 * f_2$ as given in Section 2.6. This indicates that (3.5) is indeed an extension of our previous definition of the convolution product.

Although the convolution of two distributions is always well defined when one of them has compact support, this condition is not always necessary. A bounded (measurable) function g, for example, may be convoluted with any function $f \in L^1(\mathbf{R}^n)$, since the resulting integral $\int f(y)g(x - y)\, dy$ is bounded by $M\|f\|_1$, M being the least upper bound of $|g|$. Naturally, this result holds if g is merely bounded almost everywhere in $\mathbf{R}^n$.

The linear space of complex measurable functions on Ω which are bounded almost everywhere is denoted by $L^\infty(\Omega)$. $L^\infty(\Omega)$ becomes a normed linear space if we define the norm of $g \in L^\infty(\Omega)$, denoted by $\|g\|_\infty$, to be $\inf\{M : |g(x)| \leqslant M \text{ a.e. in } \Omega\}$, called the *essential supremum* of g (see [4]). Thus we can state that if $f \in L^1(\mathbf{R}^n)$ and $g \in L^\infty(\mathbf{R}^n)$, then

$$|f * g| \leqslant \|f\|_1 \|g\|_\infty$$

which implies that $f * g \in L^\infty(\mathbf{R}^n)$ and $\|f * g\|_\infty \leqslant \|f\|_1 \|g\|_\infty$.

When f and g are both in $L^1(\mathbf{R}^n)$ it is not obvious that their convolution $(f * g)(x) = \int f(x - y)\, g(y)\, dy$ exists. At $x = 0$, for example, if we take $f(-y) = g(y)$, this integral may diverge, since not every integrable function is square integrable. But we can show that the function $F(x) = \int f(x - y)\, g(y)\, dy$ exists for almost all x in $\mathbf{R}^n$ by showing that $F = f * g \in L^1(\mathbf{R}^n)$. To that end, let

$$F_k(x) = \int_{|y| \leqslant k} f(x - y) g(y)\, dy$$

$$|F_k(x)| \leqslant \int_{|y| \leqslant k} |f(x - y) g(y)|\, dy$$

$$\int |F_k(x)|\, dx \leqslant \int \left[\int_{|y| \leqslant k} |f(x - y) g(y)|\, dy \right] dx$$

$$= \int_{|y| \leqslant k} \left[\int |f(x - y)|\, dx \right] |g(y)|\, dy \text{ (by Fubini's theorem)}$$

$$= \|f\|_1 \int_{|y| \leqslant k} |g(y)|\, dy$$

$$\leqslant \|f\|_1 \|g\|_1$$

In the limit as $k \to \infty$, we obtain $\|F\|_1 = \|f * g\|_1 \leqslant \|f\|_1 \|g\|_1$.

The next example generalizes these results from L^1 and L^∞ to L^p.

Example 3.5 Let $f \in L^1(\mathbf{R}^n)$ and $g \in L^p(\mathbf{R}^n)$, $1 \leqslant p \leqslant \infty$. We have already shown that $f * g \in L^p(\mathbf{R}^n)$ and that $\|f * g\|_p \leqslant \|f\|_1 \|g\|_p$ when $p = 1, \infty$. Now let us take $1 < p < \infty$. Since $|g|^p \in L^1(\mathbf{R}^n)$ we have

$$\int |f(x - y)|\, |g(y)|^p\, dy < \infty \qquad \text{a.e.}$$

Therefore, as a function of y, the product $|f(x - y)|^{1/p}\, |g(y)|$ lies in $L^p(\mathbf{R}^n)$ for almost all x. Furthermore, since $|f| \in L^1(\mathbf{R}^n)$, we have (for almost all x) $|f(x - y)|^{1/q} \in L^q(\mathbf{R}^n)$, where $1/p + 1/q = 1$. Hence, by Hölder's inequality, the function

$$|f(x - y)\, g(y)| = [|f(x - y)|^{1/p}\, |g(y)|]\, [|f(x - y)|^{1/q}]$$

lies in $L^1(\mathbf{R}^n)$ for almost all x. For such values of x let

$$h(x) = \int f(x - y) g(y)\, dy$$

Hölder's inequality then gives

$$
\begin{aligned}
|h(x)| &\leq \int |f(x - y)|\, |g(y)|\, dy \\
&\leq \left[\int |f(x - y)|\, |g(y)|^p\, dy\right]^{1/p} \left[\int |f(x - y)|\, dy\right]^{1/q} \\
|h(x)|^p &\leq \left[\int |f(x - y)|\, |g(y)|^p\, dy\right] \|f\|_1^{p/q} \\
\int |h(x)|^p\, dx &\leq \|f\|_1^{p/q} \int\int \left[|f(x - y)|\, |g(y)|^p\, dy\right] dx \\
&= \|f\|_1^{p/q} \int \left[\int |f(x - y)|\, dx\right] |g(y)|^p\, dy \text{ (by Fubini's theorem)} \\
&= \|f\|_1^{p/q}\, \|f\|_1\, \|g\|_p^p \\
&= \|f\|_1^p\, \|g\|_p^p
\end{aligned}
$$

Thus $h = f * g \in L^p(\mathbf{R}^n)$ and $\|f * g\|_p \leq \|f\|_1\, \|g\|_p$.

Example 3.6 Let $T \in \mathcal{D}'(\mathbf{R}^n)$. Then for any $\phi \in \mathcal{D}(\mathbf{R}^n)$

$$
\begin{aligned}
(\delta * T)(\phi) &= (\delta \otimes T)(\phi(x + y)) \\
&= T_y\, (\delta(\phi(x + y))) \\
&= T_y\, (\phi(y)) \\
&= T(\phi)
\end{aligned}
$$

Thus δ is the unit element of the product operation $*$. Furthermore

$$
\begin{aligned}
(\partial^\alpha \delta) * T(\phi) &= T_y((\partial^\alpha \delta)_x \phi(x + y)) \\
&= T_y((-1)^{|\alpha|}\, \partial^\alpha \phi(y)) \\
&= \partial^\alpha\, T(\phi)
\end{aligned}
$$

Therefore

$$(\partial^\alpha \delta) * T = \partial^\alpha T = \delta * \partial^\alpha T \tag{3.6}$$

We now list some of the basic properties of the convolution product $T_1 * T_2$, always with the assumption that at least one of the distributions in the product has compact support:

1. $\text{supp}(T_1 * T_2) \subset \text{supp}\, T_1 + \text{supp}\, T_2$

   To see this let $\text{supp}\, T_i = E_i$ and suppose, without loss of generality, that E_1 is compact and E_2 is closed. First we shall show that the set

   $$E_1 + E_2 = \{x + y : x \in E_1, y \in E_2\}$$

   is closed. Let $(x_k + y_k)$ be a sequence in $E_1 + E_2$ which converges to the point a, where $x_k \in E_1$, and $y_k \in E_2$. Since E_1 is compact (x_k) has a subsequence (x'_k) which converges to $x \in E_1$. Now both (x'_k) and the corresponding subsequence $(x'_k + y'_k)$ converge, so their difference (y'_k) also converges to some $y \in E_2$, since E_2 is closed. Thus $a = x + y$ is in $E_1 + E_2$, which must therefore be closed, and its complement $\Omega = \mathbf{R}^n - (E_1 + E_2)$ open. Now for any $(x,y) \in \text{supp}\,(T_1 \otimes T_2) = E_1 \times E_2$ we have $x + y \in E_1 + E_2$ and therefore $\text{supp}\,(T_1 \otimes T_2)$ does not intersect $\text{supp}(\phi(x + y))$ for any $\phi \in \mathcal{D}(\Omega)$. Hence $T_1 * T_2$ vanishes on $\mathcal{D}(\Omega)$ and its support must be in $E_1 + E_2$. In particular, if T_1 and T_2 both have compact support, so does $T_1 * T_2$.

2. $T_1 * (T_2 * T_3) = (T_1 * T_2) * T_3 = T_1 * T_2 * T_3$

   where $T_1, T_2, T_3 \in \mathcal{D}'(\mathbf{R}^n)$ and at least two of the three distributions have compact support. Clearly both $T_1 * T_2$ and $T_2 * T_3$ are then in $\mathcal{D}'(\mathbf{R}^n)$ and the convolution products $T_1 * (T_2 * T_3)$ and $(T_1 * T_2) * T_3$ are well defined. To show that they are equal, we note that for any $\phi \in \mathcal{D}(\mathbf{R}^n)$

   $$\begin{aligned}[T_1 * (T_2 * T_3)]\,(\phi) &= [T_1 \otimes (T_2 * T_3)]\,(\phi(x + y')) \\ &= [T_1 \otimes (T_2 \otimes T_3)]\,(\phi(x + y + z)) \\ &= [(T_1 \otimes T_2) \otimes T_3]\,(\phi(x + y + z)) \\ &= [(T_1 * T_2) * T_3]\,(\phi)\end{aligned}$$

   This associative property of $*$, together with the result of Example 3.6, implies that the linear space $\mathcal{E}'(\mathbf{R}^n)$ is a commutative and associative algebra under the convolution product, with δ as its unit element.

3. $\partial^\alpha(T_1 * T_2) = (\partial^\alpha \delta) * T_1 * T_2 = (\partial^\alpha T_1) * T_2 = T_1 * (\partial^\alpha T_2)$

   This follows directly from equation (3.6) and the commutative and associative properties of $*$.

4. If f is a function on $\mathbf{R}^n$ and h is any point in $\mathbf{R}^n$, the *translation* τ_h of f by h is the function defined on $\mathbf{R}^n$ by

$$\tau_h f(x) = f(x - h).$$

We clearly have $\tau_h \phi \in C_0^\infty(\mathbf{R}^n)$ whenever $\phi \in C_0^\infty(\mathbf{R}^n)$, and we define the translation of the distribution $T \in \mathcal{D}'(\mathbf{R}^n)$ by

$$(\tau_h T)(\phi) = T(\tau_{-h}\phi) \qquad \phi \in \mathcal{D}(\mathbf{R}^n)$$

which is again a distribution in $\mathbf{R}^n$. When the distribution T is defined by a locally integrable function $f(x)$, its translation $\tau_h T$ is clearly defined by $f(x - h)$, as expected. In the case of the Dirac measure, we have

$$\begin{aligned}\tau_h \delta(\phi) &= \delta(\tau_{-h}\phi) \qquad \phi \in \mathcal{D}(\mathbf{R}^n)\\ &= \phi(h)\\ &= \delta_h(\phi)\end{aligned}$$

which implies that

$$\tau_h \delta = \delta_h$$

More generally, for any $T \in \mathcal{D}'(\mathbf{R}^n)$

$$\begin{aligned}\tau_h T(\phi) &= T(\tau_{-h}\phi)\\ &= T_x(\phi(x + h))\\ &= T_x(\delta_h(\phi(x + y)))\\ &= (\delta_h * T)(\phi)\end{aligned}$$

or

$$\tau_h T = \delta_h * T \qquad T \in \mathcal{D}'(\mathbf{R}^n)$$

If either T_1 or T_2 has compact support, this gives

$$\begin{aligned}\tau_h(T_1 * T_2) &= \delta_h * T_1 * T_2\\ &= (\tau_h T_1) * T_2\\ &= T_1 * (\tau_h T_2)\end{aligned}$$

by the commutative and associative properties of $*$.

Even though we can sometimes define the convolution product of several distributions where more than one is without compact support, such products may not satisfy all the properties listed above, as the next example shows.

Example 3.7 Let 1 denote the distribution represented by the constant function 1 on $\mathbf{R}^n$. Then $(H * \delta') * 1$ and $H * (\delta' * 1)$ are both well defined distributions but they are not equal, for

$$(H * \delta') * 1 = (H' * \delta) * 1 = (\delta * \delta) * 1 = \delta * 1 = 1$$

whereas

$$H * (\delta' * 1) = H * (\delta * 1') = H * 0 = 0$$

The equality $\delta' * 1 = 0$ also shows that if T_1 and T_2 are two nonzero distributions, it may happen that $T_1 * T_2 = 0$. In other words the equality $S_1 * T = S_2 * T$ for some $T \neq 0$ does not necessarily imply that $S_1 = S_2$.

On the other hand, if $T \in \mathcal{D}'(\mathbf{R}^n)$ and $S_1, S_2 \in \mathcal{E}'(\mathbf{R}^n)$ are such that $S_1 * T = S_2 * T = \delta$ then, by the commutative and associative properties of $*$,

$$\begin{aligned} S_1 &= \delta * S_1 \\ &= (S_2 * T) * S_1 \\ &= S_2 * \delta \\ &= S_2 \end{aligned}$$

Another important class of distributions on which the convolution product is well defined is given in the following example:

Example 3.8 Let $\mathcal{D}'_+(\mathbf{R}) = \{T \in \mathcal{D}'(\mathbf{R}): \operatorname{supp} T \subset [0,\infty)\}$. If $T, S \in \mathcal{D}'_+(\mathbf{R})$, we can still define the convolution of T and S by the equation (3.5),

$$\langle S * T, \phi \rangle = \langle S_x, \langle T_y, \phi(x + y) \rangle \rangle$$

This can be seen by noting that, when x is fixed and $\phi \in \mathcal{D}(\mathbf{R})$, $\phi(x + y)$, as a function of y, has compact support; hence $\psi(x) = \langle T_y, \phi(x + y)\rangle$ is a well-defined function in $C^\infty(\mathbf{R})$. Moreover supp ψ is bounded from above, because $y \in \operatorname{supp} T$ and $x + y \in \operatorname{supp} \phi \subset [-M,M]$ together imply that $y \geq 0$ and $|x + y| \leq M$, and therefore $x \leq x + y \leq M$. Thus $\operatorname{supp} S \subset [0,\infty)$ intersects $\operatorname{supp} \psi \subset (-\infty, M]$ in a bounded set. Hence we can define $\langle S * T, \phi \rangle = \langle S, \psi \rangle$ as $\lim \langle S, \phi_n \psi \rangle$ where ϕ_n is a C_0^∞ function which equals 1 on $[-n,n]$.

Since

$$\operatorname{supp}(S * T) \subset \operatorname{supp} S + \operatorname{supp} T \subset [0,\infty)$$

we see that $S * T$ also lies in $\mathcal{D}'_+(\mathbf{R})$, i.e. $\mathcal{D}'_+(\mathbf{R})$ is closed under the operation $*$. $\mathcal{D}'_+(\mathbf{R})$ derives its importance in this treatment from its applicability to initial value problems in Chapter 6.

Let $T \in \mathcal{D}'(\mathbf{R}^n)$. If there exists a distribution $S \in \mathcal{D}'(\mathbf{R}^n)$ such that $S * T = \delta$ then S is called an *inverse* of T in $\mathcal{D}'(\mathbf{R}^n)$ with respect to the binary operation $*$, and denoted by T^{-1}. We have already seen that in $\mathcal{E}'$ such an inverse is unique. It is also unique in any subspace of $\mathcal{D}'$ where the convolution product is a commutative and associative algebra, such as $\mathcal{D}'_+$ (see Exercise 3.2).

Now we shall look at the possibility of inverting some simple distributions in $\mathbf{R}$.

Example 3.9 (i) Let $S * H = \delta$. Then

$$\delta' = (S * H)' = S * H' = S * \delta = S$$

Hence $H^{-1} = \delta'$. Similarly $(\delta')^{-1} = S^{-1} = H$.

(ii) Let $S * (\delta' - \lambda\delta) = \delta$. Then $S' - \lambda S = \delta$. Noting that the C^∞ function $e^{\lambda x}$ solves the corresponding homogeneous equation $S' - \lambda S = 0$, we set $S = e^{\lambda x} T$ and obtain $e^{\lambda x} T' = \delta$. This gives $T' = \delta$, and hence $T = H$. Consequently $(\delta' - \lambda\delta)^{-1} = S = e^{\lambda x}H$.

3.4 REGULARIZATION OF DISTRIBUTIONS

The fundamental result of this section is Theorem 3.5, which states that the convolution of a distribution with a smooth function is a smooth function. This leads to the possibility of approximating a distribution by a smooth function as expressed by Theorem 3.6, which is one of the most important means that we have at our disposal for applying the methods of function theory to distributions.

For any function f on $\mathbf{R}^n$ we define its *reflection* in 0 as the function $\check{f}$ defined on $\mathbf{R}^n$ by $\check{f}(x) = f(-x)$, and we extend this definition to $\mathcal{D}'(\mathbf{R}^n)$ by duality,

$$\check{T}(\phi) = T(\check{\phi}) \qquad \phi \in \mathcal{D}(\mathbf{R}^n)$$

Theorem 3.5 For all $T \in \mathcal{D}'(\mathbf{R}^n)$ and all $\psi \in C_0^\infty(\mathbf{R}^n)$, the convolution

$$(T * \psi)(x) = T(\tau_x\check{\psi}) \tag{3.7}$$

is in $C^\infty(\mathbf{R}^n)$.

Proof. For any $\phi \in \mathcal{D}(\mathbf{R})$ we have

$$(T * \psi)(\phi) = T_x(\langle\psi(y),\phi(x + y)\rangle)$$

$$\langle\psi(y),\phi(x + y)\rangle = \int \psi(y)\phi(x + y)\, dy$$

$$= \int \psi(\xi - x)\phi(\xi)\, d\xi$$
$$= \langle \psi(\xi - x), \phi(\xi) \rangle$$
$$= \langle \check{\psi}(x - \xi), \phi(\xi) \rangle$$
$$= \langle \tau_\xi \check{\psi}(x), \phi(\xi) \rangle$$

Hence

$$(T * \psi)(\phi) = T_x\,(\langle \tau_\xi \check{\psi}(x), \phi(\xi) \rangle)$$
$$= \langle T(\tau_\xi \check{\psi}), \phi(\xi) \rangle$$

which proves the equality (3.7). Furthermore,

$$(T * \psi)(x) = T(\tau_x \check{\psi}) = T_y(\psi(x - y))$$

is a $C^\infty(\mathbf{R}^n)$ function by Corollary 2 to Theorem 3.3. □

Corollary 1 $T(\phi) = (T * \check{\psi}(0)$ for every $\phi \in \mathcal{D}(\mathbf{R}^n)$ and $T \in \mathcal{D}'(\mathbf{R}^n)$.

As a consequence, if $T * \phi = 0$ for every $\phi \in \mathcal{D}(\mathbf{R}^n)$, then $T = 0$.

Corollary 2 Let $T \in \mathcal{E}'(\mathbf{R}^n)$. If $\psi \in C^\infty(\mathbf{R}^n)$ then $(T * \psi)(x) = T(\tau_x \check{\psi})$ is in $C^\infty(\mathbf{R}^n)$. If $\psi \in C_0^\infty(\mathbf{R}^n)$ then $(T * \psi)(x)$ is in $C_0^\infty(\mathbf{R}^n)$.

The first part of Corollary 2 is proved by multiplying ψ by a $C_0^\infty(\mathbf{R}^n)$ function equal to 1 on supp T. The second part follows from the observation that supp $(T * \psi) \subset$ supp $T +$ supp ψ.

If $T \in \mathcal{D}^{m'}(\mathbf{R}^n)$ and $\psi \in C_0^m(\mathbf{R}^n)$, equation (3.7) still holds and, furthermore, the convolution $T * \psi$ is then continuous in $\mathbf{R}^n$; for if (x_i) is a sequence in $\mathbf{R}^n$ which converges to x, then

$$\lim (T * \psi)(x_i) = \lim \langle T_y, \psi(x_i - y) \rangle$$
$$= \langle T_y, \lim \psi(x_i - y) \rangle$$
$$= \langle T_y, \psi(x - y) \rangle$$
$$= (T * \psi)(x)$$

where the second equality follows from the fact that T is a continuous linear functional on $\mathcal{D}^m(\mathbf{R}^n)$.

If $T \in \mathcal{D}^{m'}(\mathbf{R}^n)$ has compact support then we can take ψ in $C^m(\mathbf{R}^n)$ and reach the same conclusion. Thus,

Corollary 3 If $T \in \mathcal{D}^{m'}(\mathbf{R}^n)$ and $\psi \in C^m(\mathbf{R}^n)$, then $(T * \psi)(x) = \langle T_y, \psi(x - y) \rangle$ is a continuous function in $\mathbf{R}^n$ provided T or ψ has compact support.

The following theorem generalizes some results from Theorem 2.8. Recall that the function β_λ defined by equation (2.8) for $\lambda > 0$ is a C_0^∞ function whose support is the closed ball $\bar{B}(0,\lambda)$ in $\mathbf{R}^n$ and whose integral over $\mathbf{R}^n$ is 1.

Theorem 3.6 For any $T \in \mathcal{D}'(\mathbf{R}^n)$ the $C^\infty(\mathbf{R}^n)$ function $T * \beta_\lambda$ converges strongly to T as $\lambda \to 0$, i.e. $(T * \beta_\lambda)(\phi)$ converges to $T(\phi)$ uniformly on every bounded subset of $\mathcal{D}(\mathbf{R}^n)$.

Proof. $T * \beta_\lambda$ is in $C^\infty(\mathbf{R}^n)$ by Theorem 3.5. Let E be any bounded subset of $\mathcal{D}(\mathbf{R}^n)$. By Theorem 2.2, there is a compact subset K of $\mathbf{R}^n$ such that E is a bounded subset of $\mathcal{D}_K(\mathbf{R}^n)$. By Theorem 2.8, for every $\phi \in E$ the support of $\beta_\lambda * \phi$ lies in a λ-neighborhood of K. If $\lambda \in (0,1)$ then there is a compact set K_0 such that $K \subset K_0 \subset \mathbf{R}^n$ and

$$\text{supp}\,(\beta_\lambda * \phi) \subset K_0 \qquad \phi \in E$$

Let m be any nonnegative integer. Since $\partial^\alpha\phi(x) \in \mathcal{D}_K(\mathbf{R}^n)$ for all $|\alpha| \leq m$, there is a positive number ε, which depends on m, such that

$$\partial^\alpha\phi(x - y) \in \mathcal{D}_{K_0}(\mathbf{R}^n)$$

for every $y \in B(0,\varepsilon)$. The function $\partial^\alpha\phi(x - y)$ converges to $\partial^\alpha\phi(x)$ as $y \to 0$ uniformly on K_0 for all $|\alpha| \leq m$. We also have

$$\begin{aligned} |(\beta_\lambda * \partial^\alpha\phi - \partial^\alpha\phi)(x)| &= \left| \int \beta_\lambda(y)[\partial^\alpha\phi(x - y) - \partial^\alpha\phi(x)]\, dy \right| \\ &\leq \int \beta_\lambda(y)\, |\partial^\alpha\phi(x - y) - \partial^\alpha\phi(x)|\, dy \end{aligned}$$

For all values of λ in the open interval $(0,\varepsilon)$ we have supp $\beta_\lambda \subset \bar{B}(0,\varepsilon)$ and the integration may be performed over $B(0,\varepsilon)$. Thus the left-hand side of the inequality tends to 0 uniformly as $\lambda \to 0$ for all x in K and all $|\alpha| \leq m$.

From Corollary 1 to Theorem 3.5, we have

$$\begin{aligned} (T * \beta_\lambda - T)(\phi) &= (T * \beta_\lambda) * \check{\phi}(0) - T * \check{\phi})(0) \\ &= T * (\beta_\lambda * \check{\phi} - \check{\phi})(0) \\ &= T\,(\beta_\lambda * \phi - \phi). \end{aligned}$$

The last equality follows from the observation that $\beta_\lambda * \check{\phi} = \check{\beta}_\lambda * \check{\phi} = (\beta_\lambda * \phi)\check{}$ since β_λ is an even function. As $\lambda \to 0$, $\beta_\lambda * \phi \to \phi$ uniformly for all $\phi \in E$, and therefore $T * \beta_\lambda - T$ converges to 0 uniformly on E. □

Corollary If $T \in \mathscr{E}'(\mathbf{R}^n)$ then $T * \beta_\lambda$ converges uniformly to T on every bounded subset of $\mathscr{E}(\mathbf{R}^n)$.

Example 3.10 By setting $T = \delta$ in the above theorem, we see that β_λ converges strongly to δ in both $\mathscr{D}'(\mathbf{R}^n)$ and $\mathscr{E}'(\mathbf{R}^n)$.

In Section 2.6 the convolution of a locally integrable function f with β_λ was called a *regularization* of f, and we now extend the notion of regularization from functions to distributions. We call $T * \beta_{1/k} = T * \gamma_k$ a *regularizing sequence* of functions for the distribution $T \in \mathscr{D}'(\mathbf{R}^n)$. Thus γ_k is a regularizing sequence for δ. In consequence, if $T * \phi = 0$ for every $\phi \in \mathscr{D}(\mathbf{R}^n)$, then $T = T * \delta = \lim T * \gamma_k = 0$, as we have already observed in a different context. This is used in the next example to reestablish a result from the previous chapter, namely that T' can only vanish in $\mathbf{R}$ if T is a constant function (almost everywhere).

Example 3.11 Let $T \in \mathscr{D}'(\mathbf{R})$ satisfy $T' = 0$, and let γ_k be a regularizing sequence for δ. Then the C^∞ function $T * \gamma_k$ satisfies $(T * \gamma_k)' = T' * \gamma_k = 0$ in $\mathbf{R}$ for every k, so $T * \gamma_k = c_k$ for some constant c_k. Now $c_k = T * \gamma_k \to T$ in $\mathscr{D}'$, and it remains to show that the sequence of constants c_k also converges in $\mathbf{C}$. But for any $\phi \in \mathscr{D}(\mathbf{R})$ such that $\int \phi(x)\,dx = 1$, the sequence $c_k = \langle c_k, \phi \rangle$ converges in $\mathbf{C}$ because c_k converges in $\mathscr{D}'$; hence its limit, the constant $\lim c_k$, coincides with T.

In general, the convergence of a sequence of functions f_k to f in $\mathscr{D}'$ does not imply that its pointwise limit is f, or that it is even a function (recall the sequence $\sin kx$ which converges to 0 in $\mathscr{D}'$). But when f_k is constant, we have just shown that both assertions can be made.

As pointed out in Example 2.21, this result leads to the conclusion that, if $T^{(k)} = 0$, then T is (almost everywhere) a polynomial of degree less than k. When $k = 2$ we can also reach this conclusion by a regularization process which is given below. The significance of this method is that it can be generalized from $\mathbf{R}$ to $\mathbf{R}^n$ (see Theorem 4.11).

Example 3.12 If $T \in \mathscr{D}'(\mathbf{R})$ satisfies $T'' = 0$, we shall show that T is a linear function a.e.

For any $\phi \in \mathscr{D}(\mathbf{R})$ we know that $T * \phi$ is a C^∞ function and that $(T * \phi)'' = T'' * \phi = 0$. Therefore $T * \phi$ is a linear function of the form $(T * \phi)(x) = ax + b$.

Now let $h(x) = ax + b$, $x \in \mathbf{R}$. Recalling that β is a C^∞ function supported in $[-1,1]$ with $\int \beta(x)\,dx = 1$, we can write

$$(h * \beta)(x) = \int_{-\infty}^{\infty} h(x - y)\beta(y)\, dy$$

$$= \int_{-\infty}^{\infty} [a(x - y) + b]\beta(y)\, dy$$

$$= ax + b$$

since $\int_{-\infty}^{\infty} y\beta(y)\, dy = 0$, the integrand being an odd function. Thus $h * \beta = h$.

Let $\beta_{1/k} \in \mathfrak{D}(\mathbf{R})$ be the regularizing sequence defined by equation (2.8). Then

$$(T * \beta) * \beta_{1/k} = (T * \beta_{1/k}) * \beta$$

$$= T * \beta_{1/k}$$

In the limit as $k \to \infty$, we obtain $T = T * \beta$ a.e. Since $T * \beta$ is a linear function, so is the distribution T (almost everywhere).

Theorem 3.7 For any $T \in \mathfrak{D}'(\mathbf{R}^n)$, the linear map L from $\mathfrak{D}(\mathbf{R}^n)$ to $\mathscr{E}(\mathbf{R}^n)$ defined by

$$L(\phi) = T * \phi$$

is continuous and commutes with the translation operator τ_h, $h \in \mathbf{R}^n$. Conversely, if L is a continuous linear map from $\mathfrak{D}(\mathbf{R}^n)$ to $\mathscr{E}(\mathbf{R}^n)$ which commutes with τ_h, then there is a unique $T \in \mathfrak{D}'(\mathbf{R}^n)$ such that

$$L(\phi) = T * \phi \qquad \phi \in \mathfrak{D}(\mathbf{R}^n)$$

Proof. (i) For any sequence $\phi_k \to \phi$ in $\mathfrak{D}_K$ we have

$$\lim (T * \phi_k)(x) = \lim T(\tau_x \check{\phi}_k) \qquad x \in \mathbf{R}^n$$

$$= T(\tau_x \check{\phi})$$

$$= (T * \phi)(x)$$

because both T and τ_x are continuous.

If $T \in \mathfrak{D}'(\mathbf{R}^n)$ then, for all $\phi \in \mathfrak{D}(\mathbf{R}^n)$, Theorem 3.5 gives

$$(T * \tau_h\phi)(x) = T(\tau_x(\tau_h\phi)\check{\ })$$

$$= T(\tau_x\tau_{-h}\check{\phi})$$

$$= T(\tau_{x-h}\check{\phi})$$

$$= (T * \phi)(x - h)$$

$$= \tau_h(T * \phi)(x)$$

Thus $L\tau_h = \tau_h L$.

(ii) Suppose L is a continuous linear map from $\mathcal{D}(\mathbf{R}^n)$ to $\mathcal{E}(\mathbf{R}^n)$ which commutes with τ_h. Then the map

$$\phi \mapsto L(\check{\phi})(0)$$

is a continuous linear function on $\mathcal{D}(\mathbf{R}^n)$; so there is a $T \in \mathcal{D}'(\mathbf{R}^n)$ such that

$$L(\check{\phi})(0) = T(\phi) \qquad \phi \in \mathcal{D}(\mathbf{R}^n)$$

$$\begin{aligned} L(\phi)(x) &= \tau_{-x}L(\phi)(0) \\ &= L(\tau_{-x}\phi)(0) \\ &= T(\tau_{-x}\phi)^{\vee} \\ &= T(\tau_x\check{\phi}) \\ &= (T * \phi)(x) \end{aligned}$$

The uniqueness of T follows from the observation that $T * \phi = 0$ for all $\phi \in \mathcal{D}(\mathbf{R}^n)$ implies that $T = 0$. □

3.5 LOCAL STRUCTURE OF DISTRIBUTIONS

In Example 2.8 we saw that the Dirac distribution on $\mathbf{R}$ is the second derivative of the continuous function $x_+ = xH(x)$, and we now conclude, on the basis of Theorem 3.2, that every distribution on $\mathbf{R}$ with support $\{0\}$ is a finite linear combination of derivatives of x_+. More generally, we can show that every distribution is, locally, a derivative of some continuous function. In this sense distributions are the natural generalization of continuous functions, which is achieved by supplementing these functions with their (distributional) derivatives of all orders. To prove this important structural theorem we introduce the following notation: For $x \in \mathbf{R}^n$ and $k \in \mathbf{N}$, we define

$$(x_i)_+^k = x_i^k H(x_i) \qquad i = 1, \ldots, n$$

$$x^k = x_1^k x_2^k \cdots x_n^k$$

$$x_+^k = (x_1)_+^k (x_2)_+^k \cdots (x_n)_+^k$$

$$\partial^k = \partial_1^k \partial_2^k \cdots \partial_n^k$$

Since $1/(k-1)!\,\partial_i^k(x_i)_+^{k-1} = \delta$ is the Dirac measure on $\mathbf{R}$ for any $i = 1, \ldots, n$, we have the corresponding result in $\mathbf{R}^n$

$$\partial^k E_k = \delta \tag{3.8}$$

where

$$E_k = \frac{1}{[(k-1)!]^n} x_+^{k-1}$$

is in $C^{k-2}(\mathbf{R}^n)$ and δ is the Dirac measure on $\mathbf{R}^n$, which is the tensor product of $\delta \in \mathcal{D}'(\mathbf{R})$ with itself n times. Although we have not distinguished in our notation between these two distributions, it should be clear from the context which underlying space we are dealing with.

Theorem 3.8 If $T \in \mathcal{D}'(\mathbf{R}^n)$ and K is a compact subset of $\mathbf{R}^n$, then there is a continuous function f on $\mathbf{R}^n$ and a multi-index $\alpha \in \mathbf{N}_0^n$ such that

$$\begin{aligned} T(\phi) &= \langle \partial^\alpha f, \phi \rangle \\ &= (-1)^{|\alpha|} \langle f, \partial^\alpha \phi \rangle \end{aligned}$$

for every $\phi \in \mathcal{D}_K$.

Proof. Let $\psi \in C_0^\infty(\mathbf{R}^n)$ and $\psi = 1$ on a neighborhood of K. The distribution ψT equals T on K, has compact support and is therefore of finite order, say m. From equation (3.8) and the properties of the convolution product, we can write

$$\begin{aligned} \psi T &= \delta * \psi T \\ &= (\partial^{m+2} E_{m+2}) * \psi T \\ &= \partial^{m+2}(E_{m+2} * \psi T) \end{aligned}$$

Now $E_{m+2} \in C^m(\mathbf{R}^n)$ and the distribution ψT, being of order m, may be extended to a continuous linear functional on $C_0^m(\mathbf{R}^n)$ in the topology of $\mathcal{D}^m(\mathbf{R}^n)$. Since ψT has compact support, the convolution $E_{m+2} * \psi$ is a continuous function on $\mathbf{R}^n$, by Corollary 3 to Theorem 3.5, and represents the desired function f. □

When T has compact support this result takes a global form:

Corollary If $T \in \mathcal{E}'(\mathbf{R}^n)$, then there is a continuous function f on $\mathbf{R}^n$ and a multi-index α such that $T = \partial^\alpha f$.

Proof. if supp $T = K$ is compact then T is of finite order, say m. By Theorem 3.8, $T = \partial^{m+2}f$, where $f = E_{m+2} * T \in C^0(\mathbf{R}^n)$. □

Example 3.13 Let $T \in \mathscr{E}'(\mathbf{R})$ be a distribution of compact support, and hence of finite order m. Then

$$E_{m+2} = \frac{1}{(m+1)!} x_+^{m+1} \in C^m(\mathbf{R})$$

$$E_{m+2}^{(m+2)} = \delta$$

Now if $f = T * E_{m+2}$ then $T = T * \delta = T * E_{m+2}^{(m+2)} = f^{(m+2)}$. Since $T \in \mathscr{E}'(\mathbf{R})$ is of order m, it can be extended to a bounded linear functional on $C^m(\mathbf{R})$. Hence, using equation (3.7), we can write

$$f(x) = \langle T_y, E_{m+2}(x - y)\rangle$$

which is clearly continuous.

In this example, even though T has compact support, the continuous function f which satisfies $T = f^{(m+2)}$ may not have compact support. In fact, when $T = \delta$ then $f = E_{m+2}$, which has support $[0,\infty)$. This example also indicates the relation between the order of differentiation of f which is needed to represent T, namely $m + 2$, and the order of T.

It is of course evident that the representation $T = \partial^\alpha f$ in the statement of both Theorem 3.8 and its corollary is not unique. The choice $\alpha = (\alpha_1, \ldots, \alpha_n) = (m + 2, \ldots, m + 2)$ always works when f is chosen to be $E_{m+2} * T$, but obviously there are other possibilities. The second point worth noting is that this representation remains valid whether K is taken in $\mathbf{R}^n$ or in any of its open subsets. Hence Theorem 3.8 and its corollary still hold if $\mathbf{R}^n$ is replaced by Ω. More significantly, the corollary remains valid if the distribution T is merely of finite order. The proof of this stronger result relies essentially on a partition of unity in Ω.

Let Ω be any open set in $\mathbf{R}^n$. An open covering $\{\Omega_i: i \in \mathbf{N}\}$ of Ω is called *locally finite* if every compact subset of Ω intersects at most a finite number of Ω_i. Following the procedure outlined in Section 2.6, we can construct a sequence of functions ψ_i in $C_0^\infty(\Omega)$ such that, for each $i \in \mathbf{N}$, supp $\psi_i \subset \Omega_i$, $0 \leq \psi_i \leq 1$, and

$$\sum_1^\infty \psi_i(x) = 1 \qquad \text{for every } x \in \Omega$$

Since any $x \in \Omega$ lies in at most a finite number of the sets Ω_i, this sum has only a finite number of nonzero terms. The collection $\{\psi_i\}$ is called a *locally finite partition of unity* in Ω subordinate to the cover $\{\Omega_i\}$.

Theorem 3.9 If $T \in \mathcal{D}'(\Omega)$ is of finite order, then there exists a continuous function f in Ω and a multi-index α such that $T = \partial^\alpha f$ in Ω.

Proof. Let $T \in \mathcal{D}'(\Omega)$ be of order m, and let $\{\psi_i\}$ be a locally finite partition of Ω subordinate to the (locally finite) cover $\{\Omega_i\}$. We can then write

$$T = \sum \psi_i T = \sum T_i$$

where $T_i = \psi_i T$ is a distribution with compact support in Ω_i, and of order $m_i \leqslant m$, since the order of $\psi_i T$ cannot exceed the order of T. By the corollary to Theorem 3.8, it is represented in Ω_i by

$$T_i = \partial^{m_i+2}(E_{m_i+2} * T_i) = \partial^{m+2}(E_{m+2} * T_i)$$

where the convolution of E_{m+2} and T_i is well defined because $T_i = \psi_i T$ can be extended as 0 into $\mathbf{R}^n - \Omega_i$. Now $E_{m+2} * T_i = f_i$ is a continuous function in Ω, and T is represented by the sum

$$T = \sum_i T_i = \sum_i \partial^{m+2} f_i = \partial^{m+2} \sum_i f_i$$

Since any compact set in Ω intersects the supports of at most a finite number of the functions f_i, this sum over i is finite, and therefore the function $g = \Sigma_i f_i$ is continuous in Ω. □

If the distribution T is not of finite order, the representation $T = \Sigma \partial^{m+2} f_i$ is still valid, and we obtain a global version of Theorem 3.8:

Corollary For every $T \in \mathcal{D}'(\Omega)$ there exist continuous functions f_i in Ω and multi-indices $\alpha_i \in \mathbf{N}_0^n$ such that $T = \Sigma\, \partial^{\alpha_i} f_i$, in the sense that

$$\langle T,\phi\rangle = \sum_{i=1}^{N} (-1)^{|\alpha_i|} \langle f_i, \partial^{\alpha_i}\phi\rangle \qquad \text{for all } \phi \in \mathcal{D}(\Omega)$$

where the (finite) integer N depends on supp ϕ.

3.6 APPLICATIONS TO DIFFERENTIAL EQUATIONS

Because not every function is differentiable, the treatment of differential equations is severely limited when the solutions are assumed to be functions.

In Section 3.5 we have seen that distributions are the smallest extension of continuous functions which is closed under differentiation. For that reason they are well suited to the treatment of differential equations, and in fact, the theory of distributions was developed largely with that purpose in mind [7].

As pointed out in Section 2.8, for a given $T \in \mathcal{D}'(\Omega)$, the distribution S which satisfies $\partial_k S(\phi) = T(\phi)$ for every $\phi \in \mathcal{D}(\Omega)$ is called a primitive of T. We shall establish the existence of a primitive when $\Omega = \mathbf{R}$.

Theorem 3.10 Any distribution in $\mathcal{D}'(\mathbf{R})$ has a primitive distribution which is unique up to an additive constant.

Proof. Let $T \in \mathcal{D}'(\mathbf{R})$. We wish to determine a distribution S such that

$$S'(\phi) = -S(\phi') = T(\phi)$$

for every $\phi \in \mathcal{D}(\mathbf{R})$. This determines S on the space

$$\mathcal{D}_0(\mathbf{R}) = \{\psi \in \mathcal{D}(\mathbf{R}) : \psi = \phi' \text{ for some } \phi \in \mathcal{D}(\mathbf{R})\}$$

We have already seen in Example 2.21 that $\psi \in \mathcal{D}_0(\mathbf{R})$ if and only if

$$\int_{-\infty}^{\infty} \psi(x)\, dx = 0$$

Let ϕ_0 be a fixed function in $\mathcal{D}(\mathbf{R})$ such that $\langle 1,\phi_0\rangle = 1$. For any $\phi \in \mathcal{D}(\mathbf{R})$ we can write

$$\begin{aligned} \phi(x) &= \phi(x) - \langle 1,\phi\rangle\phi_0(x) + \langle 1,\phi\rangle\phi_0(x) \\ &= \psi(x) + \langle 1,\phi\rangle\phi_0(x) \end{aligned} \tag{3.9}$$

where $\psi = \phi - \langle 1,\phi\rangle\phi_0 \in \mathcal{D}_0(\mathbf{R})$, and (3.9) provides a decomposition of ϕ into its components in $\mathcal{D}_0(\mathbf{R})$ and $\mathcal{D}(\mathbf{R}) - \mathcal{D}_0(\mathbf{R})$. We first define S on $\mathcal{D}_0(\mathbf{R})$ by

$$S(\psi) = -T(\chi)$$

where $\chi(x) = \int_{-\infty}^{x} \psi(t)\, dt$ lies in $\mathcal{D}(\mathbf{R})$. Then we extend the definition to $\mathcal{D}(\mathbf{R})$ by

$$S(\phi) = -T(\chi) + \langle c,\phi\rangle$$

where c is an arbitrary complex constant. If S is a distribution, then

$$S'(\phi) = -S(\phi') = -\left[-T\left(\int_{-\infty}^{x} \phi'(t)\,dt\right) + \langle c,\phi'\rangle\right] = T(\phi) + 0$$

for all $\phi \in \mathcal{D}(\mathbf{R})$, which means that S is a primitive of T.

To show that S is in $\mathcal{D}'(\mathbf{R})$ let (ϕ_k) be any sequence in $\mathcal{D}(\mathbf{R})$ which converges to 0. This implies that supp ϕ_k is in some fixed compact set $K \subset \mathbf{R}$ for all k, and that $\partial^\alpha\phi_k \to 0$ uniformly on K, so that $\langle 1,\phi_k\rangle \to 0$. Therefore

$$\psi_k(x) = \phi_k(x) - \langle 1,\phi_k\rangle\phi_0(x) \to 0$$

$$\chi_k(x) = \int_{-\infty}^{x} \psi_k(t)\,dt \to 0$$

in $\mathcal{D}(\mathbf{R})$, and hence

$$S(\phi_k) = -T(\chi_k) + \langle c,\phi_k\rangle \to 0$$

which proves that S is a distribution in $\mathbf{R}$.

If S_1 and S_2 are two primitives of T, then for any $\phi \in \mathcal{D}(\mathbf{R})$,

$$\begin{aligned}(S_1 - S_2)'(\phi) &= S_1'(\phi) - S_2'(\phi)\\ &= T(\phi) - T(\phi)\\ &= 0\end{aligned}$$

By Example 2.21 we conclude that $S_1 - S_2$ must be a constant. □

Note that the proof above works equally well when $\mathbf{R}$ is replaced by any open interval in $\mathbf{R}$.

Let L be a *linear partial differential operator* of order $m \geq 1$ of the form

$$L = \sum_{|\alpha| \leq m} c_\alpha(x)\partial^\alpha \tag{3.10}$$

where $\alpha \in \mathbf{N}_0^n$ and c_α are C^∞ functions on $\mathbf{R}^n$. L clearly maps $\mathcal{D}'(\Omega)$ into $\mathcal{D}'(\Omega)$. The corresponding equation

$$Lu = f$$

where f is generally given as a distribution in $\Omega \subset \mathbf{R}^n$, is called a *linear partial differential equation* of order m. The restriction to *linear* differential equations is, of course, necessary because we cannot define multiplication in $\mathcal{D}'$ as a natural extension of multiplication of functions [3].

We should start by explaining what we mean by a "solution" to the differential equation $Lu = f$ in Ω. In the classical theory, that means a *function* which is *differentiable* up to order m in Ω and which satisfies the equation in the sense of equality between functions. We shall demand a little more smoothness and call u a *strong* solution of $Lu = f$ in Ω if $u \in C^m(\Omega)$ and the (continuous) function Lu equals f in Ω. A *weak* solution of $Lu = f$ is a distribution $u \in \mathfrak{D}'(\Omega)$ which satisfies $Lu = f$ in the sense of distributions, i.e., in the sense that

$$\langle Lu,\phi\rangle = \langle f,\phi\rangle$$

for all $\phi \in \mathfrak{D}(\Omega)$.

Since any continuous function defines a distribution in $\mathfrak{D}'$ and all its continuous derivatives coincide with its corresponding distributional derivatives, every strong solution of $Lu = f$ is also a weak solution. The question is: are there weak solutions of the equation $Lu = f$ which are not strong solutions? The answer is yes, as the next example shows.

Example 3.14 The ordinary differential equation $xu' = 0$ on $\mathbf{R}$ has the strong solution $u = c_1$. But the function $u(x) = c_2H(x)$ satisfies the equation as a distribution, because $u' = c_2\delta$ and

$$\langle xu',\phi\rangle = c_2\langle\delta,x\phi\rangle = 0 \qquad \text{for all } \phi \in \mathfrak{D}(\mathbf{R})$$

Hence $u = c_1 + c_2H$ is a weak solution of $xu' = 0$.

This solution, incidentally, violates the (classical) rule that an ordinary differential equation of order 1 has a general solution with just one arbitrary constant; so it would seem that this "rule" no longer holds when distributions are admitted to the class of solutions.

The next result gives a criterion by which we can decide when a distributional solution of $Lu = f$ is a strong solution.

Theorem 3.11 Let L be a linear differential operator of order m and u be a weak solution of $Lu = f$ in Ω. If $u \in C^m(\Omega)$ and $f \in C^0(\Omega)$, then u is also a strong solution of the equation.

Proof. As a weak solution of $Lu = f$, u satisfies

$$\langle Lu,\phi\rangle = \langle f,\phi\rangle \qquad \phi \in \mathfrak{D}(\Omega)$$

which can also be written as

$$\int_{\Omega} (Lu - f)\phi = 0$$

for all $\phi \in \mathcal{D}(\Omega)$. This implies that $Lu - f = 0$ on Ω, because otherwise there would be a point $x \in \Omega$ where $Lu(x) - f(x) \neq 0$. Since $Lu - f$ is continuous, there is a neighborhood U of x where $Lu - f$ does not vanish. Now we can choose $\phi \in \mathcal{D}(\Omega)$ to be a positive function supported in U. For such a choice, we would have

$$\int_{\Omega} (Lu - f)\phi = \int_{U} (Lu - f)\phi \neq 0$$

in contradiction to the equality above. □

Example 3.15 Suppose f and g are continuous functions on the open interval $I = (a,b)$. We shall show that, if T is a distribution satisfying the differential equation $T' + fT = g$, then T is a C^1 function which, consequently, is a strong solution of the equation.

First we choose a function $\phi \in C^1(I)$ such that $\phi' = f$. The function $u(x) = ce^{-\phi(x)}$, where c is constant, then satisfies the equation $u' + fu = 0$. Using the method of variation of parameters to construct a solution of $u' + fu = g$, we now assume that c is a function of x. In that case the equation $u' + fu = g$ is satisfied if

$$c'e^{-\phi} - ce^{-\phi}\phi' + ce^{-\phi} f = g$$

or

$$c'e^{-\phi} = g$$

$$c(x) = \int_{x_0}^{x} e^{\phi(t)} g(t)\, dt$$

where x_0 is a fixed point in (a,b). Since $c \in C^1(I)$, the function $u(x) = c(x) \exp(-\phi(x))$ is in $C^1(I)$ and satisfies $u' + fu = g$.

Suppose now that T is any distribution on (a,b) such that $T' + fT = g$. With $\phi' = f$, the distribution $S = e^{\phi}(T - u)$ satisfies

$$\begin{aligned} S' &= e^{\phi}\phi'(T - u) + e^{\phi}(T' - u') \\ &= e^{\phi}[(T' + fT) - (u' + fu)] \\ &= 0 \end{aligned}$$

Therefore S is a constant, say λ, and $T = u + \lambda e^{-\phi} \in C^1(I)$ is a strong solution of $T' + fT = g$.

Example 3.16 Let $L = (d^m/dx^m) + c_1(d^{m-1}/dx^{m-1}) + \cdots + c_m$ be a differential operator in $\mathbf{R}$ with constant coefficients. If $\lambda_1, \lambda_2, \ldots, \lambda_m$ are the roots of the polynomial $P(x) = x^m + c_1x^{m-1} + \cdots + c_m$, we shall show that

$$u = He^{\lambda_1 x} * He^{\lambda_2 x} * \cdots * He^{\lambda_m x}$$

is a solution of the ordinary differential equation $Lu = \delta$.

Since $P(x) = (x - \lambda_1)(x - \lambda_2) \cdots (x - \lambda_m)$, we can write

$$L = \left(\frac{d}{dx} - \lambda_1\right)\left(\frac{d}{dx} - \lambda_2\right) \cdots \left(\frac{d}{dx} - \lambda_m\right)$$

$$\begin{aligned} L\delta &= \left(\frac{d}{dx} - \lambda_1\right)\left(\frac{d}{dx} - \lambda_2\right) \cdots \left(\frac{d}{dx} - \lambda_m\right)\delta \\ &= \left(\frac{d}{dx} - \lambda_1\right)\left(\frac{d}{dx} - \lambda_2\right) \cdots \left(\frac{d}{dx} - \lambda_m\right)\delta * \delta * \cdots * \delta \\ &= \left(\frac{d}{dx} - \lambda_1\right)\delta * \left(\frac{d}{dx} - \lambda_2\right)\delta * \cdots * \left(\frac{d}{dx} - \lambda_m\right)\delta \\ &= (\delta' - \lambda_1\delta) * (\delta' - \lambda_2\delta) * \cdots * (\delta' - \lambda_m\delta) \end{aligned}$$

By writing

$$Lu = Lu * \delta = u * L\delta$$

we see that u satisfies $Lu = \delta$ if $u * L\delta = \delta$, i.e. if $u = (L\delta)^{-1}$. Referring to Example 3.9, and using the property that $(v * w)^{-1} = v^{-1} * w^{-1}$, we conclude that

$$\begin{aligned} u &= [(\delta' - \lambda_1\delta) * (\delta' - \lambda_2\delta) * \cdots * (\delta' - \lambda_m\delta)]^{-1} \\ &= (\delta' - \lambda_1\delta)^{-1} * (\delta' - \lambda_2\delta)^{-1} * \cdots * (\delta' - \lambda_m\delta)^{-1} \\ &= He^{\lambda_1 x} * He^{\lambda_2 x} * \cdots * He^{\lambda_m x}. \end{aligned}$$

When $c_1 = c_2 = \cdots = c_m = 0$, then $\lambda_1 = \lambda_2 = \cdots = \lambda_m = 0$ and we retrieve the solution of $(d^m/dx^m)u = \delta$ as given by

$$H * H * \cdots * H = \frac{1}{(m-1)!} x_+^{m-1} = \frac{1}{(m-1)!} Hx^{m-1}.$$

If v is any other solution then $(d^m/dx^m)(v - u) = 0$. This implies, in view of Example 2.21, that $v - u$ is a polynomial of degree $\leq m - 1$. Hence

$$u = \frac{1}{(m-1)!}Hx^{m-1} + b_1x^{m-1} + b_2x^{m-2} + \cdots + b_m$$

is the general solution of $d^mu/dx^m = \delta$, where $b_1, \ldots, b_m$ are arbitrary constants. These constants may be evaluated by imposing conditions on u and its derivatives at one or more points in $\mathbf{R}$.

The uniqueness of the solution to the equation $Lu = f$ in Ω generally requires additional information about u in the form of *boundary conditions*. These are usually equations involving u and its derivatives of order less than m, to be satisfied on the boundary $\partial\Omega$ or on some other hypersurface in Ω of dimension $m - 1$. Since the solutions we obtain by using the techniques of distribution theory are invariably weak solutions, the question of extending such solutions in Ω to $\partial\Omega$ and of imposing additional conditions in a lower dimension is quite complicated when u does not have the required degree of smoothness, and may not even be possible for some domains, as we shall see in Section 6.1. In practice, however, the solutions that we shall have occasion to deal with at this stage will be regular distributions, represented by functions which are smooth enough, except possibly for some isolated singular points in the interior of the domain Ω. Under such circumstances it makes sense to impose boundary conditions on the solution. For our present purposes, therefore, the theory of distributions does not so much add to the known solutions of differential equations as it justifies the methods used in obtaining them.

Because of their special role in formulating the fundamental laws of physics, second order differential equations have historically been the subject of extensive study. Indeed, the classification and much of the basic theory of linear differential equations is modeled on that of second order equations. We shall therefore devote the rest of this section to some representative examples of such equations. In $\mathbf{R}$ the second order linear differential operator

$$L = c_1(x)\frac{d^2}{dx^2} + c_2(x)\frac{d}{dx} + c_3(x) \qquad c_1 \neq 0 \tag{3.11}$$

gives rise to the general ordinary differential equation $Lu = 0$ of order 2.

Example 3.17 Let $E \in \mathcal{D}'(\mathbf{R})$ satisfy the differential equation

$$\frac{d^2}{dx^2}E = \delta \tag{3.12}$$

where δ is the Dirac distribution on $\mathbf{R}$. In Example 2.8 we found that one solution to equation (3.12) is given by

$$x_+ = xH(x)$$

Any other solution E will satisfy the homogeneous equation

$$\frac{d^2}{dx^2}[E - xH(x)] = 0$$

and must, therefore, have the form

$$E(x) = xH(x) + ax + b \tag{3.13}$$

which is a continuous function on $\mathbf{R}$. The arbitrary constants a and b may be determined by imposing boundary conditions on E. If we require, for example, that

$$E(0) = 0 \qquad E(1) = 1$$

then $a = b = 0$ and (3.13) becomes

$$E(x) = xH(x) \tag{3.14}$$

Example 3.18 The solution in $(0,1)$ of the differential equation

$$u'' = f \tag{3.15}$$

for some given $f \in L^1(0,1)$ may now be constructed by using the result of Example 3.17. If f is extended into $\mathbf{R}$ by setting $f = 0$ outside $(0,1)$, then $f \in L^1(\mathbf{R})$ and we can write

$$\begin{aligned}(f * E)'' &= f * E'' \\ &= f * \delta \\ &= f\end{aligned} \tag{3.16}$$

Thus one solution of equation (3.15) is given by

$$\begin{aligned}u(x) &= (f * E)(x) \\ &= \int_0^1 (x - \xi)H(x - \xi)f(\xi)\, d\xi\end{aligned}$$

$$= \int_0^x (x - \xi)f(\xi)\, d\xi \qquad 0 \leq x \leq 1$$

and the general solution is therefore

$$u(x) = \int_0^x (x - \xi)f(\xi)\, d\xi + ax + b \tag{3.17}$$

In terms of the initial values

$$u(0) = b \qquad u'(0) = a$$

the solution (3.17) becomes

$$u(x) = \int_0^x (x - \xi)f(\xi)\, d\xi + u'(0)x + u(0)$$

Alternatively, the constants a and b in equation (3.17) can be expressed in terms of the boundary values of u at $x = 0$ and $x = 1$ from the pair of equations

$$u(0) = b$$

$$u(1) = \int_0^1 (1 - \xi)f(\xi)\, d\xi + a + b$$

The requirement that f be integrable on (0,1) in Example 3.18 is, of course, not necessary and was only made in order to allow us to express the convolution $f * g$ as an integral. We could equally well have assumed that f is a distribution on (0,1) which can be extended to a distribution in **R** with compact support in [0,1]. In fact, equation (3.12) is really a special case of equation (3.15) where we chose f to be δ. The resulting solution E is called a *fundamental solution* of the differential operator d^2/dx^2. The function $He^{\lambda_1 x} * \cdots * He^{\lambda_m x}$, which was shown in example (3.16) to satisfy the mth order equation

$$\left(\frac{d^m}{dx^m} + \cdots + c_m\right)u = \delta$$

is a fundamental solution of the operator $d^m/dx^m + \cdots + c_m$.

In general, E is a fundamental solution of the differential operator L, given by (3.10), if $E \in \mathcal{D}'(\mathbf{R}^n)$ satisfies $LE = \delta$. The importance of the fundamental solution in the theory of differential equations lies in the fact that it allows us to solve the more general equation $Lu = f$, as shown by example (3.18). When E is a fundamental solution of L and f is a distribution with compact support in $\Omega \subset \mathbf{R}^n$, then

$$\begin{aligned} L(f * E) &= f * LE \\ &= f * \delta \\ &= f \end{aligned}$$

and $f * E$ is therefore a solution of the differential equation

$$Lu = f.$$

Example 3.19 To determine a fundamental solution for the operator

$$L = \frac{d^2}{dx^2} + \omega^2$$

where ω is a (nonzero) constant, we note that the solution of $Lu = 0$ is a linear combination of $\cos \omega x$ and $\sin \omega x$. Following the method of Example 2.13 with $f_1 = a \cos \omega x$ and $f_2 = b \sin \omega x$, we shall assume that one solution of $LE = \delta$ is given by

$$E(x) = \begin{cases} a \cos \omega x & x \leq 0 \\ b \sin \omega x & x > 0 \end{cases}$$

The condition that E be continuous at $x = 0$ implies that $f_1(0) = f_2(0)$, so $a = 0$. For E' to have a unit jump discontinuity at $x = 0$ we must have $f_2'(0) - f_1'(0) = 1$, or $b\omega = 1$. Therefore, with $a = 0$ and $b = 1/\omega$, we obtain

$$E(x) = \frac{1}{\omega} H(x) \sin \omega x$$

Note that this result can also be obtained by substituting $\lambda_1 = i$, $\lambda_2 = -i$ into the formula of Example 3.16.

The solution of the differential equation $Lu = f$, where $f \in \mathcal{E}'(\mathbf{R})$, is now given by

$$u = f * E$$

When $\operatorname{supp} f \subset [0,1]$ and f is integrable, we can write

$$u(x) = \frac{1}{\omega} \int_0^1 f(\xi) H(x - \xi) \sin \omega(x - \xi) \, d\xi$$

$$= \frac{1}{\omega} \int_0^x f(\xi) \sin \omega(x - \xi) \, d\xi \qquad 0 \leqslant x \leqslant 1.$$

The general solution of $Lu = f$ is therefore

$$u(x) = \frac{1}{\omega} \int_0^x f(\xi) \sin \omega(x - \xi) \, d\xi + c_1 \cos \omega x + c_2 \sin \omega x,$$

where c_1 and c_2 are arbitrary constants which may be determined by imposing appropriate boundary conditions on u.

Example 3.20 The general linear ordinary differential operator of order 2 with constant coefficients is given by

$$L = c_1 \frac{d^2}{dx^2} + c_2 \frac{d}{dx} + c_3 \qquad c_1 \neq 0$$

Let w_1 and w_2 be two linearly independent solutions of $Lu = 0$, which we know from the classical theory are C^∞ functions. Assuming that a fundamental solution of L has the form

$$E(x) = \begin{cases} aw_1(x) & x \leqslant 0 \\ bw_2(x) & x > 0 \end{cases}$$

and following the procedure of Example 2.13, we see that in order to satisfy $LE = \delta$ we must have

$$bw_2(0) - aw_1(0) = 0$$

$$bw_2'(0) - aw_1'(0) = 1/c_1$$

Hence

$$a = \frac{w_2(0)}{c_1 W(0)} \qquad b = \frac{w_1(0)}{c_1 W(0)}$$

where $W(x) = w_1(x)w_2'(x) - w_1'(x)w_2(x)$ is the Wronskian of the two linearly independent solutions w_1 and w_2 of $Lu = 0$, which therefore cannot vanish.

If L is the general linear differential operator given in (3.10), then we have for any $T \in \mathcal{D}'(\Omega)$

$$\begin{aligned}\langle LT,\phi\rangle &= \left\langle \sum_{|\alpha|\le m} c_\alpha(x)\partial^\alpha T,\phi\right\rangle \qquad \phi\in\mathcal{D}(\Omega)\\ &= \left\langle T, \sum_{|\alpha|\le m}(-1)^{|\alpha|}\partial^\alpha(c_\alpha(x)\phi)\right\rangle\\ &= \langle T,L^*\phi\rangle\end{aligned}$$

where the operator L^*, defined by

$$L^*\phi = \sum_{|\alpha|\le m}(-1)^{|\alpha|}\partial^\alpha(c_\alpha(x)\phi),$$

is known as the *formal adjoint* of L. We always have $(L^*)^* = L$, and when $L^* = L$ we say that L is *formally self-adjoint*. The reason for using the qualification "formal" is that we are only looking at the differential equation $Lu = f$, and have not taken account of any boundary conditions on u.

The general form of the linear ordinary differential operator of order 2 which is formally self-adjoint may be obtained by solving the equation

$$(c_1\phi)'' - (c_2\phi)' + c_3\phi = c_1\phi'' + c_2\phi' + c_3\phi$$

for the functions c_1, c_2, and c_3, where ϕ is an arbitrary test function. The equality is clearly satisfied if and only if $c_2 = c_1'$. Thus we arrive at

$$L = \frac{d}{dx}\left(p\frac{d}{dx}\right) + q$$

where $p = c_1 \neq 0$ and $q = c_3$ are C^∞ functions on $\mathbf{R}$, as representing the general formally self-adjoint linear differential operator of order 2 on $\mathbf{R}$. If w_1 and w_2 are any two linearly independent solutions of the homogeneous differential equation

$$Lu = (pu')' + qu = 0$$

then a fundamental solution of L can still be represented by

$$\begin{aligned}E(x) &= \begin{cases} aw_1(x) & x\le 0\\ bw_2(x) & x>0\end{cases}\\ &= aw_1(x) + [bw_2(x) - aw_1(x)]H(x)\end{aligned}$$

Hence

$$\begin{aligned}pE' &= p[aw_1' + (bw_2 - aw_1)'H + (bw_2 - aw_1)\delta]\\ &= p[aw_1' + (bw_2' - aw_1')H]\end{aligned}$$

if we assume that $aw_1(0) = bw_2(0)$. With $Lw_1 = Lw_2 = 0$, we have

$$\begin{aligned} LE &= (pE')' + qE \\ &= a(pw_1')' + aqw_1 + [p(bw_2' - aw_1')H]' + q(bw_2 - aw_1)H \\ &= p(0)\,[bw_2'(0) - aw_1'(0)]\delta \end{aligned}$$

Thus, if E is to be a fundamental solution, a and b must also satisfy

$$p(0)[bw_2'(0) - aw_1'(0)] = 1$$

Consequently $a = w_2(0)/p(0)W(0)$, $b = w_1(0)/p(0)W(0)$, and

$$E(x) = \frac{1}{p(0)W(0)}\{w_2(0)w_1(x) + [w_1(0)w_2(x) - w_2(0)w_1(x)]H(x)\}$$

The general solution in $(0,1)$ of the differential equation

$$(pu')' + qu = f$$

with f integrable on $(0,1)$, is therefore

$$\begin{aligned} u(x) &= f * E(x) + c_1w_1(x) + c_2w_2(x) \\ &= \frac{1}{p(0)W(0)}\left[w_1(0)\int_0^x f(\xi)w_2(x - \xi)d\xi + w_2(0)\int_x^1 f(\xi)w_1(x - \xi)d\xi\right] \\ &\quad + c_1w_1(x) + c_2w_2(x) \qquad 0 \leqslant x \leqslant 1 \end{aligned}$$

where c_1 and c_2 may be determined from the boundary conditions on u.

The method that we have used for constructing a solution to the differential equation $Lu = f$ by taking the convolution of f with a fundamental solution of L works equally well when L is a partial differential operator. Recalling the results of Examples 2.14 and 2.15,

$$\Delta \log |x| = (\partial_1^2 + \partial_2^2)\log |x| = 2\pi\delta$$

$$\Delta \frac{1}{|x|} = (\partial_1^2 + \partial_2^2 + \partial_3^2)(x_1^2 + x_2^2 + x_3^2)^{-1/2} = -4\pi\delta$$

we may now conclude that $(1/2\pi)\log|x|$ and $-1/4\pi|x|$ are fundamental solutions of Δ in $\mathbf{R}^2$ and $\mathbf{R}^3$, respectively.

Example 3.21 In $\mathbf{R}^3$ the partial differential equation

$$\Delta u = f \tag{3.18}$$

also known as the nonhomogenous *Laplace equation*, or the *Poisson equation*, has a solution which is given by

$$u = f * \left(-\frac{1}{4\pi|x|} \right)$$

when this convolution is well defined. The solution of equation (3.18) may be interpreted physically as the potential generated by f, a typical example being the Newtonian gravitational potential due to a mass density distribution given by f. When f is an integrable function with compact support, u is represented by the function

$$u(x) = -\frac{1}{4\pi} \int_{\mathbb{R}^3} \frac{f(\xi)}{|x - \xi|} d\xi \tag{3.19}$$

which clearly tends to 0 as $|x| \to \infty$, i.e., as we move away from the mass distribution. If u is to satisfy other boundary conditions then the solution (3.19) has to be supplemented by a solution of the homogeneous equation $\Delta u = 0$ before such boundary conditions can be met. We shall return to this topic in Section 6.2.

A somewhat different problem arises if the nonhomogeneous term appears in the boundary condition rather than in the differential equation, as the next two examples indicate.

Example 3.22 The temperature distribution u on a slender, infinite conducting bar as a function of time t and position x may be described by the *initial value problem*

$$\partial_t u = \partial_x^2 u \qquad (x,t) \in (-\infty,\infty) \times (0,\infty)$$

$$u(x,0) = g(x) \qquad x \in (-\infty,\infty)$$

where the differential equation governs the heat flow along the bar for all $t > 0$, and g describes the initial temperature distribution at $t = 0$.

Here the "fundamental solution" that we need to construct $u = f * E$, would have to satisfy the equation

$$(\partial_t - \partial_x^2)E(x,t) = 0 \tag{3.20}$$

on the upper half-plane $(-\infty,\infty) \times (0,\infty)$, and the condition

$$E(x,0) = \delta_{(x,0)} \tag{3.21}$$

on the boundary $t = 0$, $-\infty < x < \infty$. Such an E is given by

$$E(x,t) = \frac{1}{\sqrt{4\pi t}} \exp(-x^2/4t) \tag{3.22}$$

and in the next chapter it will be shown how this is obtained. Note that equation (3.20) is satisfied and, in order to satisfy (3.21), it suffices to show that $E(x,t)$ is a delta-convergent sequence as $t \to 0^+$ in the terminology of Section 2.4. But this has already been shown in Example 2.19, where we can choose $n = 1$ and $\lambda = 4t$.

In its dependence on, x, $E(x,t)$ is a C^∞ function which decays exponentially as $|x| \to \infty$ for every $t > 0$; so the convolution

$$u(x,t) = (g * E)(x,t) = \frac{1}{\sqrt{4\pi t}} \int_{-\infty}^{\infty} g(\xi) \exp(-(x - \xi)^2/4t)\, d\xi \qquad (3.23)$$

is well defined for a wide class of functions, including all locally integrable functions $g(x)$ whose growth as $|x| \to \infty$ is no faster than some power of x. It represents the temperature distribution u in $(-\infty,\infty) \times (0,\infty)$. Equation (3.23) clearly shows that $u \to 0$ as $t \to \infty$ or as $|x| \to \infty$.

Example 3.23 The motion of an infinite vibrating string is described by the wave equation

$$\partial_t^2 u = \partial_x^2 u \qquad (3.24)$$

where $-\infty < x < \infty$ and $0 < t < \infty$. If the string is released with initial shape u_0 and initial velocity u_1, then we have the *initial conditions*

$$u = u_0 \qquad \partial_t u = u_1 \qquad \text{at } t = 0 \qquad -\infty < x < \infty.$$

Let

$$E_0 = \tfrac{1}{2}[H(x + t) - H(x - t)]$$

$$E_1 = \partial_t E_0 = \tfrac{1}{2}[\delta(x + t) + \delta(x - t)]$$

Then the two differential equations

$$(\partial_t^2 - \partial_x^2)E_0 = 0$$

$$(\partial_t^2 - \partial_x^2)E_1 = 0$$

clearly hold in the upper half-plane $t > 0$, $-\infty < x < \infty$. Furthermore, when $t = 0$ we have

$$E_0 = 0$$

$$E_1 = \delta$$

$$\partial_t E_1 = 0$$

Consequently,

$$u = u_0 * E_1 + u_1 * E_0$$

where the convolution is taken with respect to x, is a solution of the boundary value problem. When $u_0 \in C^2(\mathbf{R})$ and $u_1 \in C^1(\mathbf{R}) \cap L^1(\mathbf{R})$ we can write this in the form

$$\begin{aligned} u(x,t) &= \tfrac{1}{2}[u_0(x - t) + u_0(x + t)] \\ &\quad + \tfrac{1}{2}\int_{-\infty}^{\infty} u_1(\xi)[H(x + t - \xi) - H(x - t - \xi)]\, d\xi \\ &= \tfrac{1}{2}[u_0(x - t) + u_0(x + t)] + \tfrac{1}{2}\int_{x-t}^{x+t} u_1(\xi)\, d\xi \end{aligned} \tag{3.25}$$

It is straightforward to verify that the expression (3.25) satisfies equation (3.24) and that $u(x,t) \to u_0(x)$ and $\partial_t u(x,t) \to u_1(x)$ as $t \to 0^+$. If the string is released from rest then $u_1 = 0$ and the solution is the average of the two traveling waves $u_0(x - t)$ and $u_0(x + t)$, both having the same shape u_0 but traveling in opposite directions with velocities ± 1.

The differential equation $\partial_t^2 u = c^2 \partial_x^2 u$, with c a positive constant, only differs from (3.24) in a change of scale of the time coordinate defined by $t \mapsto ct$. Its solution

$$u(x,t) = \tfrac{1}{2}[u_0(x - ct) + u_0(x + ct)] + \frac{1}{2c}\int_{x-ct}^{x+ct} u_1(\xi)\, d\xi \tag{3.26}$$

describes a traveling wave whose amplitude u at (x,t) depends only on the values of u_0 and u_1 in the interval $(x - ct, x + ct)$. When $u_1 = 0$, we again have a traveling wave of velocity $\pm c$. Its wave front is moving along the rays $x = \pm ct$. The contribution of the integral term, when $u_1 \neq 0$, to $u(x,t)$ depends only on the values of u_1 in the interval $(x - ct, x + ct)$ so its velocity cannot exceed c, and the position of its wave front satisfies the inequality $-ct \leqslant x \leqslant ct$. In the (x,t) plane the wave is therefore propagating in the so-called *forward wave cone* shown in Figure 3.2, and corresponding to $t \geqslant 0$. The region bounded by the lines $x = \pm ct$ with $t \leqslant 0$ is called the *backward wave cone*.

The formula (3.26) indicates that the initial disturbances u_0 and u_1 propagate in different ways. Whereas u_0 splits into two halves, each moving along an edge of the (forward) cone with speed c, u_1 propagates in a diffusive manner over the whole range of velocities within the cone. This is clearly illustrated in the limiting case when $u_0 = \delta$ and $u_1 = \delta$, for then

$$
\begin{aligned}
u &= u_0 * E_1 + u_1 * E_0 \\
&= E_1 + E_0 \\
&= \tfrac{1}{2}[\delta(x + ct) + \delta(x - ct)] + \frac{1}{2c}[H(x + ct) - H(x - ct)]
\end{aligned}
$$

Examples (3.21), (3.22), and (3.23) are typical of the three classical types of second order partial differential equations with constant coefficients: the *elliptic, parabolic* and *hyperbolic* equations. In its homogeneous form, any second order partial differential equation with constant coefficients may be transformed, by an appropriate change of coordinates, to one of the following forms:

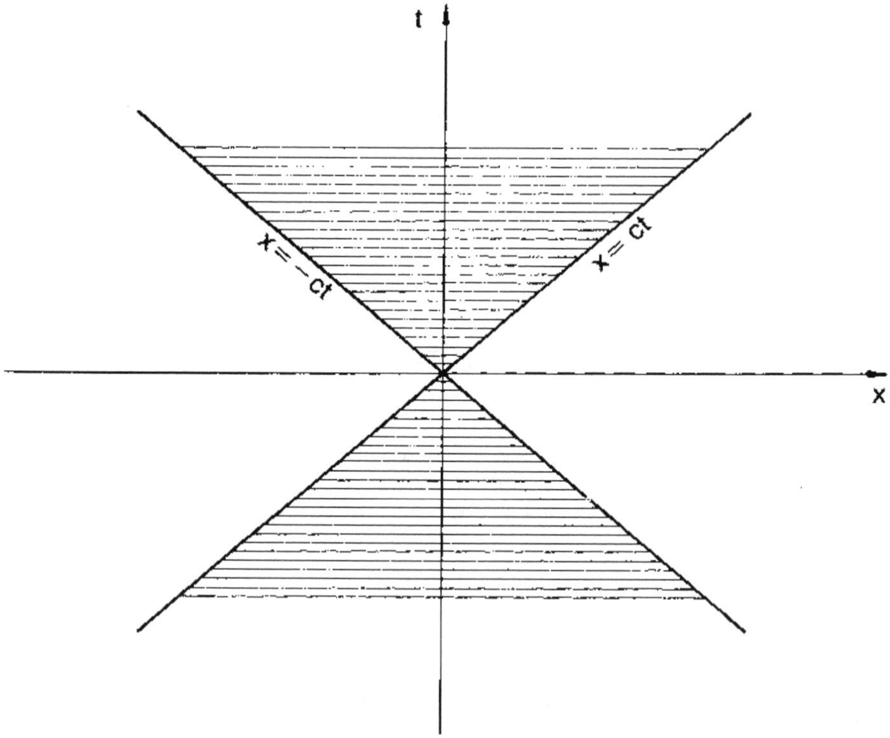

Figure 3.2

$\Delta u = 0$	Laplace's equation
$(\partial_t - \Delta)u = 0$	heat equation
$(\partial_t^2 - \Delta)u = 0$	wave equation

where Δ is the Laplacian operator in $\mathbf{R}^n$. If we replace ∂_t by τ and ∂_k by ξ_k in the above operators, the Laplacian Δ becomes a polynomial in $\xi_1, \ldots, \xi_n$ whose level surfaces are spherical or, up to a change of scale, elliptical. The heat operator corresponds to the parabolic surface $\tau - |\xi|^2 = 0$ and the wave operator to the hyperbolic surface $\tau^2 - |\xi|^2 = 0$. We shall return to these equations in Chapter 6.

EXERCISES

3.1 Give an example of a sequence of functions in $\mathcal{D}(\mathbf{R})$ which converges to 0 in $\mathcal{E}(\mathbf{R})$ but not in $\mathcal{D}(\mathbf{R})$, and one of a sequence in $\mathcal{E}'(\mathbf{R})$ which converges to 0 in $\mathcal{D}'(\mathbf{R})$ but not in $\mathcal{E}'(\mathbf{R})$.

3.2 Prove that $\mathcal{D}'_+(\mathbf{R})$ is a commutative and associative algebra under the convolution operation $*$.

3.3 Compute the following convolutions of distributions in $\mathcal{D}'(\mathbf{R})$:

(a) $x_+ * (H \sin x)$

(b) $\delta' * x_+^{-1/2}$

(c) $H * x^{-1}$

3.4 Show that if $\phi \in C^\infty(\Omega_1 \times \Omega_2)$ has compact support as a function of x and y separately, then $T_1(\phi) \in C_0^\infty(\Omega_2)$ for every $T_1 \in \mathcal{E}'(\Omega_1)$.

3.5 If P is a polynomial of degree $\leq m$ and T is a distribution with compact support, show that $P * T$ is a polynomial of degree $\leq m$.

3.6 Prove that

$$\psi_k(x) = \frac{k}{\pi^{n/2}}\left(1 - \frac{|x|^2}{k}\right)^{k^3} \qquad x \in \mathbf{R}^n$$

is a delta-convergent sequence. Deduce from this that every distribution with compact support in $\mathbf{R}^n$ is the limit of a sequence of polynomials.

3.7 Let T be a distribution in $\mathbf{R}$ such that $T' \in L^2(\mathbf{R})$. Show that $T \in L^2_{\text{loc}}(\mathbf{R})$.

3.8 Find all $u \in \mathcal{D}'(\mathbf{R})$ such that $x^m u = \delta$, $m \in \mathbf{N}$.

3.9 Find all $u \in \mathcal{D}'(\mathbf{R}^2)$ such that $(x + iy)u = 0$.

3.10 Obtain the general (weak) solution of the differential equation

(a) $xu'' = 0$

(b) $x^m u' = 0 \qquad m \in \mathbf{N}$

3.11 Solve the differential equation $xu' + \lambda u = 0$ in $\mathcal{D}'(\mathbf{R})$, where λ is a nonnegative integer. Discuss the possibility of solving the equation for other values of λ.

3.12 Solve the equation $(\sin x)u = 0$ in $\mathcal{D}'(\mathbf{R})$.

3.13 Use the result of Example 3.9 to solve $(a\delta' - b\delta) * v = f$, where f is a given distribution in $\mathbf{R}$ with compact support.

3.14 Obtain the general solution of $x^n T^{(m)} = 0$ in $\mathbf{R}$.

3.15 If $T \in \mathcal{D}'(\mathbf{R}^2)$, show that $xT = 0$ if and only if $T = \delta \otimes u$, where $u \in \mathcal{D}'(\mathbf{R})$ and δ is the Dirac distribution with respect to the variable x on $\mathbf{R}$.

3.16 Given $E_j = [(j - 1)!]^{-n} x_+^{j-1}$, $x \in \mathbf{R}$, prove that $E_j * E_k = E_{j+k}$ for all $j,k \in \mathbf{N}$. Generalize this to $E_\lambda * E_\mu = E_{\lambda+\mu}$ for all $\lambda, \mu \in \mathbf{C}$ and Re λ, Re μ both positive.

3.17 Determine a fundamental solution of the Helmholtz operator $\Delta + k^2$ in $\mathbf{R}^3$.

3.18 Show that the solution (3.23) of the initial value problem for the heat equation is a C^∞ function in $\mathbf{R} \times (0,\infty)$ when $g \in L^1(\mathbf{R})$.

4

Fourier Transforms and Tempered Distributions

4.1 THE CLASSICAL FOURIER TRANSFORMATION IN L^2

We start by presenting a brief outline of the classical theory of the Fourier transformation, with a view to extending the definitions and results from functions on $\mathbf{R}^n$ to distributions. Throughout this chapter the underlying open set Ω will be $\mathbf{R}^n$ and we shall write L^p, $\mathcal{D}$, $\mathcal{D}'$, etc. for $L^p(\mathbf{R}^n)$, $\mathcal{D}(\mathbf{R}^n)$, $\mathcal{D}'(\mathbf{R}^n)$, etc.

The *Fourier transform* of a function $f \in L^1$ is a function $\mathcal{F}(f) = \hat{f}$ on $\mathbf{R}^n$ defined by

$$\hat{f}(\xi) = \int e^{-i\langle x,\xi\rangle} f(x)\, dx \qquad \xi \in \mathbf{R}^n \tag{4.1}$$

where $\langle x,\xi\rangle = \sum_{j=1}^n x_j\xi_j$. The *Fourier transformation* is the mapping $\mathcal{F}: f \mapsto \hat{f}$ defined, so far, on L^1.

From equation (4.1) we have

$$|\hat{f}(\xi)| \leq \int |f(x)|\, dx = \|f\|_1$$

Furthermore, if ξ_k is a sequence in $\mathbf{R}^n$ which converges to ξ, then since

$$|\hat{f}(\xi_k) - \hat{f}(\xi)| \leq \int |f(x)|\, |e^{-i\langle x,\xi_k\rangle} - e^{-i\langle x,\xi\rangle}|\, dx$$

and $|e^{-i\langle x,\xi_k\rangle} - e^{-i\langle x,\xi\rangle}| \to 0$ as $\xi_k \to \xi$, we conclude that $\hat{f}(\xi_k) \to \hat{f}(\xi)$ by Lebesgue's convergence theorem. Thus $\hat{f}$ is a bounded continuous function on $\mathbf{R}^n$. From the Riemann-Lebesgue lemma [4], it follows that $\hat{f}(\xi) \to 0$ as $|\xi| \to \infty$. But in general $\hat{f}$ may not be integrable, as may be seen by taking the Fourier transform of the function in $\mathbf{R}$ which equals 1 on the open interval $(0,1)$ and 0 otherwise. A simple calculation gives the result as $2 \sin \xi/\xi$, which is not in L^1.

When $\hat{f} \in L^1$,

$$f(x) = (2\pi)^{-n} \int e^{i\langle \xi,x\rangle}\, \hat{f}(\xi)\, d\xi \tag{4.2}$$

almost everywhere [4]. Since the right-hand side is continuous the equality holds everywhere provided f, which was assumed to be integrable, is also continuous.

For any two functions $f,g \in L^1$ we obviously have the linearity property $\mathcal{F}(af + bg) = a\hat{f} + b\hat{g}$, $a,b \in \mathbf{C}$. Consequently, if C^0_∞ is the Banach space of continuous functions on $\mathbf{R}^n$ which tend to 0 at ∞, equipped with the norm $\|f\| = |f|_0 = \sup\{|f(x)| : x \in \mathbf{R}^n\}$, then the Fourier transformation $\mathcal{F}$ satisfies the inequality $\|\mathcal{F}(f)\| = |\hat{f}|_0 \leq \|f\|_1$ and is therefore an injective, continuous linear map from L^1 to C^0_∞.

When $f,g \in L^1$ we also have $f\hat{g} \in L^1$, since $\hat{g}$ is bounded, and

$$\begin{aligned}\int f(x)\,\hat{g}(x)\,dx &= \int f(x) \int g(\xi) e^{-i\langle x,\xi\rangle}\, d\xi\, dx \\ &= \int g(\xi) \int f(x) e^{-i\langle x,\xi\rangle}\, dx\, d\xi\end{aligned}$$

by Fubini's theorem. Therefore

$$\int f(x)\hat{g}(x)\,dx = \int g(\xi)\hat{f}(\xi)\,d\xi \tag{4.3}$$

The relation (4.3) points the way to extending the definition of the Fourier transformation from L^1 to $\mathcal{D}'$ by taking one of the two functions as a test function and the other as a distribution, but the extension is not as simple as it might seem. If we take g in $\mathcal{D}$ then (4.3) may be written as

$$\langle \hat{f}, g\rangle = \langle f, \hat{g}\rangle \tag{4.4}$$

which would allow an extension of the Fourier transformation to $\mathcal{D}'$ provided $\hat{g}$ is in $\mathcal{D}$. But in fact $\hat{g}$ is analytic (see Section 4.5) when $g \in \mathcal{D}$ and cannot have compact support unless it is identically zero. This indicates that $\mathcal{D}$ is too small as a space of test functions, or, equivalently, that $\mathcal{D}'$ is too large for the purpose of extending the Fourier transformation.

On the other hand, if g is taken in $\mathcal{E}$ it may not be integrable and, as a consequence, its Fourier transform may not exist. So it would seem that $\mathcal{E}$ is too big as a space of test functions, and what we should seek is a new space of test functions larger than $\mathcal{D}$ and smaller than $\mathcal{E}$ which would allow an extension of the Fourier transformation through the relation (4.4). Such a test function space, call it X, should meet certain conditions in order to serve our purpose:

(i) X should be a subspace of C^∞ in order that the distributions in X' have derivatives of all orders.
(ii) The Fourier transformation should be "well behaved" on X, in the sense that it maps X onto itself.
(iii) Since $\partial_k \mathcal{F}(\phi) = -i\mathcal{F}(x_k\phi)$, X should be closed under multiplication by polynomials.

With these conditions, we should also choose X as small as possible, in order that X' be as large as possible. That is what we propose to do in the next section.

4.2 TEMPERED DISTRIBUTIONS

A function $\phi \in C^\infty$ is said to be *rapidly decreasing* if

$$\sup_{x\in\mathbf{R}^n} |x^\alpha \partial^\beta \phi(x)| < \infty$$

for all pairs of multi-indices α and β. This is equivalent to the condition that

$$\lim_{|x|\to\infty} |x^\alpha \partial^\beta \phi(x)| = 0$$

It is also equivalent to the condition that

$$\sup_{|\beta|\leq m}\ \sup_{x\in\mathbf{R}^n} (1 + |x|^2)^m\, |\partial^\beta \phi(x)| < \infty$$

for all $m \in \mathbf{N}_0$. We shall use $\mathcal{S}$ to denote the set of all rapidly decreasing functions, which is clearly a linear space under the usual operations of addition and multiplication by scalars. A function in $\mathcal{S}$ approaches 0 as $|x| \to \infty$ faster than any power of $1/|x|$. An example of such a function is $\exp(-|x|)$.

For any $\phi \in \mathcal{S}$ we define the seminorms

$$p_{\alpha\beta}(\phi) = \sup_{x \in \mathbf{R}^n} |x^\alpha \partial^\beta \phi(x)|$$

with $\alpha, \beta \in \mathbf{N}_0^n$. The countable family $\{p_{\alpha\beta}\}$ defines a Hausdorff, locally convex, topology on $\mathcal{S}$ which is metrizable and complete. With this topology, $\mathcal{S}$ is therefore a Fréchet space, and a sequence (ϕ_k) converges to 0 in $\mathcal{S}$ if and only if

$$x^\alpha \partial^\beta \phi_k(x) \to 0$$

uniformly on $\mathbf{R}^n$ as $k \to \infty$. Furthermore, $x^\alpha \partial^\beta \phi$ is clearly in $\mathcal{S}$ whenever ϕ is in $\mathcal{S}$ for any pair, α, $\beta \in \mathbf{N}_0^n$.

Theorem 4.1 The topological vector spaces $\mathcal{D}$, $\mathcal{S}$ and $\mathcal{E}$ are related by $\mathcal{D} \subset \mathcal{S} \subset \mathcal{E}$, with continuous injection. Moreover, $\mathcal{D}$ is a dense subspace of $\mathcal{S}$ and $\mathcal{S}$ is a dense subspace of $\mathcal{E}$.

Proof. The inclusion relations clearly hold between $\mathcal{D}$, $\mathcal{S}$, and $\mathcal{E}$ as sets. For any sequence (ϕ_k) in $\mathcal{D}$ which converges to 0 there is a compact set $K \subset \mathbf{R}^n$ such that (ϕ_k) lies in $\mathcal{D}_K$ and converges to 0 in $\mathcal{D}_K$, by Theorem 2.3. Hence $\phi_k \to 0$ in $\mathcal{S}$. On the other hand, if the sequence (ϕ_k) in $\mathcal{S}$ converges to 0 then, for any $\alpha \in \mathbf{N}_0^n$, $\partial^\alpha \phi_k \to 0$ uniformly on every compact subset of $\mathbf{R}^n$. This means that (ϕ_k) converges to 0 in $\mathcal{E}$, and the first part of the theorem is proved.

The second part follows from the simple observation that $\mathcal{D}$ is dense in $\mathcal{E}$, as we have seen in the proof of Theorem 3.1. □

Theorem 4.2 $\mathcal{S}$ is a dense subspace of L^p, $1 \leq p < \infty$, with the identity map from $\mathcal{S}$ into L^p continuous.

Proof. Let $\phi \in \mathcal{S}$. Since $(1 + |x|^2)^m \phi$ is in $\mathcal{S}$ for any positive integer m, we obviously have $\phi \in L^p$. If we now let $\phi_k \to 0$ in $\mathcal{S}$, then

$$\sup_{x \in \mathbf{R}^n} (1 + |x|^2)^m |\phi_k(x)|^p \to 0$$

for every m as $k \to \infty$. When $m > (1/2)\, n$, $(1 + |x|^2)^{-m}$ is integrable and we then have

$$\begin{aligned} \|\phi_k\|_p^p &= \int (1 + |x|^2)^m |\phi_k(x)|^p (1 + |x|^2)^{-m}\, dx \\ &\leq M \sup_{x \in \mathbf{R}^n} (1 + |x|^2)^m |\phi_k(x)|^p \end{aligned}$$

Therefore $\phi_k \to 0$ in L^p. Since $\mathcal{D}$ is dense in L^p, by Theorem 2.11, so is $\mathcal{S}$. □

This theorem shows that the convolution $\phi * \psi$ of any pair of functions ϕ, ψ in $\mathcal{S}$ is well defined in $\mathbf{R}^n$ and is in fact an $\mathcal{S}$ function. To see this we note that the integral

$$(\phi * \psi)(x) = \int \phi(x - y)\, \psi(y)\, dy$$

is uniformly convergent in $\mathbf{R}^n$. Therefore we can write

$$\begin{aligned}\sup_{x\in\mathbf{R}^n} |x^\alpha \partial^\beta(\phi * \psi)(x)| &\leq \int \sup_{x\in\mathbf{R}^n} |x^\alpha\, \partial^\beta\, \phi(x - y)|\, |\psi(y)|\, dy \\ &\leq M \int |\psi(y)|\, dy \\ &< \infty\end{aligned}$$

Theorem 4.1 implies that the relation

$$\mathscr{E}' \subset \mathcal{S}' \subset \mathscr{D}'$$

must hold between the topological dual spaces with the identity maps from $\mathscr{E}'$ to $\mathcal{S}'$ and from $\mathcal{S}'$ to $\mathscr{D}'$ continuous. We have seen that every locally integrable function f on $\mathbf{R}^n$ defines a distribution in $\mathscr{D}'$ by

$$\phi \mapsto \int f\phi \qquad \phi \in \mathscr{D}$$

For the function f to define a distribution in $\mathcal{S}'$ by this mapping, with $\phi \in \mathcal{S}$, it must, in addition to being locally integrable, satisfy a growth condition at ∞ : f cannot grow faster than some power of x as $|x| \to \infty$, otherwise the integral $\int f\phi$ will not be defined. The exponential function $e^{|x|}$, for example, does not define a distribution in $\mathcal{S}'$. Loosely speaking, we can say that the elements of $\mathcal{S}'$ are the distributions of polynomial growth as $|x| \to \infty$ (see, however, Example 4.1). Hence they are called *tempered distributions*.

Now we list some observations regarding the structure and properties of $\mathcal{S}$ and its dual space $\mathcal{S}'$:

(i) Any polynomial function f on $\mathbf{R}^n$ defines a tempered distribution by the formula

$$\langle f, \phi\rangle = \int f(x)\phi(x)\, dx \qquad \phi \in \mathcal{S}$$

This follows from

$$\begin{aligned}|\langle f,\phi\rangle| &\leq \int |f(x)\, \phi(x)|\, dx \\ &\leq M \sup_{x\in\mathbf{R}^n} (1 + |x|^2)^m\, |\phi(x)|\end{aligned}$$

for m > (1/2) $(n + k)$, with $M = \|f(x)\,(1 + |x|^2)^{-m}\|_1$ and k equal to the degree of the polynomial f.

(ii) The same definitions and properties of convergence, differentiation, translation and reflection in the origin which were given in $\mathcal{D}'$ apply to the elements of $\mathcal{S}'$. Since $\mathcal{S}$ is closed under multiplication by polynomials, we can define the product of a polynomial P on $\mathbf{R}^n$ with a tempered distribution by

$$PT(\phi) = T(P\phi) \qquad \phi \in \mathcal{S}$$

This definition clearly extends to any C^∞ function f with polynomial growth at ∞, i.e. an $f \in C^\infty$ for which there is a positive integer m such that $|x|^{-m}|\partial^\alpha f(x)|$ remains bounded as $|x| \to \infty$ for all $\alpha \in \mathbf{N}_0^n$. Thus the linear space of multipliers of $\mathcal{D}'$, which is C^∞, is also "tempered" by a growth condition before it can serve as a linear space of multipliers of $\mathcal{S}'$.

(iii) If $1 \leqslant p < \infty$ and $\phi \in \mathcal{S}$, then for any positive integer m

$$|\phi(x)| = (1 + |x|^2)^{-m}\,(1 + |x|^2)^m|\phi(x)| \leqslant M(1 + |x|^2)^{-m}$$

where $M = \sup\{(1 + |x|^2)^m\,|\phi(x)| : x \in \mathbf{R}^n\}$. Now $|\phi|^p$ is integrable if $m > (1/2)\,n/p$. Hence $\mathcal{S} \subset L^p$. Since any $\phi \in \mathcal{S}$ is bounded on $\mathbf{R}^n$, we also have $\mathcal{S} \subset L^\infty$. Thus $\mathcal{S}$ is a subspace of L^p for $1 \leqslant p \leqslant \infty$.

(iv) $L^p \subset \mathcal{S}'$ for $1 \leqslant p \leqslant \infty$, because if $f \in L^p$ and ϕ is any C^∞ function with compact support K, then

$$\begin{aligned}|\langle f,\phi\rangle| &= \left|\int_K f(x)\phi(x)\,dx\right| \\ &= \left|\int \phi(x)I_K(x)f(x)\,dx\right| \\ &\leqslant M\,|\phi|_0\,\|f\|_p\end{aligned}$$

by Hölder's inequality. Thus f defines a continuous linear functional on C_0^∞ in the topology induced by $\mathcal{S}$. But since C_0^∞ is dense in $\mathcal{S}$, f can be extended to a continuous linear functional of $\mathcal{S}$.

More generally, any locally integrable function f such that $|x|^{-m}|f(x)|$ is bounded (almost everywhere) as $|x| \to \infty$, for some positive integer m, defines a distribution in $\mathcal{S}'$. On the other hand, consider the following example:

Example 4.1 The function $f(x) = e^x \sin(e^x)$, $x \in \mathbf{R}$, cannot be dominated at ∞ by a polynomial because there is no positive integer m such that $x^{-m}|f(x)| = x^{-m}\,e^x|\sin(e^x)|$ remains bounded as $x \to \infty$. However, if $\phi \in \mathcal{S}(\mathbf{R})$, then

$$\left|\int f(x)\phi(x)\,dx\right| = \left|\int e^x \sin(e^x)\,\phi(x)\,dx\right|$$
$$= \left|\int \phi(x)\,d(-\cos(e^x))\right|$$
$$= \left|\int \cos(e^x)\,\phi'(x)\,dx\right|$$
$$\leq \int |\phi'(x)|dx$$
$$= \int (1 + x^2)|\phi'(x)|\,\frac{1}{1 + x^2}\,dx$$
$$\leq M \sup(1 + x^2)|\phi'(x)|$$

Thus f defines a distribution in $\mathcal{S}'(\mathbf{R})$.

(v) Since $\mathcal{D}_F \subset \mathcal{S}$ and convergence in $\mathcal{D}_F$ implies convergence in $\mathcal{S}$, we therefore have the inclusion $\mathcal{S}' \subset \mathcal{D}_F'$. Thus every tempered distribution is of finite order. From Theorem 3.9, we conclude that every tempered distribution is a derivative of some continuous function of polynomial growth. Note that the tempered distribution $e^x \sin(e^x)$ in the above example is the first derivative of the bounded function $-\cos(e^x)$. The various powers x_+^λ, x_-^λ and $|x|^\lambda$ that were discussed in Section 2.8 are examples of tempered distributions, each being dominated at $\pm\infty$ by $|x|^m$ whenever $m \geq \operatorname{Re}\lambda$.

4.3 FOURIER TRANSFORMATION IN $\mathcal{S}$

Since $\mathcal{S} \subset L^1$ the Fourier transform $\hat{\phi}$ of any $\phi \in \mathcal{S}$ exists, and moreover

$$\partial_k\hat{\phi}(\xi) = \partial_k \int e^{-i\langle x,\xi\rangle}\,\phi(x)\,dx$$
$$= \int \frac{\partial}{\partial\xi_k}\,e^{-i\langle x,\xi\rangle}\,\phi(x)\,dx$$
$$= -i\int e^{-i\langle x,\xi\rangle}\,x_k\phi(x)\,dx$$
$$= -i\mathcal{F}(x_k\phi)$$

The second equality, where differentiation is carried inside the integral, is justified by the uniform convergence of the integral as a function of ξ.

On the other hand, we also have

$$\mathcal{F}(\partial_k\phi)(\xi) = \int e^{-i(x,\xi)}\, \partial_k\phi(x)\, dx$$
$$= i\xi_k \int e^{-i(x,\xi)}\phi(x)\, dx$$
$$= i\xi_k\hat{\phi}\ (\xi)$$

where the second equality follows from integrating by parts. Using the notation $D_k = -i\partial_k$, we therefore arrive at the relations

$$\mathcal{F}(D_k\ \phi) = \xi_k\mathcal{F}(\phi)$$
$$\mathcal{F}\ (x_k\ \phi) = -D_k\mathcal{F}(\phi)$$

This process may be repeated any number of times, and with respect to any index, so we conclude that

$$\mathcal{F}(D^\alpha\ \phi) = \xi^\alpha\mathcal{F}(\phi) \tag{4.5}$$
$$\mathcal{F}(x^\alpha\phi) = (-1)^{|\alpha|}D^\alpha\mathcal{F}(\phi) \tag{4.6}$$

where $D^\alpha = (-\mathrm{i})^{|\alpha|}\partial^\alpha$ and $\alpha = (\alpha_1, \ldots, \alpha_n)$

Theorem 4.3 The Fourier transformation is a continuous linear map from $\mathcal{S}$ into $\mathcal{S}$.

Proof. For any $\phi \in \mathcal{S}$ and $\alpha,\beta \in \mathbf{N}_0^n$ the relations (4.5) and (4.6) imply that

$$\xi^\alpha D^\beta\ \hat{\phi}(\xi) = \xi^\alpha(-1)^{|\beta|}\ \mathcal{F}(x^\beta\ \phi)$$
$$= \mathcal{F}(D^\alpha\ (-x)^\beta\ \phi)$$
$$= \int e^{-i(x,\xi)}\ D^\alpha[(-x)^\beta\ \phi(x)]\ dx$$
$$|\xi^\alpha D^\beta\ \hat{\phi}(\xi)| \leq \int |D^\alpha[x^\beta\ \phi(x)]|\ dx$$
$$= \int (1 + |x|^2)^{-m}\ (1 + |x|^2)^m\ |D^\alpha\ [x^\beta\ \phi(x)]|\ dx$$

We can choose m so that $\int (1 + |x|^2)^{-m}\ dx = M < \infty$, and then

$$|\xi^\alpha D^\beta\ \hat{\phi}(\xi)| \leq \sup_{x\in\mathbf{R}^n} (1 + |x|^2)^m\ |D^\alpha[x^\beta\ \phi(x)]|M$$

Since ϕ is in $\mathcal{S}$, the right-hand side of this inequality is finite and hence $\hat{\phi}$ is also in $\mathcal{S}$. Now $\mathcal{F}$ is linear and $\hat{\phi} \to 0$ as $\phi \to 0$ in $\mathcal{S}$, so $\mathcal{F}$ is continuous on $\mathcal{S}$. □

We have already indicated that the inversion relation (4.2),

$$f(x) = (2\pi)^{-n} \int e^{i(\xi,x)} \hat{f}(\xi)\, d\xi$$

holds (almost everywhere) whenever f and $\hat{f}$ are integrable. If we denote the right-hand side by $\mathcal{F}^{-1}(\hat{f})(x)$ we see that $\mathcal{F}^{-1}$ is of the same form as the map $\mathcal{F}$. By Theorem 4.3 $\mathcal{F}^{-1}$ is therefore a continuous map from $\mathcal{S}$ into $\mathcal{S}$. We shall now prove (4.2) for any function in $\mathcal{S}$, independently of the classical result. This is all we need in order to extend $\mathcal{F}$ to $\mathcal{S}'$. The proof utilizes the relation

$$\mathcal{F}(\exp(-\tfrac{1}{2}|x|^2)) = (2\pi)^{n/2} \exp(-\tfrac{1}{2}|\xi|^2)$$

which is derived in the following example.

Example 4.2 Let $\gamma(x) = \exp(-\tfrac{1}{2}|x|^2)$, $x \in \mathbf{R}^n$. For $n = 1$, γ satisfies the differential equation

$$\gamma'(x) + x\gamma(x) = 0 \qquad x \in \mathbf{R}$$

Taking the Fourier transform and using (4.5) and (4.6) we obtain

$$\xi\hat{\gamma}(\xi) + (\hat{\gamma})'(\xi) = 0 \qquad \xi \in \mathbf{R}$$

whose solution is given by

$$\hat{\gamma}(\xi) = c \exp(-\tfrac{1}{2}\xi^2)$$

with $c = \hat{\gamma}(0) = \int_{-\infty}^{\infty} \exp((-1/2)\, x^2)\, dx = (2\pi)^{1/2}$. When $n \geq 1$ we can write

$$\begin{aligned}
\hat{\gamma}(\xi) &= \int_{\mathbf{R}^n} \prod_{k=1}^{n} \exp(-ix_k\, \xi_k) \exp(-\tfrac{1}{2}x_k^2)\, dx \\
&= \prod_{k=1}^{n} \int_{-\infty}^{\infty} \exp(-ix_k\, \xi_k - \tfrac{1}{2}x_k^2)\, dx_k \\
&= \prod_{k=1}^{n} \hat{\gamma}(\xi_k) \\
&= (2\pi)^{(1/2)n} \exp(-\tfrac{1}{2}|\xi|^2)
\end{aligned}$$

Theorem 4.4 If $\phi \in \mathcal{S}$, then

$$\phi(x) = \mathcal{F}^{-1}(\hat{\phi})(x) = (2\pi)^{-n} \int e^{i(x,\xi)}\, \hat{\phi}(\xi)\, d\xi \tag{4.7}$$

Proof. For any $\phi,\psi \in \mathcal{S}$ we have, by Fubini's theorem,

$$
\begin{aligned}
\int \hat{\phi}(x)\, \psi(x)\, e^{i\langle\xi,x\rangle}\, dx &= \int \left[\int e^{-i\langle y,x\rangle}\, \phi(y)\, dy \right] \psi(x)\, e^{i\langle\xi,x\rangle}\, dx \\
&= \int \phi(y) \left[\int e^{-i\langle y-\xi,x\rangle}\, \psi(x)\, dx \right] dy \\
&= \int \phi(y)\, \hat{\psi}(y - \xi)\, dy \\
&= \int \phi(\xi + y)\, \hat{\psi}(y)\, dy \qquad (4.8)
\end{aligned}
$$

Furthermore, when $\psi \in \mathcal{S}$ and $\varepsilon > 0$,

$$
\begin{aligned}
\mathcal{F}(\psi(\varepsilon x))(y) &= \int e^{-i\langle y,x\rangle}\, \psi(\varepsilon x)\, dx \\
&= \int e^{-i\langle y,\xi/\varepsilon\rangle}\, \psi(\xi)\, d\xi/\varepsilon^n \\
&= \varepsilon^{-n}\, \hat{\psi}\,(y/\varepsilon)
\end{aligned}
$$

Using this in equation (4.8),

$$
\begin{aligned}
\int \hat{\phi}(x)\, \psi(\varepsilon x) e^{i\langle\xi,x\rangle}\, dx &= \int \phi(\xi + y)\, \mathcal{F}(\psi(\varepsilon x))(y)\, dy \\
&= \int \phi(\xi + y)\, \hat{\psi}(y/\varepsilon)\, dy/\varepsilon^n \\
&= \int \phi(\xi + y)\, \hat{\psi}(y/\varepsilon)\, d(y/\varepsilon) \\
&= \int \phi(\xi + \varepsilon y)\, \hat{\psi}(y)\, dy
\end{aligned}
$$

Since these integrals are uniformly convergent, we can take the limit as $\varepsilon \to 0$ inside the integral sign. The result is

$$\psi(0) \int \hat{\phi}(x)\, e^{i\langle\xi,x\rangle}\, dx = \phi(\xi) \int \hat{\psi}(y)\, dy$$

If we choose $\psi(x) = \exp\left(-\tfrac{1}{2}|x|^2\right)$, then $\psi(0) = 1$ and

$$\int \hat{\psi}(y)\, dy = (2\pi)^{\frac{1}{2}n} \int \exp\left(-\tfrac{1}{2}|y|^2\right) dy = (2\pi)^n$$

The equality (4.7) immediately follows. □

From Example 4.2 and this theorem, we obtain the equation

$$\mathcal{F}^{-1}\left(e^{-(1/2)|\xi|^2}\right) = (2\pi)^{-(1/2)n}\, e^{-(1/2)|x|^2}$$

Theorems (4.3) and (4.4) together imply that the Fourier transformation defines a *topological isomorphism* from $\mathcal{S}$ onto $\mathcal{S}$, i.e., a bijection from $\mathcal{S}$ to $\mathcal{S}$ which preserves the algebraic properties of the linear space $\mathcal{S}$ because it is linear, and the topological properties of $\mathcal{S}$ because it is a homeomorphism. The next theorem exhibits some fundamental relations involving the Fourier transformation in $\mathcal{S}$. Recall that $\phi\psi$ and $\phi * \psi$ are both in $\mathcal{S}$ when $\phi, \psi \in \mathcal{S}$.

Theorem 4.5 If $\phi, \psi \in \mathcal{S}$ then

$$\int \hat{\phi}\, \psi = \int \phi\, \hat{\psi} \tag{4.9}$$

$$\int \phi\, \overline{\psi} = (2\pi)^{-n} \int \hat{\phi}\, \bar{\hat{\psi}} \tag{4.10}$$

$$\mathcal{F}(\phi * \psi) = \hat{\phi}\, \hat{\psi} \tag{4.11}$$

$$\mathcal{F}(\phi\psi) = (2\pi)^{-n}\, \hat{\phi} * \hat{\psi} \tag{4.12}$$

Equation (4.10) is known as *Parseval's relation*.

Proof. Equation (4.9) follows from equation (4.8) by setting $\xi = 0$. To prove (4.10) we note that, in view of (4.7),

$$\begin{aligned}
\mathcal{F}(\bar{\hat{\psi}})(\xi) &= \int e^{-i\langle \xi, x\rangle}\, \bar{\hat{\psi}}(x)\, dx \\
&= \text{c.c.} \int e^{i\langle \xi, x\rangle}\, \hat{\psi}(x)\, dx \\
&= (2\pi)^n \overline{\psi}(\xi)
\end{aligned}$$

where c.c. denotes complex conjugation. Equation (4.10) now follows if ψ is replaced by $(2\pi)^{-n}\, \bar{\hat{\psi}}$ in (4.9).

To prove (4.11) we use Fubini's theorem:

$$\begin{aligned}
\mathcal{F}(\phi * \psi)(\xi) &= \int e^{-i\langle \xi, x\rangle} \left[\int \phi(y)\, \psi(x - y)\, dy\right] dx \\
&= \int \phi(y) \left[\int e^{-i\langle \xi, x\rangle}\, \psi(x - y)\, dx\right] dy \\
&= \int \phi(y) \left[\int e^{-i\langle \xi, y + \eta\rangle}\, \psi(\eta)\, d\eta\right] dy \\
&= \int e^{-i\langle \xi, y\rangle}\, \phi(y)\, dy \int e^{-i\langle \xi, \eta\rangle}\, \psi(\eta)\, d\eta \\
&= \hat{\phi}(\xi)\, \hat{\psi}(\xi)
\end{aligned}$$

Finally, the inversion formula (4.7) implies

$$\phi(x) = (2\pi)^{-n} \int e^{i(x,\xi)}\, \hat{\phi}(\xi)\, d\xi = (2\pi)^{-n}\, \hat{\hat{\phi}}(-x)$$
$$\hat{\hat{\phi}}\,(x) = (2\pi)^{n}\, \phi(-x) \tag{4.13}$$

Now we use equation (4.11) to obtain

$$\begin{aligned}\hat{\phi} * \hat{\psi}\,(\xi) &= \mathcal{F}^{-1}\,(\hat{\hat{\phi}}\,\hat{\hat{\psi}})\,(\xi)\\ &= (2\pi)^{-n} \int e^{i(\xi,x)}\, \hat{\hat{\phi}}(x)\, \hat{\hat{\psi}}(x)\, dx\\ &= (2\pi)^{n} \int e^{i(\xi,x)}\, \phi(-x)\, \psi(-x)\, dx\\ &= (2\pi)^{n} \int e^{-i(\xi,x)}\, \phi(x)\, \psi(x)\, dx\\ &= (2\pi)^{n}\, \mathcal{F}(\phi\, \psi)(\xi)\end{aligned}$$

□

Example 4.3 Equation (4.11) can be used to construct two nonzero functions $\phi, \psi \in \mathcal{S}$ such that $\phi * \psi = 0$.

Let ϕ_0 and ψ_0 be two nonzero functions in $\mathcal{D}$ such that supp $\phi_0 \cap$ supp $\psi_0 = \emptyset$, and define $\phi = \mathcal{F}^{-1}(\phi_0)$ and $\psi = \mathcal{F}^{-1}(\psi_0)$. Since $\phi_0, \psi_0 \in \mathcal{S}$ and $\mathcal{F}$ is bijective, ϕ and ψ are nonzero functions in $\mathcal{S}$. Equation (4.11) now gives

$$\mathcal{F}(\phi * \psi) = \mathcal{F}(\phi)\mathcal{F}(\psi) = \phi_0\psi_0 = 0$$

which implies that $\phi * \psi = 0$.

However, if $\phi \in \mathcal{S}$ and $\phi * \phi = 0$, then $0 = \mathcal{F}(\phi * \phi) = [\mathcal{F}(\phi)]^2$. Hence $\mathcal{F}(\phi) = 0$ and $\phi = 0$.

4.4 FOURIER TRANSFORMATION IN $\mathcal{S}'$

As with the other operations on distributions, we define the Fourier transformation on $\mathcal{S}'$ by duality.

Definition For any $T \in \mathcal{S}'$ the Fourier transform $\mathcal{F}(T) = \hat{T}$ is defined by

$$\hat{T}(\phi) = T(\hat{\phi}) \qquad \phi \in \mathcal{S}$$

Since $\hat{\phi} \in \mathcal{S}$ for every $\phi \in \mathcal{S}$, and since the Fourier transformation is continuous on $\mathcal{S}$, it is clear that $\hat{T} \in \mathcal{S}'$ for every $T \in \mathcal{S}'$.

Now $\mathcal{S}$ can be considered a subspace of $\mathcal{S}'$, so if $\psi \in \mathcal{S}$ corresponds to $T_\psi \in \mathcal{S}'$, then

$$\hat{T}_\psi(\phi) = T_\psi(\hat{\phi}) = T_{\hat{\psi}}(\phi)$$

in view of (4.9); hence $\hat{T}_\psi = T_{\hat{\psi}}$.

Since $\mathcal{F} : \mathcal{S} \to \mathcal{S}$ is continuous, the same is true of $\mathcal{F} : \mathcal{S}' \to \mathcal{S}'$ in the (weak) topology of $\mathcal{S}'$, i.e., if $T_k \to T$ in $\mathcal{S}'$ then $\hat{T}_k \to \hat{T}$ in $\mathcal{S}'$. For if $\phi \in \mathcal{S}$ then $\hat{T}_k(\phi) = T_k(\hat{\phi}) \to T(\hat{\phi}) = \hat{T}(\phi)$.

When f is an L^1 function, $\hat{f}$ is a C^0_∞ function and therefore $T_{\hat{f}} \in \mathcal{S}'$. Hence, for any $\phi \in \mathcal{S}$,

$$
\begin{aligned}
T_{\hat{f}}(\phi) &= \int \hat{f}(\xi)\phi(\xi)\, d\xi \\
&= \int\left[\int e^{-i(\xi,x)} f(x)dx\right]\phi(\xi)\, d\xi \\
&= \int f(x)\left[\int e^{-i(x,\xi)}\, \phi(\xi)\, d\xi\right] dx \\
&= \int f(x)\hat{\phi}(x)\, dx \\
&= T_f(\hat{\phi})
\end{aligned}
$$

which shows that $\hat{T}_f = T_{\hat{f}}$ and that the Fourier transform of T, as a distribution, coincides with its transform as an L^1 function. In other words, the above definition of the Fourier transform is an extension of the one given by equation (4.1).

Similarly we define the inverse Fourier transform of $T \in \mathcal{S}'$ by

$$\mathcal{F}^{-1}(T)(\phi) = T(\mathcal{F}^{-1}(\phi)) \qquad \phi \in \mathcal{S} \tag{4.14}$$

and we see that $\mathcal{F}^{-1}$ is also a continuous map from $\mathcal{S}'$ into $\mathcal{S}'$ which satisfies $\mathcal{F}^{-1}(\hat{T}) = T$. Moreover, equation (4.13) gives

$$\hat{\hat{T}}(\phi) = T(\hat{\hat{\phi}}) = (2\pi)^n T(\check{\phi}) = (2\pi)^n\, \check{T}(\phi) \qquad \phi \in \mathcal{S}$$

or

$$\hat{\hat{T}} = (2\pi)^n \check{T} \qquad T \in \mathcal{S}' \tag{4.15}$$

Thus we have proved

Theorem 4.6 The Fourier transformation $\mathcal{F}$ from $\mathcal{S}'$ to $\mathcal{S}'$ with the inversion formula (4.15) is a topological isomorphism.

The definition of the Fourier transform of a tempered distribution by duality carries the properties of the Fourier transformation in $\mathcal{S}$, as discussed in Section 4.3, into $\mathcal{S}'$. In particular equations (4.5) and (4.6) imply

$$\mathcal{F}(D^\alpha T) = \xi^\alpha \mathcal{F}(T) \tag{4.16}$$

$$\mathcal{F}(x^\alpha T) = (-1)^{|\alpha|} D^\alpha \mathcal{F}(T) \tag{4.17}$$

for every $T \in \mathcal{S}'$, where multiplication of T by any polynomial P has already been defined by $PT(\phi) = T(P\phi)$ for all $\phi \in \mathcal{S}$.

Example 4.4 For any $\phi \in \mathcal{S}$ we have

$$\langle \hat{\delta},\phi\rangle = \langle \delta,\hat{\phi}\rangle = \hat{\phi}(0) = \langle 1,\phi\rangle$$

Hence $\hat{\delta} = 1$. Since $\hat{\hat{\delta}} = (2\pi)^n\check{\delta} = (2\pi)^n\delta$, we also obtain $\hat{1} = (2\pi)^n\delta$. Using equations (4.16) and (4.17), these results may be generalized to

$$\mathcal{F}(D^\alpha\delta) = \xi^\alpha$$
$$\mathcal{F}(x^\alpha) = (-1)^{|\alpha|}(2\pi)^n D^\alpha\delta \qquad \alpha \in \mathbf{N}^n$$

A distribution $T \in \mathcal{D}'$ is said to be *even* if $\check{T} = T$, in the sense that $T(\check{\phi}) = T(\phi)$ for every $\phi \in \mathcal{D}$, and *odd* if $\check{T} = -T$. When T is an even distribution in $\mathcal{S}'$ then, for any $\phi \in \mathcal{S}$,

$$\begin{aligned}\hat{T}(\check{\phi}) &= T(\check{\phi})\hat{}\\ &= T(\hat{\phi})\check{}\\ &= T(\hat{\phi}) \qquad \text{because } T \text{ is even}\\ &= \hat{T}(\phi)\end{aligned}$$

Therefore $\hat{T}$ is even. Conversely, if $\hat{T}$ is even, we can also show that T is even.

Similarly, T is odd if and only if $\hat{T}$ is odd. It is also worth noting that the relation (4.15) implies $\mathcal{F}(T) = (2\pi)^n\mathcal{F}^{-1}(T)$ when $T \in \mathcal{S}'$ is even, and $\mathcal{F}(T) = -(2\pi)^n\mathcal{F}^{-1}(T)$ when T is odd.

Example 4.5 Let $T = pv(1/x)$, $x \in \mathbf{R}$.

(i) First we shall establish that T is odd. If $\phi \in \mathcal{D}(\mathbf{R})$, then

$$\begin{aligned}\langle T,\check{\phi}\rangle &= \lim_{\varepsilon\to 0}\int_{|x|\geq\varepsilon}\frac{1}{x}\phi(-x)\,dx\\ &= -\lim_{\varepsilon\to 0}\int_{|x|\geq\varepsilon}\frac{1}{x}\phi(x)\,dx\\ &= -\langle T,\phi\rangle\end{aligned}$$

(ii) Now we shall compute $\hat{T}$. Noting that

$$\begin{aligned}\langle xT,\phi\rangle &= \langle T,x\phi\rangle \\ &= \lim_{\varepsilon\to 0}\int_{|x|\geq\varepsilon} \phi(x)\,dx \\ &= \int \phi(x)\,dx \\ &= \langle 1,\phi\rangle\end{aligned}$$

we conclude that $xT = 1$. Therefore

$$\mathcal{F}(xT) = \hat{1} = 2\pi\delta$$

But $\mathcal{F}(xT) = -D\hat{T} = i(d\hat{T}/d\xi)$, and hence $d\hat{T}/d\xi = -2\pi i\delta$. This implies that $\hat{T} = -2\pi iH + c$ for some constant c. Since $\hat{T}$ is odd, this constant satisfies $-2\pi i + c = -c$. Thus

$$\hat{T} = -2\pi iH + \pi i$$

(iii) The expressions for $\hat{H}$ and $\mathcal{F}^{-1}(H)$ can now be derived:

$$\begin{aligned}\hat{\hat{T}} &= -2\pi i\hat{H} + \pi i\hat{1} \\ &= -2\pi i\hat{H} + 2\pi^2 i\delta \\ &= 2\pi\check{T} \qquad \text{by formula (4.15)} \\ &= -2\pi T \qquad \text{since } T \text{ is odd.}\end{aligned}$$

Hence

$$\hat{H} = \pi\delta - ipv\,\frac{1}{x}$$

On the other hand,

$$\begin{aligned}T &= \mathcal{F}^{-1}(\hat{T}) \\ &= -2\pi i\mathcal{F}^{-1}(H) + \pi i\mathcal{F}^{-1}(1) \\ &= -2\pi i\mathcal{F}^{-1}(H) + \pi i\delta\end{aligned}$$

Therefore

$$\mathcal{F}^{-1}(H) = \tfrac{1}{2}\delta - \frac{1}{2\pi i}\,pv\,\frac{1}{x}$$

Though we have extended the definition of the Fourier transformation from L^1 to $\mathcal{S}'$, its important properties as given by the equations (4.9) to (4.12), which hold in $\mathcal{S}$, do not extend to all of $\mathcal{S}'$. However, they can be extended to appropriate subspaces of $\mathcal{S}'$. This will be the subject of the next two sections.

4.5 FOURIER TRANSFORMATION IN L^2

For any open subset Ω of $\mathbf{R}^n$, $L^2(\Omega)$ is the Banach space of (Lebesgue) square integrable complex functions on Ω under the norm

$$\|f\|_2 = \left[\int_\Omega |f(x)|^2\,dx\right]^{1/2}$$

From the Schwarz inequality we have

$$\left|\int_\Omega f(x)\bar{g}(x)\,dx\right| \leq \|f\|_2\,\|g\|_2$$

for any pair of functions $f,g \in L^2(\Omega)$, and consequently the complex number

$$(f,g) = \int_\Omega f(x)\bar{g}(x)\,dx$$

is always finite. It is called the *inner product* of f and g in L^2. In particular

$$(f,f) = \|f\|_2^2$$

In line with our convention, L^2 will denote $L^2(\mathbf{R}^n)$. Since L^2 is not a subspace of L^1 the definition of the Fourier transform by (4.1) does not apply to all L^2 functions. But when $f \in L^1 \cap L^2$ then it turns out that $\hat{f}$ is also in L^2 and Parseval's relation (4.10) gives

$$\|f\|_2 = (2\pi)^{-n/2}\,\|\hat{f}\|_2$$

The resulting symmetry would seem to indicate that there is a clear advantage in considering the Fourier transformation on L^2.

Parseval's relation (4.10), which was proved in $\mathcal{S}$, will now be shown to hold in L^2 as a subspace of $\mathcal{S}'$. This will follow from the next result, known as *Plancherel's theorem*.

Theorem 4.7 If $f \in L^2$ then $\hat{f} \in L^2$ and

$$\|\hat{f}\|_2 = (2\pi)^{n/2}\,\|f\|_2 \tag{4.18}$$

Proof. When we set $\psi = \phi$ in Parseval's relation (4.10) we obtain

$$\|\phi\|_2 = (2\pi)^{-n/2}\|\hat{\phi}\|_2 \qquad \phi \in \mathcal{S} \tag{4.19}$$

Since C_0^∞ is dense in L^2 and $C_0^\infty \subset \mathcal{S} \subset L^2$, we conclude that $\mathcal{S}$ is also dense in L^2. Now equations (4.19) may be extended to L^2, since convergence in $\mathcal{S}$ implies convergence in L^2. □

Using the *parallelogram law*

$$\|f + g\|_2^2 = (f + g, f + g) = \|f\|_2^2 + 2\text{Re}(f,g) + \|g\|_2^2$$

which holds for any pair $f,g \in L^2$, and equation (4.18), we obtain the L^2 version of Parseval's relation:

Corollary $(\hat{f},\hat{g}) = (2\pi)^n(f,g)$ for all $f,g \in L^2$.

Example 4.6 Suppose $f \in \mathcal{S}'$ satisfies the differential equation $(-\Delta + c)f = g$ in $\mathbf{R}^n$, where $c > 0$. If $g \in L^2$ then we can show that $f \in L^2$ and, more generally, $D_k^m f \in L^2$ for all $0 \leq m \leq 2$, $1 \leq k \leq n$.

$$\begin{aligned}\mathcal{F}[(-\Delta + c)f] &= \mathcal{F}[(D_1^2 + \cdots + D_n^2 + c)f] \\ &= (\xi_1^2 + \cdots + \xi_n^2 + c)\hat{f}\end{aligned}$$

Since, by hypothesis, $(-\Delta + c)f \in L^2$ we have $(|\xi|^2 + c)\hat{f} \in L^2$ by Theorem 4.7. Hence

$$(|\xi|^2 + 1)\hat{f} = \frac{|\xi|^2 + 1}{|\xi|^2 + c}(|\xi|^2 + c)\hat{f} \in L^2$$

With $\xi = (\xi_1, \ldots, \xi_n) \in \mathbf{R}^n$ it is straightforward to verify that

$$|\xi_k|^m \leq |\xi|^2 + 1$$

for all $0 \leq m \leq 2$ and $1 \leq k \leq n$. This implies that $\mathcal{F}(D_k^m f) = \xi_k^m \hat{f} \in L^2$, hence $D_k^m f \in L^2$.

4.6 FOURIER TRANSFORMATION IN $\mathcal{E}'$

With equation (4.11) in mind, it is natural to consider now the Fourier transform of a convolution product of two tempered distributions. This requires a closer look at the Fourier transforms of distributions with compact support. It turns out that these enjoy certain regularity properties, which are given in the next theorem. First we shall define what it means for a function of several complex variables to be *analytic*, based on the corresponding definition for a function of a single complex variable.

Let f be defined on an open connected set Ω in $\mathbf{C}^n$. f is analytic in Ω if for each $k \in \{1, \ldots, n\}$, with $z_1, \ldots, z_{k-1}, z_{k+1}, \ldots, z_n$ all fixed, the function

$$f_k(z_k) = f(z_1, \ldots, z_k, \ldots, z_n)$$

of the single variable z_k is analytic on $\{z_k \in \mathbf{C} : z = (z_1, \ldots, z_k, \ldots, z_n) \in \Omega\}$. As in the single variable theory this means that the function has a power series expansion about every point $c \in \Omega$,

$$f(z) = \sum_{\alpha} a_\alpha (z - c)^\alpha \tag{4.20}$$

which is valid for every point z in the open ball

$$B(c,r) = \left\{ z \in \Omega : |z - c| = \left[\sum_{k=1}^{n} |z_k - c_k|^2 \right]^{1/2} < r \right\}$$

for some positive number r. When f is analytic in $\mathbf{C}^n$ it is called *entire*. The summation index α in (4.20) runs through $\mathbf{N}_0^n$, and a_α are the Taylor coefficients

$$a_\alpha = \frac{1}{\alpha!} \partial_z^\alpha f(c)$$

When $z_k = x_k + iy_k$, we shall use the notation

$$\partial_{z_k} = \tfrac{1}{2}(\partial_{x_k} - i\partial_{y_k})$$

$$\bar{\partial}_{z_k} = \partial_{\bar{z}_k} = \tfrac{1}{2}(\partial_{x_k} + i\partial_{y_k}) \qquad k = 1, \ldots, n$$

and the *Cauchy-Riemann equations* take the form [9]

$$\bar{\partial}_{z_k} f = \frac{1}{2}\left[\frac{\partial f}{\partial x_k} + i \frac{\partial f}{\partial_{y_k}} \right] = 0 \qquad k = 1, \ldots, n$$

When Ω is an open subset of $\mathbf{R}^n$, we shall say that f is (real) analytic in Ω if it has a power series expansion of the form (4.20) about every point $c \in \Omega$, with z replaced by $x \in B(c,r) \subset \mathbf{R}^n$. Clearly, this is so if and only if the function f can be extended to an open neighborhood of Ω in $\mathbf{C}^n$ where f is (complex) analytic.

Theorem 4.8 The Fourier transform of $T \in \mathscr{E}'$ is an analytic function in $\mathbf{R}^n$ which is given by

$$\hat{T}(\xi) = T_x(e^{-i\langle x,\xi\rangle}) \tag{4.21}$$

Furthermore, the right-hand side may be extended as an analytic function to $\mathbf{C}^n$, known as the *Fourier–Laplace transform* of T.

Proof. As a function of ξ, $T_x(e^{-i\langle x,\xi\rangle})$ is in C^∞ by Corollary 1 to Theorem 3.3, and it remains to show that equation (4.21) holds in $\mathscr{S}'$.

For any $\phi \in \mathscr{D}$ we have $\hat{T}(\phi) = T(\hat{\phi})$. If we now consider ϕ as an element in $\mathscr{E}'$ then $\hat{\phi}(\xi) = \langle \phi(x), e^{-i\langle x,\xi\rangle}\rangle$, and

$$\begin{aligned} T(\hat{\phi}) &= \langle T_x \otimes \phi_\xi, e^{-i\langle x,\xi\rangle}\rangle \\ &= \int T_x(e^{-i\langle x,\xi\rangle})\phi(\xi)\, d\xi \end{aligned}$$

by Theorem 3.4, as it applies to distributions with compact support. But since $\mathscr{D}$ is dense in $\mathscr{S}$, equation (4.21) holds in $\mathscr{S}'$.

By replacing ξ by $\zeta = \xi + i\eta$, $\hat{T}$ may be extended into $\mathbf{C}^n$, where it is also a C^∞ function of ζ, and the derivatives $\partial_{\zeta_k}\hat{T}$ and $\bar{\partial}_{\zeta_k}\hat{T}$ may be computed by differentiating $\exp(-i\langle x,\zeta\rangle)$. Since the exponential function is entire, so is $\hat{T}(\zeta)$. Hence $\hat{T}$ is analytic in $\mathbf{R}^n$. □

Example 4.7 Let T be a distribution in $\mathbf{R}$ such that $T^{(m)} = \delta$ for some positive integer m. Applying the Fourier transformation we get $(i\xi)^m\hat{T} = 1$. Hence

$$\hat{T} = 1/(i\xi)^m$$

Since $\hat{T}$ is singular at $\xi = 0$, Theorem 4.8 implies that T cannot have compact support. In other words, any fundamental solution of the operator d^m/dx^m in $\mathbf{R}$ cannot have compact support. Compare this to the result of Example 3.16.

Example 4.8 Suppose T is a distribution with compact support such that $\langle T_x, x^\alpha\rangle = 0$ for every $\alpha \in \mathbf{N}_0^n$. We shall prove that $T = 0$, and thereby conclude that the set of all polynomials in $\mathbf{R}^n$ with constant coefficients is dense in C^∞.

(i) Since $T \in \mathscr{E}'$, $\hat{T}(\xi)$ can be extended as an analytic function $f(\zeta)$ in $\mathbf{C}^n$ such that $f(\zeta) = T_x(e^{-i\langle x,\zeta\rangle})$. For any $\alpha \in \mathbf{N}_0^n$,

$$\partial^\alpha f(\zeta) = T_x(\partial_\zeta^\alpha e^{-i\langle x,\zeta\rangle})$$
$$= (-i)^{|\alpha|}T_x(x^\alpha e^{-i\langle x,\zeta\rangle})$$

At $\zeta = 0$,

$$\partial^\alpha f(0) = (-i)^{|\alpha|}T_x(x^\alpha) = 0 \qquad \alpha \in \mathbf{N}_0^n$$

Since f is an entire function in $\mathbf{C}^n$, it is represented by the power series

$$f(\zeta) = \sum_\alpha \frac{1}{\alpha!}\partial^\alpha f(0)\zeta^\alpha = 0$$

for any $\zeta \in \mathbf{C}^n$. Thus f, and therefore $\hat{T}$, vanishes identically. Consequently $T = 0$, since the Fourier transformation is injective in $\mathcal{S}'$.

(ii) Let $\mathcal{P}$ be the set of all polynomials in $\mathbf{R}^n$ with constant coefficients. If $\overline{\mathcal{P}}$ is a proper subset of C^∞, then, by the Hahn-Banach theorem [4], there exists a nonzero continuous linear functional T on C^∞ such that $\langle T,P\rangle = 0$ for every $P \in \overline{\mathcal{P}}$. This implies, in particular, that T is a nonzero distribution with compact support which satisfies $\langle T,x^\alpha\rangle = 0$ for every $\alpha \in \mathbf{N}^n$, thereby contradicting (i).

The next theorem gives the distributional version of equation (4.11).

Theorem 4.9 If $T_1 \in \mathcal{S}'$ and $T_2 \in \mathcal{E}'$ then $T_1 * T_2 \in \mathcal{S}'$ and

$$\mathcal{F}(T_1 * T_2) = \mathcal{F}(T_2)\mathcal{F}(T_1) \tag{4.22}$$

where the right-hand side is a well-defined distribution because $\mathcal{F}(T_2)$ is in C^∞.

Proof. Let $\phi \in \mathcal{D}$. From the properties of the convolution and Corollary 1 to Theorem 3.5, we have

$$(T_1 * T_2)(\phi) = (T_1 * T_2 * \check{\phi})(0)$$

and $(T_2 * \check{\phi})(x) = T_2(\tau_x\phi)$ is a C_0^∞ function by Corollary 2 to Theorem 3.5. On the other hand,

$$(T_1 * T_2 * \check{\phi})(0) = \langle T_{1_y}, T_2 * \check{\phi})(-y)\rangle$$
$$= \langle T_{1_y}, T_2(\tau_{-y}\phi)\rangle$$
$$= \langle T_{1_y}, \check{T}_2(\tau_y\check{\phi})\rangle$$
$$= \langle T_{1_y}, \check{T}_2 * \phi(y)\rangle$$

Therefore

$$(T_1 * T_2)(\phi) = T_1(\check{T}_2 * \phi) \tag{4.23}$$

If we now take ϕ in $\mathcal{S}$, it is clear that

$$(\check{T}_2 * \phi)(x) = \check{T}_2(\tau_x\check{\phi}) = T_2(\tau_{-x}\phi)$$

will also be in $\mathcal{S}$; for if T_2 is of order m then

$$\sup_{\substack{x\in\mathbb{R}^n\\|\alpha+\beta|\le k}} |x^\alpha\partial^\beta(\check{T}_2 * \phi)(x)| \le M_k \sup_{\substack{\mathbb{R}^n\\|\alpha+\beta|\le k+m}} |x^\alpha\partial^\beta\phi(x)|$$

Thus $T_1 * T_2$ is a continuous linear functional on $\mathcal{S}$. To compute its Fourier transform, let $\phi \in \mathcal{S}$ so that $\hat{\phi}$ is also in $\mathcal{S}$ and equation (4.23) gives

$$\begin{aligned}(T_1 * T_2)(\hat{\phi}) &= T_1(\check{T}_2 * \hat{\phi})\\ (\check{T}_2 * \hat{\phi})(x) &= T_2(\tau_{-x}\hat{\phi})\\ &= T_{2_y}(\tau_{-x}\hat{\phi}(y))\\ &= T_{2_y}(\hat{\phi}(x + y))\end{aligned}$$

If $\phi \in \mathcal{D}$, then we can write

$$\begin{aligned}T_{2_y}(\hat{\phi}(x+y)) &= T_{2_y}\left(\int e^{-i(x+y,\xi)}\,\phi(\xi)\,d\xi\right)\\ &= \int T_{2_y}(e^{-i(y,\xi)})\phi(\xi)e^{-i(x,\xi)}\,d\xi\\ &= \int \hat{T}_2(\xi)\phi(\xi)e^{-i(x,\xi)}\,d\xi\end{aligned}$$

by Theorem 4.8. Similarly,

$$\begin{aligned}(T_1 * T_2)(\hat{\phi}) &= T_{1_x}(T_{2_y}(\hat{\phi}(x+y)))\\ &= T_{1_x}\left(\int \hat{T}_2(\xi)\phi(\xi)e^{-i(x,\xi)}\,d\xi\right)\\ &= \int \hat{T}_1(\xi)\hat{T}_2(\xi)\phi(\xi)\,d\xi\\ &= \hat{T}_1\hat{T}_2(\phi) \qquad \phi\in\mathcal{D}\end{aligned} \tag{4.24}$$

Since $\mathcal{D}$ is dense in $\mathcal{S}$, equation (4.24) holds for all $\phi \in \mathcal{S}$. But since $(T_1 * T_2)(\hat{\phi}) = (T_1 * T_2)^\wedge(\phi)$, $\phi \in \mathcal{S}$, we conclude that $(T_1 * T_2)^\wedge = \hat{T}_1\hat{T}_2$. □

Example 4.9 (i) Let $T_a = (1/2)(\delta_a + \delta_{-a})$ for some real number a. In order to find the Fourier transform of T_a we shall first compute $\hat{\delta}_a$:

$$
\begin{aligned}
\langle \hat{\delta}_a, \phi \rangle &= \langle \delta_a, \hat{\phi} \rangle \\
&= \hat{\phi}(a) \\
&= \int e^{-ixa} \phi(x)\, dx \\
&= \langle e^{-iax}, \phi \rangle \qquad \phi \in \mathcal{D}(\mathbf{R})
\end{aligned}
$$

Hence $\hat{\delta}_a(\xi) = e^{-ia\xi}$ and $\hat{T}_a(\xi) = (1/2)(e^{-ia\xi} + e^{ia\xi}) = \cos a\xi$.

(ii) To verify that $\mathcal{F}(T_a * T_b) = \mathcal{F}(T_a)\mathcal{F}(T_b)$, we use the property $\delta_a * \delta_b = \tau_a\delta_b = \delta_{a+b}$ to write

$$
\begin{aligned}
T_a * T_b &= \tfrac{1}{4}\,(\delta_{a+b} + \delta_{-(a+b)} + \delta_{a-b} + \delta_{-(a-b)}) \\
\mathcal{F}(T_a * T_b) &= \tfrac{1}{4}\,[2\cos(a+b)\xi + 2\cos(a-b)\xi] \\
&= \cos a\xi \cos b\xi \\
&= \mathcal{F}(T_a)\,\mathcal{F}(T_b)
\end{aligned}
$$

(iii) Now the Fourier transforms of $\sin x$ and $\cos x$ may be computed using the results above:

$$
\begin{aligned}
\mathcal{F}(\cos x) &= \hat{\hat{T}}_1 = 2\pi\check{T}_1 = \pi(\check{\delta}_1 + \check{\delta}_{-1}) = \pi(\delta_{-1} + \delta_1) \\
\mathcal{F}(\sin x) &= \mathcal{F}(-iD\cos x) \\
&= -i\xi\mathcal{F}(\cos x) \\
&= -i\pi\xi(\delta_1 + \delta_{-1}) \\
&= i\pi(\delta_{-1} - \delta_1)
\end{aligned}
$$

The next result, known as the *Paley-Wiener-Schwartz theorem,* describes the behavior at ∞ of the Fourier-Laplace transform of a distribution with compact support.

Theorem 4.10 (i) If $T \in \mathscr{E}'$ and $\operatorname{supp} T \subset \{x \in \mathbf{R}^n : |x| \le r\} = \overline{B}(0,r)$ then there is a constant M and a nonnegative integer N such that

$$
|\hat{T}(\zeta)| \le M(1 + |\zeta|)^N \exp(r|\operatorname{Im}\zeta|) \qquad \zeta \in \mathbf{C}^n \tag{4.25}
$$

(ii) Conversely, every entire function in $\mathbf{C}^n$ satisfying (4.25) is the Fourier-Laplace transform of a distribution with support contained in $\overline{B}(0,r)$.

(iii) If $T \in C_0^\infty$ and supp $T \subset \overline{B}(0,r)$, then for every integer $m \geqslant 0$ there is a constant M_m such that

$$|\hat{T}(\zeta)| \leqslant M_m(1 + |\zeta|)^{-m} \exp(r|\text{Im}\,\zeta|) \qquad \zeta \in \mathbf{C}^n \tag{4.26}$$

(iv) Conversely, every entire function in $\mathbf{C}^n$ satisfying (4.26) for every $m \in \mathbf{N}_0$ is the Fourier-Laplace transform of a C_0^∞ function with support contained in $\overline{B}(0,r)$.

Proof. (i) Let K = supp $T \subset \overline{B}(0,r)$ and let ψ be a C_0^∞ function which equals 1 on a neighborhood of K. Then we have $T(\phi) = T(\psi\phi)$ for all $\phi \in \mathscr{E}$. Now $\psi\phi$ is in $\mathscr{D}$. By Theorem 2.10, T is of finite order on $\mathscr{D}$, so there is an integer $N \geqslant 0$ and a constant M_1 such that

$$|T(\phi)| = |T(\psi\phi)| \leqslant M_1|\psi\phi|_N \tag{4.27}$$

Assuming supp $\psi = K_0 \supset \mathring{K} \supset K$, and using Leibnitz' formula,

$$|\psi\phi|_N \leqslant M_2 \sup \{|\partial^\alpha\phi(x)| : x \in K_0\ |\alpha| \leqslant N\} \tag{4.28}$$

for some constant M_2. Since (4.28) is true for every K_0 such that $\mathring{K}_0 \supset K$, it is also true for K. With $\phi(x) = e^{-i(x,\zeta)}$ and $\zeta = \xi + i\eta$, we obtain

$$\begin{aligned}\sup \{|\partial^\alpha\phi(x)| : x \in K, |\alpha| \leqslant N\} &\leqslant \sup \{|\zeta|^{|\alpha|} \exp((x,\eta)) : |x| \leqslant r,\ |\alpha| \leqslant N\}\\ &\leqslant (1 + |\zeta|)^N \exp(r|\eta|) \end{aligned}\tag{4.29}$$

Applying the inequalities (4.27), (4.28), and (4,29) to $|\hat{T}(\zeta)| = |T_x(e^{-i(x,\zeta)})|$ gives (4.25).

(ii) When T is a C_0^∞ function we can use (4.5) to write

$$\zeta^\alpha\hat{T}(\zeta) = \int e^{-i(x,\zeta)} D^\alpha T(x)\, dx$$

for any $\alpha \in \mathbf{N}_0^n$. With supp T in $\overline{B}(0,r)$ this gives the estimate

$$|\zeta^\alpha\hat{T}(\zeta)| \leqslant M \exp(r|\eta|)$$

for some constant M, from which (4.26) follows.

(iii) Conversely, if the inequality (4.26) holds for every nonnegative integer m, then the integral

$$(2\pi)^{-n} \int \hat{T}(\xi)e^{i(x,\xi)}\, d\xi$$

is absolutely convergent on $\mathbf{R}^n$, and clearly defines the inverse Fourier transform of $\hat{T}(\xi)$, which is $T(x)$. Thus

$$\partial^\alpha T(x) = (-i)^{|\alpha|}(2\pi)^{-n}\int \hat{T}(\xi)\xi^\alpha e^{i\langle x,\xi\rangle}\,d\xi \qquad \alpha \in \mathbf{N}_0^n$$

is also absolutely convergent, and we conclude that T is in C^∞. To show that T has compact support we note that the above integrand extends to an entire function on $\mathbf{C}^n$, so we can use Cauchy's theorem with each variable $\zeta_1, \ldots, \zeta_n$ to shift the integration from $\mathbf{R}^n$ into $\mathbf{C}^n$. For any fixed point $\eta \in \mathbf{R}^n$, therefore,

$$T(x) = (2\pi)^{-n}\int \hat{T}(\xi + i\eta)\, e^{i\langle x,\xi+i\eta\rangle}\,d\xi$$

Using the inequality (4.26) with $m = n + 1$,

$$\begin{aligned}|T(x)| &\leq (2\pi)^{-n}M_{n+1}\exp(-\langle x,\eta\rangle + r|\eta|)\int (1 + |\xi|)^{-n-1}\,d\xi\\ &\leq M\exp(r|\eta| - \langle x,\eta\rangle)\end{aligned}$$

Taking $\eta = tx$ and letting $t \to \infty$, we see that $T(x) = 0$ for all $x \in \mathbf{R}^n$ with $|x| > r$. Therefore the support of T must lie in $\overline{B}(0,r)$.

(iv) Let $\hat{T}(\zeta)$ be an entire function which satisfies the inequality (4.25). It then follows that $\hat{T}(\xi)$ is a function of polynomial growth at ∞ and therefore lies in $\mathcal{S}'$. Its inverse Fourier transform T must also be in $\mathcal{S}'$. To prove that supp T is compact, we regularize T using the C^∞ functions β_λ defined by equation (2.8) for all $\lambda > 0$, and satisfying supp $\beta_\lambda \subset \overline{B}(0,\lambda)$. Now

$$T_\lambda = T * \beta_\lambda$$

is in C^∞, and its Fourier transform, according to Theorem 4.9, is given by

$$\hat{T}_\lambda = \hat{\beta}_\lambda\hat{T}$$

For each $\lambda > 0$, $\hat{T}_\lambda(\xi)$ extends to an analytic function on $\mathbf{C}^n$. Since $\hat{T}$ satisfies (4.25) and $\hat{\beta}_\lambda$ satisfies (4.26), $\hat{T}_\lambda$ must satisfy

$$|\hat{T}_\lambda(\zeta)| \leq MM_m(1 + |\zeta|)^{N-m}\exp((r + \lambda)\,|\mathrm{Im}\,\zeta|)$$

for $m = 0,1,2, \ldots$ and $\zeta \in \mathbf{C}^n$. By choosing m greater than N we see that $\hat{T}_k$ satisfies an estimate of the type (4.26) with r replaced by $r + \lambda$. By part (iii) we therefore conclude that supp $T_\lambda \subset \overline{B}(0,r + \lambda)$. Since $T_\lambda \to T$ as $\lambda \to 0$, we must have

$$\operatorname{supp} T \subset \cap\,\{\overline{B}(0,r + \lambda) : \lambda > 0\} = \overline{B}(0,r)$$ □

4.7 APPLICATIONS TO DIFFERENTIAL EQUATIONS

The fundamental property of the Fourier transformation which makes it important in the treatment of differential equations is contained in equation

(4.16): $\mathcal{F}(D^\alpha T) = \xi^\alpha \mathcal{F}(T)$. In other words $D^\alpha T$ in $\mathcal{S}'$ is transformed to $\xi^\alpha \hat{T}$, so that differentiation is replaced by multiplication by a polynomial. This can result in considerable simplification of the differential equation. Once the expression for $\hat{T}$ is known, T may then be obtained, in principle, by taking the inverse transform $\mathcal{F}^{-1}(\hat{T})$. This last step is usually the more difficult part of solving the differential equation.

With some differential equations we shall find it necessary to apply the Fourier transformation with respect to some, but not all, of the variables in $\mathbf{R}^n$. Thus if $T \in \mathcal{S}'(\mathbf{R}^{n_1} \times \mathbf{R}^{n_2})$, with $n_1 + n_2 = n$, then the Fourier transform of T with respect to $x \in \mathbf{R}^{n_1}$, denoted by $\mathcal{F}_1(T)$, is defined by

$$\langle \mathcal{F}_1(T), \phi \rangle = \langle T, \mathcal{F}_1(\phi) \rangle$$

for all $\phi \in \mathcal{S}(\mathbf{R}^{n_1} \times \mathbf{R}^{n_2})$. $\mathcal{F}_1(\phi)$ is well defined by the integral formula

$$\mathcal{F}_1(\phi(\cdot,y))(\xi) = \int_{\mathbf{R}^{n_l}} e^{-i(x,\xi)}\, \phi(x,y)\, dx \qquad \xi \in \mathbf{R}^{n_1}, y \in \mathbf{R}^{n_2}$$

It is also denoted by $\hat{\phi}(\xi,y)$ and lies in $\mathcal{S}(\mathbf{R}^{n_1} \times \mathbf{R}^{n_2})$. Thus $\mathcal{F}_1(T) \in \mathcal{S}'(\mathbf{R}^{n_1} \times \mathbf{R}^{n_2})$ and, if ∂_y^α is a partial differential operator in $y \in \mathbf{R}^{n_2}$, then

$$\mathcal{F}_1(\partial_y^\alpha T) = \partial_y^\alpha \mathcal{F}_1(T). \tag{4.30}$$

This follows from

$$\begin{aligned}
\langle \mathcal{F}_1(\partial_y^\alpha T), \phi \rangle &= \langle \partial_y^\alpha T, \mathcal{F}_1(\phi) \rangle \\
&= (-1)^{|\alpha|} \langle T, \partial_y^\alpha \mathcal{F}_1(\phi) \rangle \\
&= (-1)^{|\alpha|} \langle T, \mathcal{F}_1(\partial_y^\alpha \phi) \rangle \\
&= (-1)^{|\alpha|} \langle \mathcal{F}_1(T), \partial_y^\alpha \phi \rangle \\
&= \langle \partial_y^\alpha \mathcal{F}_1(T), \phi \rangle \qquad \phi \in \mathcal{S}(\mathbf{R}^{n_1} \times \mathbf{R}^{n_2})
\end{aligned}$$

Note that the commutation of $\mathcal{F}_1$ with ∂_y^α on $\mathcal{S}(\mathbf{R}^{n_1} \times \mathbf{R}^{n_2})$ is based on the linearity and continuity of $\mathcal{F}_1$.

Example 4.10 Let $L = \sum_{k=0}^{m} c_k D^k$ be a differential operator in $\mathbf{R}$ of order m with constant coefficients. If $u \in \mathcal{E}'(\mathbf{R})$ satisfies $Lu = 0$, then, upon transformation,

$$0 = \mathcal{F}(Lu) = \sum_0^m c_k \xi^k \hat{u}$$

Hence $\hat{u}(\xi) = 0$ except possibly at the zeros of the polynomial $c_0 + c_1\xi +$

$\cdots + c_m\xi^m$. But since u has compact support, $\hat{u}$ is continuous and must therefore vanish in all $\mathbf{R}$.

Thus the ordinary differential equation $Lu = 0$ has only the trivial solution in $\mathscr{E}'$.

Example 4.11 Let $L = \sum_{|\alpha|\leq m} c_\alpha D^\alpha$ be a differential operator in $\mathbf{R}^n$ of order m with constant coefficients, and $u \in \mathscr{S}'$ be a solution of $Lu = 0$.

The application of the Fourier transformation gives

$$0 = \mathscr{F}\left(\sum c_\alpha D^\alpha u\right) = \left(\sum c_\alpha \xi^\alpha\right)\hat{u} = P(\xi)\hat{u}$$

where $P(\xi)$ is the polynomial $\sum_{|\alpha|\leq m} c_\alpha \xi^\alpha$. If $P(\xi) = 0$ only when $\xi = 0$, then supp $\hat{u} \subset \{0\}$ and hence, by Theorem 3.2, $\hat{u} = \sum_{|\alpha|\leq k} a_\alpha \partial^\alpha \delta$ for some k. By taking the inverse Fourier transform, we obtain

$$u = \sum_{|\alpha|\leq k} b_\alpha x^\alpha$$

Thus the only solution of $Lu = 0$ in $\mathscr{S}'$ for this type of operator is a polynomial. In other words, the fundamental solution of L in $\mathscr{S}'$ is unique up to an additive polynomial.

The *Cauchy-Riemann operator* in $\mathbf{R}^2$,

$$\bar{\partial} = \tfrac{1}{2}(\partial_1 + i\partial_2)$$

is an example of such an operator, for the polynomial

$$P(i\xi) = \tfrac{1}{2}i(\xi_1 + i\xi_2)$$

vanishes only at $\xi = 0$. Consequently, its fundamental solution in $\mathscr{S}'(\mathbf{R}^2)$ is unique up to an additive polynomial. Since every entire function f satisfies $\bar{\partial} f = 0$ in $\mathbf{R}^2$, the fundamental solution of $\bar{\partial}$ in $\mathscr{D}'(\mathbf{R}^2)$ is unique up to an additive entire function.

Example 4.12 We shall show in this example that $1/\pi z = 1/\pi(x + iy)$ is a fundamental solution of the Cauchy-Riemann operator in the plane.

Since $1/|z| = 1/r \in L^1_{\text{loc}}(\mathbf{R}^2)$, $1/z$ defines a distribution in $\mathbf{R}^2$. For any $\phi \in \mathscr{D}(\mathbf{R}^2)$,

$$\left\langle \bar{\partial}\frac{1}{z}, \phi \right\rangle = -\left\langle \frac{1}{z}, \bar{\partial}\phi \right\rangle = -\tfrac{1}{2}\int_{\mathbf{R}^2} \frac{1}{x + iy}\left(\frac{\partial\phi}{\partial x} + i\frac{\partial\phi}{\partial y}\right) dxdy$$

Changing to polar coordinates,

$$\frac{\partial}{\partial x} = \cos\theta \frac{\partial}{\partial r} - \frac{\sin\theta}{r}\frac{\partial}{\partial \theta}$$

$$\frac{\partial}{\partial y} = \sin\theta \frac{\partial}{\partial r} + \frac{\cos\theta}{r}\frac{\partial}{\partial \theta}$$

Hence

$$\left\langle \bar{\partial}\frac{1}{z},\phi\right\rangle = -\tfrac{1}{2}\int_0^{2\pi}\int_0^{\infty}\frac{1}{re^{i\theta}}\left[e^{i\theta}\frac{\partial\bar{\phi}}{\partial r} + \frac{i}{r}e^{i\theta}\frac{\partial\bar{\phi}}{\partial\theta}\right] r\,dr\,d\theta$$

where $\bar{\phi}(r,\theta) = \phi(x,y)$. By Fubini's theorem,

$$\begin{aligned}\left\langle \bar{\partial}\frac{1}{z},\phi\right\rangle &= -\tfrac{1}{2}\int_0^{2\pi}\int_0^{\infty}\frac{\partial\bar{\phi}}{\partial r}\,dr\,d\theta - \tfrac{1}{2}i\int_0^{\infty}\frac{1}{r}\int_0^{2\pi}\frac{\partial\bar{\phi}}{d\theta}\,d\theta\,dr \\ &= -\tfrac{1}{2}[-2\pi\bar{\phi}(0)] - 0 \qquad \text{since } \bar{\phi}(r,2\pi) = \bar{\phi}(r,0) \\ &= \pi\phi(0)\end{aligned}$$

Therefore

$$\bar{\partial}(1/\pi z) = \delta$$

and any fundamental solution E of $\bar{\partial}$ in $\mathcal{D}'(\mathbf{R}^2)$ is of the form

$$E(z) = \frac{1}{\pi z} + h(z)$$

where h is an entire function in $\mathbf{C}$.

In Section 3.6 it was shown that the solution of the differential equation $Lu = f$ may be obtained from the solution of $LE = \delta$ by forming the convolution $f * E$. In this section, therefore, we shall seek the fundamental solution E of the operator L for some typical classes of operators with constant coefficients. With the Fourier transform now at our disposal, we can hope to extend some of the results of Section 3.6 to $\mathbf{R}^n$, keeping in mind, of course, that the solutions we obtain by this method lie in $\mathcal{S}'$.

(i) The Heat Equation: We shall start with the first order differential operator $d/dx + a$, where $x \in \mathbf{R}$ and a is a complex constant. According to Example 3.16, a particular solution of the equation

$$E' + aE = \delta \tag{4.31}$$

has the form

$$E(x) = e^{-ax}H(x)$$

and therefore $e^{-ax}H(x)$ is a fundamental solution of $d/dx + a$ on $\mathbf{R}$.

A fundamental solution of the *heat*, or *evolution, operator* $\partial_t - \partial_x^2$ with $(x,t) \in \mathbf{R}^2$ is a solution of

$$(\partial_t - \partial_x^2)E = \delta_{(x,t)} = \delta_x \otimes \delta_t \tag{4.32}$$

in $\mathcal{S}'(\mathrm{R}^2)$. Taking the Fourier transform of both sides of (4.32) with respect to x, and using (4.30), yields

$$(\partial_t + \xi^2)\hat{E} = \delta_t$$

which has the same form as equation (4.31). Thus

$$\hat{E}(\xi,t) = \exp(-\xi^2 t) H(t)$$

and, with $t > 0$,

$$\begin{aligned} E(x,t) &= \mathcal{F}^{-1} \exp(-\xi^2 t) \\ &= (2\pi)^{-1} \int_{-\infty}^{\infty} \exp(ix\xi - \xi^2 t)\, d\xi \\ &= (2t)^{-1/2} (2\pi)^{-1} \int_{-\infty}^{\infty} \exp\left(i \frac{x}{\sqrt{2t}} \eta - \tfrac{1}{2} \eta^2\right) d\eta \\ &= (2t)^{-1/2} f(x/\sqrt{2t}) \end{aligned}$$

where $f(x) = \mathcal{F}^{-1}(\exp(-\eta^2/2)(x)) = (2\pi)^{-1/2} \exp(-x^2/2)$ from Example 4.2. Hence

$$E(x,t) = (4\pi t)^{-1/2} \exp(-x^2/4t)$$

as given in equation (3.22).

It is a simple exercise to generalize this result to x in $\mathbf{R}^n$. In that case, the result of applying the Fourier transformation with respect to x to the equation

$$(\partial_t - \Delta_x)E = \delta$$

is

$$(\partial_t + |\xi|^2)\hat{E} = \delta_t$$

Hence

$$\hat{E}(\xi,t) = \exp(-|\xi|^2 t)H(t)$$

$$E(x,t) = \mathcal{F}^{-1}(\exp(-|\xi|^2 t)H(t))$$

$$= (4\pi t)^{-n/2}H(t)\exp(-|x|^2/4t) \tag{4.33}$$

(ii) The Wave Equation: The general wave operator

$$\partial_t^2 - \Delta_x$$

differs from the heat operator in that the derivative with respect to t is of second order, instead of first order. But this apparently small discrepancy has very significant consequences for the solutions of the corresponding differential equations. While the heat equation describes a diffusion process in which the solution decays as we move away from the energy source in both space and time, the wave equation describes a phenomenon of wave propagation with a fixed velocity.

Applying the Fourier transformation with respect to x to the equation

$$(\partial_t^2 - \Delta_x)E = \delta_x \otimes \delta_t \qquad (x,t) \in \mathbf{R}^n \times (0,\infty) \tag{4.34}$$

we arrive at

$$(\partial_t^2 + |\xi|^2)\hat{E} = \delta_t \tag{4.35}$$

The solution of equation (4.35) has already been obtained in Example 3.19. It is given by

$$\hat{E}(\xi,t) = H(t)\,|\xi|^{-1}\sin|\xi|t$$

which is a bounded, continuous function of $\xi \in \mathbf{R}^n$ and $t \in [0,\infty)$; so the fundamental solution of the wave operator in $\mathcal{S}'(\mathbf{R}^n \times (0,\infty))$ is given by

$$E = \mathcal{F}^{-1}[H(t)\,|\xi|^{-1}\sin|\xi|t] \tag{4.36}$$

When $n = 1$ and $t > 0$, it is a simple matter to verify that

$$E(x,t) = \begin{cases} \frac{1}{2} & |x| < t \\ 0 & |x| > t \end{cases}$$

$$= \tfrac{1}{2}[H(x + t) - H(x - t)]$$

which coincides with the function E_0 of Example 3.23.

For $n = 2$, the representation (4.36) gives

$$E(x,t) = \begin{cases} (2\pi)^{-1}(t^2 - |x|^2)^{-1/2} & |x| < t \\ 0 & |x| > t \end{cases}$$

$$= H(t - |x|)/2\pi\sqrt{t^2 - |x|^2}$$

and for $n = 3$,

$$E = \frac{1}{4\pi t}\,\delta_{|x|-t}$$

This last fundamental solution is an example of the Dirac distribution supported on a "surface" in $\mathbf{R}^2$, namely the half-lines $\{(x,t) \in \mathbf{R}^2 : |x| = t,\ t \geqslant 0\}$.

Though we shall not do so, a general expression for E can be derived from the representation (4.36) (see [10], for example). It turns out that this involves higher derivatives of $\delta_{|x|-t}$ or higher powers of $(t^2 - |x|^2)^{-1/2}$, depending on whether n is odd or even.

(iii) Laplace's Equation: The Laplacian operator

$$\Delta = \sum_{k=1}^{n} \partial_k^2$$

is undoubtedly the most important of all differential operators. It is already a component of both the heat and the wave operators. Laplace's equation is usually taken to be the homogeneous equation

$$\Delta u = 0$$

whose solutions in C^2 are called *harmonic functions,* while its solutions in $\mathcal{D}'$ are called *harmonic distributions.*

Harmonic functions on an open set $\Omega \subset \mathbf{R}^n$ possess a fundamental mean-value property [4], which we shall simply state here: A continuous function h defined on Ω is harmonic if and only if, for any ball B with center x and $\overline{B} \subset \Omega$,

$$h(x) = \frac{1}{\text{volume } B}\int_B h(y)dy = \frac{1}{\text{area } \partial B}\int_{\partial B} h(\sigma)\, d\sigma$$

The next theorem shows that if u is harmonic in the distributional sense, then it is also harmonic in the classical sense.

Theorem 4.11 If u is a harmonic distribution in $\Omega \subset \mathbf{R}^n$, then u is a C^∞ harmonic function almost everywhere.

Proof. (i) Let us first consider the case where $\Omega = \mathbf{R}^n$.

For any C_0^∞ function ϕ, $u * \phi$ is a C^∞ function which satisfies $\Delta(u * \phi) = \Delta u * \phi = 0$ and is therefore harmonic.

Let ψ be a fixed function in $\mathcal{D}(\mathbf{R}^n)$ which is *spherically symmetric,* in the sense that $\psi(x)$ depends only on $|x|$, and such that $\int\psi(x)\,dx = 1$. We shall denote the unit sphere in $\mathbf{R}^n$ by $S_{n-1} = \{x \in \mathbf{R}^n : |x| = 1\}$, and use polar coordinates to write $x = (|x|,\sigma) = (r,\sigma)$ with σ on S_{n-1}. Then

$$1 = \int_{\mathbf{R}^n} \psi(x)dx = \int_0^\infty \int_{S_{n-1}} \psi(r)r^{n-1}\, dr\, d\sigma$$

$$= \omega_{n-1} \int_0^\infty \psi(r)r^{n-1}\, dr$$

where $\omega_{n-1} = \int_{S_{n-1}} d\sigma$ = area of S_{n-1}.

Now for any harmonic function v, $v * \psi$ coincides with v, for

$$v * \psi(x) = \int v(x - y)\, \psi(y)\, dy$$

$$= \int_0^\infty \int_{S_{n-1}} v(x - y)\psi(r)r^{n-1}\, dr\, d\sigma$$

$$= \omega_{n-1}v(x) \int_0^\infty \psi(r)r^{n-1}\, dr$$

$$= v(x)$$

since $v(x - y)$ is harmonic as a function of y and, by the mean-value property, $\int_{S_{n-1}} v(x - y)d\sigma = \omega_{n-1}v(x)$.

Let γ_k be a sequence in $\mathcal{D}$ such that $\gamma_k \to \delta$. Since $u * \gamma_k$ is harmonic, we have $(u * \gamma_k) * \psi = u * \gamma_k$. Hence

$$\begin{aligned} u &= \lim u * \gamma_k \\ &= \lim(u * \gamma_k) * \psi \\ &= \lim u * (\gamma_k * \psi) \\ &= u * \psi \end{aligned}$$

Since $u * \psi$ is a C^∞ harmonic function, we see that the same is true of u almost everywhere in $\mathbf{R}^n$.

(ii) Now assume that u is a harmonic distribution in any open set $\Omega \subset \mathbf{R}^n$.

Let ω be an open set in Ω such that $\bar{\omega} \subset \Omega$, and choose $\phi \in \mathcal{D}(\mathbf{R}^n)$ such that supp $\phi \subset \Omega$, $0 \leq \phi \leq 1$, and $\phi = 1$ in a neighborhood of $\bar{\omega}$. Thus $v = \phi u$ is a distribution in $\mathbf{R}^n$ which coincides with u, and is therefore harmonic, in ω. Its regularizing sequence $v * \gamma_k$ is harmonic in ω if k is large enough. Therefore $v * \psi$ is a C^∞ harmonic function in ω and $v * \psi = u$ a.e. in ω.

ω being an arbitrary open set in Ω, we conclude that u is a C^∞ harmonic function a.e. in Ω. □

If u is in $\mathcal{S}'$ then the Fourier transform of Laplace's equation $\Delta u = 0$ is $|\xi|^2 \hat{u} = 0$ which, in view of Example 4.11, implies that supp $\hat{u} \subset \{0\}$ and that, therefore, u is a harmonic polynomial. Thus no solution of $\Delta u = 0$ in $\mathcal{S}'$ can remain bounded as $|x| \to \infty$ unless it is a constant, and no solution can approach 0 as $|x| \to \infty$ unless it vanishes identically in $\mathbf{R}^n$. This, in fact, is true of *any* distributional solution of $\Delta u = 0$, based on the (classical) properties of harmonic functions and Theorem 4.11.

In $\mathbf{R}^2$ we know from the classical theory that the real (or imaginary) part of any analytic function of the complex variable $x + iy$, such as $e^x \cos y$, is harmonic, and yet it is not a polynomial. The reason, of course, is that $e^x \cos y$ is not a tempered distribution, and we should expect that $\Delta u = 0$ has solutions other than polynomials if we move from $\mathcal{S}'$ to $\mathcal{D}'$. Since every harmonic function in $\Omega \subset \mathbf{R}^n$ is analytic (see Exercise 6.4), we conclude that every solution of $\Delta u = 0$ in $\mathcal{D}'(\Omega)$ is in fact analytic (almost everywhere) in Ω. Based on the known growth properties of analytic functions, we see that, even in $\mathcal{D}'$, no solution of $\Delta u = 0$ in Ω can remain bounded as $|x| \to \infty$ unless it is a constant, and no solution can approach 0 as $|x| \to \infty$ unless it vanishes identically in Ω.

The corresponding nonhomogeneous equation

$$\Delta u = f$$

describes potential functions in $\mathbf{R}^n$ that can be associated with vector fields of gravity, electromagnetism, fluid motion, etc. It is a static equation which describes a spatial distribution u arising from the source distribution f. Our main concern, of course, is to solve

$$\Delta E = \delta$$

which, after transformation, becomes

$$\hat{E}(\xi) = -|\xi|^{-2} \qquad \xi \in \mathbf{R}^n - \{0\}$$

When $n = 2$, the function $|\xi|^{-2}$ is not locally integrable in $\mathbf{R}^2$ and does not represent a distribution. Thus we cannot obtain E in $\mathbf{R}^2$ by the Fourier transform method. But we have already found in Example 2.14 that, in this case,

$$E(x) = \frac{1}{2\pi} \log|x|$$

which is a tempered distribution because it is locally integrable and $|E(x)| \leq |x|$ when $|x| > 1$.

When $n > 2$, $|\xi|^{-2}$ is locally integrable and lies in $\mathcal{S}'$. Let

$$E_R(x) = -(2\pi)^{-n} \int_{|\xi| \leq R} e^{i\langle x,\xi\rangle} |\xi|^{-2}\, d\xi \tag{4.37}$$

For an arbitrary function $\phi \in \mathcal{S}$, we have

$$\langle E - E_R,\phi\rangle = \langle \hat{E},\mathcal{F}^{-1}(\phi)\rangle - \langle E_R,\phi\rangle$$

By Fubini's theorem,

$$\langle E_R,\phi\rangle = -\int_{|\xi| \leq R} |\xi|^{-2}\mathcal{F}^{-1}(\phi)d\xi$$

and hence

$$\langle E - E_R,\phi\rangle = -\int_{|\xi| \geq R} |\xi|^{-2}\mathcal{F}^{-1}(\phi)d\xi$$

which tends to 0 as $R \to \infty$ because $\mathcal{F}^{-1}(\phi)$ is in $\mathcal{S}$. Thus

$$\langle E,\phi\rangle = \lim_{R\to\infty} \langle E_R,\phi\rangle \qquad \phi \in \mathcal{S}$$

Now the integral (4.37) can also be expressed as

$$E_R(x) = -(2\pi)^{-n} \int_{|\xi| \leq R} e^{i|x||\xi| \cos\theta} |\xi|^{-2}\, d\xi$$

where θ is the angle between $x = (x_1, \ldots, x_n)$ and $\xi = (\xi_1, \ldots, \xi_n)$ defined by $\cos\theta = \langle x,\xi\rangle/|x||\xi|$. If we set $|x| = r$ and $|\xi| = \rho$, then

$$E_R(x) = -(2\pi)^{-n} \int_{\rho=0}^{R} \int_{|\xi|=\rho} e^{ir\rho\cos\theta} \rho^{-2}\, d\sigma\, d\rho \tag{4.38}$$

where $d\sigma$ is the element of area on the surface of the ball $B(0,\rho)$ in $\mathbf{R}^n$.

Let

$$S_{n-1}(\rho) = \{\xi \in \mathbf{R}^n : |\xi| = \rho\} = \partial B(0,\rho)$$

be the $(n - 1)$-dimensional sphere of radius ρ and center 0 in $\mathbf{R}^n$, and $\omega_{n-1}(\rho)$ be the area of $S_{n-1}(\rho)$. $\omega_{n-1}(\rho)$ is related to $\omega_{n-1}(1) = \omega_{n-1}$ by

$$\omega_{n-1}(\rho) = \rho^{n-1}\,\omega_{n-1}$$

and ω_{n-1} is given by

$$\omega_{n-1} = \frac{2\pi^{(1/2)n}}{\Gamma(\frac{1}{2}n)}$$

where Γ is the *gamma function*, which is defined on $(0,\infty)$ by

$$\Gamma(t) = \int_0^\infty e^{-s}s^{t-1}\,ds$$

and satisfies the recursion formula

$$\Gamma(t+1) = t\Gamma(t) \qquad t > 0$$

Thus, with $d\sigma = \rho^{n-1}\,d\omega$, equation (4.38) becomes

$$E_R(x) = -(2\pi)^{-n}\int_{\rho=0}^{R}\int_{S_{n-1}} e^{ir\rho\cos\theta}\,\rho^{n-3}\,d\omega\,d\rho$$

$$= -(2\pi)^{-n}r^{2-n}\int_{p=0}^{rR}\int_{S_{n-1}} e^{ip\cos\theta}\,p^{n-3}\,d\omega\,dp \tag{4.39}$$

where $p = r\rho$.

Taking the limit in (4.39) as $R \to \infty$,

$$E(x) = c_n|x|^{2-n}$$

where c_n is a constant which depends on the dimension n, and may be evaluated by the following procedure:

Let $\phi \in C_0^\infty$ be a spherically symmetric function, i.e., there is a $\tilde{\phi} \in C_0^\infty(\mathbf{R})$ such that $\phi(x) = \tilde{\phi}(|x|) = \tilde{\phi}(r)$ for all $x \in \mathbf{R}^n$. Then

$$\Delta\phi(x) = \sum_{k=1}^{n} \partial_k^2\,\tilde{\phi}(r)$$

$$= \sum_{k=1}^{n} \partial_k\left[\tilde{\phi}'(r)\,\frac{x_k}{r}\right]$$

$$= \sum_{k=1}^{n}\left[\frac{1}{r}\,\tilde{\phi}'(r) + \frac{x_k^2}{r^2}\,\tilde{\phi}''(r) - \frac{x_k^2}{r^3}\,\tilde{\phi}'(r)\right]$$

$$= \frac{n}{r}\tilde{\phi}'(r) + \tilde{\phi}''(r) - \frac{1}{r}\tilde{\phi}'(r)$$

$$= \frac{n-1}{r}\tilde{\phi}'(r) + \tilde{\phi}''(r)$$

Hence

$$\begin{aligned}
\langle \Delta E, \phi \rangle &= c_n \langle r^{2-n}, \Delta\phi \rangle \\
&= c_n \int_0^\infty r^{2-n}\left(\tilde{\phi}'' + \frac{n-1}{r}\tilde{\phi}'\right) r^{n-1}\,\omega_{n-1}\,dr \\
&= c_n[\tilde{\phi}(0) - (n-1)\tilde{\phi}(0)]\omega_{n-1} \\
&= -(n-2)\,\omega_{n-1}c_n\phi(0)
\end{aligned}$$

If E is to be a fundamental solution of Δ, then we must have $c_n = -1/(n-2)\omega_{n-1}$. Thus any fundamental solution of Δ is a spherically symmetric function which is given, to within an additive harmonic function in $\mathbf{R}^n$, by

$$E(x) = \begin{cases} \dfrac{1}{2\pi}\log|x| & n = 2 \\[2ex] -\dfrac{1}{(n-2)\omega_{n-1}}|x|^{2-n} & n > 2 \end{cases} \tag{4.40}$$

for all $x \neq 0$. E is singular at $x = 0$, but the singularity is such that both E and its first order derivatives are locally integrable in $\mathbf{R}^n$.

When $x \neq 0$, E is harmonic, as it should be since $\Delta E = \delta$ has its support at the origin. Since E is negative in the neighborhood of 0, in fact for all $x \neq 0$ when $n > 2$, there is an advantage in considering the operator $-\Delta$ rather than Δ, as some authors do. When $n = 3$, $E(x) = -1/4\pi|x|$, in agreement with the result of Example 2.15.

Finally, it is worth noting that, since harmonic functions possess the mean-value property, any spherically symmetric harmonic function in $\mathbf{R}^n$, $n \geq 2$, is necessarily a constant. Consequently, any spherically symmetric fundamental solution of Δ in $\mathbf{R}^n$, $n \geq 2$, is of the form E + constant, where E is given by (4.40).

4.8 FOURIER TRANSFORMS AND HOMOGENEOUS DISTRIBUTIONS

Let Λ be a linear mapping from $\mathbf{R}^n$ to $\mathbf{R}^n$ and $F(\mathbf{R}^n)$ be the linear space of complex functions on $\mathbf{R}^n$. We define the map

$$\Lambda^* : F(\mathbf{R}^n) \to F(\mathbf{R}^n)$$

by

$$\Lambda^* f(x) = f(\Lambda x) \qquad f \in F(\mathbf{R}^n)$$

Λ^* is also linear because

$$\begin{aligned}\Lambda^*(af + bg)(x) &= (af + bg)(\Lambda x)\\ &= a\Lambda^* f(x) + b\Lambda^* g(x)\\ &= (a\Lambda^* f + b\Lambda^* g)(x)\end{aligned}$$

for all $f,g \in F(\mathbf{R}^n)$ and $a,b \in \mathbf{C}$.

Λ may be represented by a real n $\times$ n matrix, determined by the basis that we choose for $\mathbf{R}^n$. It is nonsingular if the null space of Λ is $\{0\} \subset \mathbf{R}^n$. In that case the determinant of the matrix representing Λ, det Λ, is nonzero, and the inverse map Λ^{-1} exists and is also a linear map from $\mathbf{R}^n$ to $\mathbf{R}^n$.

If Λ is nonsingular, then Λ^* maps $\mathcal{S}$ continuously onto $\mathcal{S}$. Let ϕ,ψ be functions in $\mathcal{S}$. Then

$$\begin{aligned}\langle \Lambda^*\psi,\phi\rangle &= \int \psi(\Lambda x)\, \phi(x)dx\\ &= \int \psi(y)\, \phi(\Lambda^{-1}y)\, |\det \Lambda|^{-1}dy\\ &= \int \psi(y)\, |\det \Lambda|^{-1}\, \Lambda^{-1*}\, \phi(y)dy\end{aligned}$$

or

$$\langle \Lambda^*\psi,\phi\rangle = \langle \psi, |\det \Lambda|^{-1}\, \Lambda^{-1*}\, \phi\rangle \tag{4.41}$$

Now $|\det \Lambda|^{-1}\, \Lambda^{-1*}\phi$ is in $\mathcal{S}$ whenever ϕ is in $\mathcal{S}$, so the function ψ in equation (4.41) may be extended by continuity from $\mathcal{S}$ to $\mathcal{S}'$.

Since $f(x) = f(\Lambda^{-1}\, \Lambda x) = \Lambda^*\, \Lambda^{-1*}f(x)$ for every $f \in F(\mathbf{R}^n)$, we must have

$$\Lambda^{-1*} = \Lambda^{*-1}$$

We also have, for any $\phi \in \mathcal{S}$,

$$\begin{aligned}\mathcal{F}(\Lambda^*\phi)(\xi) &= \int e^{-i\langle\xi,x\rangle}\,\phi(\Lambda x)\,dx \\ &= \int e^{-i\langle\xi,\Lambda^{-1}y\rangle}\,\phi(y)|\det\Lambda|^{-1}dy \\ &= \int e^{-i\langle\Lambda^{-1T}\xi,y\rangle}\phi(y)\,|\det\Lambda|^{-1}dy \\ &= |\det\Lambda|^{-1}\,\hat{\phi}(\Lambda^{-1T}\xi)\end{aligned}$$

where Λ^T is the transpose of Λ. Thus

$$(\Lambda^*\phi)\,\hat{}\; = |\det\Lambda|^{-1}\,(\Lambda^{-1T})^*\hat{\phi} \qquad \phi\in\mathcal{S} \tag{4.42}$$

The operators $\mathcal{F}\Lambda^*$ and $|\det\Lambda|^{-1}\,(\Lambda^{-1T})^*\mathcal{F}$ are equal and continuous on $\mathcal{S}$, so they may be extended by continuity to $\mathcal{S}'$, and we obtain

$$(\Lambda^*T)\,\hat{}\; = |\det\Lambda|^{-1}\,(\Lambda^{-1T})^*\hat{T} \qquad T\in\mathcal{S}'. \tag{4.43}$$

The reflection operator $x \mapsto -x$ introduced in Section 3.4 is an example of a (nonsingular) linear transformation on $\mathbf{R}^n$, which induces the transformation $T \to \check{T}$ in $\mathcal{S}'$. A more general example is the transformation

$$\Lambda_t x = tx \qquad x\in\mathbf{R}^n$$

which is linear and continuous for any $t \in \mathbf{R}$, but singular when $t = 0$. Thus, in accordance with (4.41), if $T \in \mathcal{D}'$ and $t \neq 0$ then Λ_t^*T is the distribution defined by

$$\langle\Lambda_t^*T,\phi\rangle = \langle T,|\det\Lambda_t|^{-1}\,\Lambda_t^{-1*}\phi\rangle \qquad \phi\in\mathcal{D}$$

But since $\det\Lambda_t = t^n$ and $\Lambda^{-1} = \Lambda_{1/t}$, this becomes

$$\langle\Lambda_t^*T,\phi\rangle = \langle T,t^{-n}\Lambda_{1/t}^*\phi\rangle$$

We say that a distribution T is *homogeneous* of degree d, where d is a complex number, if $\Lambda_t^*T = t^dT$ for any $t > 0$. Recalling that a function f defined on $\mathbf{R}^n$ is homogeneous of degree d if $f(tx) = t^df(x)$, we shall now show that the two definitions coincide when the function is locally integrable in $\mathbf{R}^n$, in the sense that $\Lambda_t^*f = t^df$ if and only if $f(tx) = t^df(x)$ a.e. Using the definition above,

$$\begin{aligned}\langle\Lambda_t^*f,\phi\rangle &= \langle f,t^{-n}\Lambda_{1/t}^*\phi\rangle \qquad \phi\in\mathcal{D} \\ &= \int f(x)\,t^{-n}\phi(x/t)dx \\ &= \int f(ty)\phi(y)dy\end{aligned}$$

Thus if $f(tx) = t^d f(x)$ a.e. then $\langle \Lambda_t^* f,\phi\rangle = \int t^d f(y)\phi(y)dy = \langle t^d f,\phi\rangle$ and $\Lambda_t^* f = t^d f$. Conversely, if $\Lambda_t^* f = t^d f$, then for any $\phi \in \mathfrak{D}$,

$$\int f(ty)\phi(y)\,dy = \int t^d f(y)\phi(y)\,dy$$

and, by Example 2.24, $f(ty) = t^d f(y)$ a.e.

Example 4.13 (i) Let $\{T_1, \ldots ,T_m\}$ be a set of nonzero distributions in $\mathbf{R}^n$ such that T_k, $1 \leqslant k \leqslant m$, is homogeneous of real degree d_k and $d_k \neq d_j$ if $k \neq j$. We shall show that the set is linearly independent over $\mathbf{C}$.

Let $a_1T_1 + \cdots + a_mT_m = 0$. Without loss of generality, assume that $d_1 > d_2 > \cdots > d_m$. For any $\phi \in \mathfrak{D}$, we have

$$0 = \left\langle \Lambda_t^* \sum_1^m a_kT_k,\phi \right\rangle = \sum_1^m a_k \langle \Lambda_t^* T_k,\phi\rangle = \sum_1^m a_k t^{d_k} \langle T_k,\phi\rangle$$

If the coefficients a_k do not all vanish, let $i \geqslant 1$ be the smallest integer for which $a_i \neq 0$. If $i = m$ then $\langle T_m,\phi\rangle = 0$ and hence $T_m = 0$, contradicting the hypothesis. If $1 \leqslant i < m$, then $a_i\langle T_i,\phi\rangle + \Sigma_i^m a_k t^{d_k - d_i} \langle T_k,\phi\rangle = 0$ for all $t > 0$ and $\phi \in \mathfrak{D}$. Now if we let $t \to \infty$, we obtain $a_i\langle T_i,\phi\rangle = 0$. Since $a_i \neq 0$ this implies that $T_i = 0$, again contradicting the hypothesis.

(ii) Now we shall show that $\partial^\alpha\delta$ is homogeneous of degree $-n - |\alpha|$.

$$\begin{aligned}
\langle \Lambda_t^* \partial^\alpha\delta,\phi\rangle &= \langle \partial^\alpha\delta, t^{-n}\Lambda_{1/t}^*\phi\rangle \\
&= t^{-n}\langle \partial^\alpha\delta,\phi(x/t)\rangle \\
&= (-1)^{|\alpha|}t^{-n}\langle \delta,\partial^\alpha\phi(x/t)\rangle \\
&= (-1)^{|\alpha|}t^{-n}t^{-|\alpha|}(\partial^\alpha\phi)(0) \\
&= t^{-n-|\alpha|}\langle \partial^\alpha\delta,\phi\rangle \qquad \phi \in \mathfrak{D}
\end{aligned}$$

Therefore $\Lambda_t^*\partial^\alpha\delta = t^{-n-|\alpha|}\partial^\alpha\delta$. In view of (i), we conclude that the distributions $\delta, \delta', \ldots ,\delta^{(m)}$ on $\mathbf{R}$ are linearly independent.

Example 4.14 For $\lambda \geqslant 0$ we shall show that $x_+^\lambda = x^\lambda H$, $x \in \mathbf{R}$, is homogeneous of degree λ.

$$\begin{aligned}
\langle \Lambda_t^* x_+^\lambda,\phi\rangle &= \langle x_+^\lambda, t^{-1}\Lambda_{1/t}^*\phi\rangle \\
&= t^{-1}\int_0^\infty x^\lambda\phi(x/t)dx
\end{aligned}$$

$$= t^{-1} \int_0^\infty t^\lambda y^\lambda \, \phi(y) t \, dy$$

$$= \langle t^\lambda x_+^\lambda, \phi \rangle$$

Hence $\Lambda_t^* x_+^\lambda = t^\lambda x_+^\lambda$.

Let $\phi \in \mathcal{S}$ be homogeneous of degree d and t be a positive number. Then, by the chain rule,

$$\partial_k[\phi(tx)] = t(\partial_k \phi)(tx)$$

Hence

$$\begin{aligned} \Lambda_t^*(\partial_k \phi)(x) &= (\partial_k \phi)(tx) \\ &= t^{-1} \partial_k[\phi(tx)] \\ &= t^{d-1} (\partial_k \phi)(x) \end{aligned}$$

which means that $\partial_k \phi$ is homogeneous of degree $d - 1$.

To obtain the analogous result for $T \in \mathcal{S}'$, given that the degree of T is d, we first note that for all $\phi \in \mathcal{S}$

$$\begin{aligned} \partial_k(\Lambda_t^* \phi)(x) &= \partial_k[\phi(tx)] \\ &= t(\partial_k \phi)(tx) \qquad (4.44) \\ &= t\Lambda_t^* (\partial_k \phi)(x) \end{aligned}$$

Now, with $\Lambda_t^{-1} = \Lambda_{1/t}$, equations (4.41) and (4.44) imply

$$\begin{aligned} \Lambda_t^* \partial_k T(\phi) &= \partial_k T(|\det \Lambda_t|^{-1} \Lambda_t^{-1*} \phi) \\ &= -T(|\det \Lambda_{1/t}| \partial_k |\Lambda_{1/t}^* \phi) \\ &= -t^{-1} T(|\det \Lambda_{1/t}| \Lambda_{1/t}^* \partial_k \phi) \\ &= -t^{-1} \Lambda_t^* T(\partial_k \phi) \\ &= t^{-1} \partial_k \Lambda_t^* T(\phi) \\ &= t^{d-1} \partial_k T(\phi) \end{aligned}$$

Thus $\partial_k T$ has degree $d - 1$.

Using the relations $\det \Lambda_t = t^n$ and $\Lambda_t^T = \Lambda_t$ in equation (4.43) gives

$$(\Lambda_t^* T)^\wedge = t^{-n} \Lambda_{1/t}^* \hat{T} \qquad T \in \mathcal{S}' \qquad (4.45)$$

If T is homogeneous of degree d, then (4.45) implies that

$$t^d\hat{T} = t^{-n}\Lambda^*_{1/t}\hat{T}$$

or

$$\Lambda^*_t\hat{T} = t^{-n-d}\hat{T}$$

Thus we have proved

Theorem 4.12 If $T \in \mathcal{S}'$ is homogeneous of degree d, then $\partial_k T$ is homogeneous of degree $d - 1$ and $\hat{T}$ is homogeneous of degree $-n - d$.

Example 4.15 The function $f(z) = 1/z = 1/(x + iy)$ is locally integrable in the plane and $|f(z)| < 1$ when $|z| > 1$; hence f defines a tempered distribution in $\mathbf{R}^2$, and we shall now compute its Fourier transform:

$$\mathcal{F}(zf) = \mathcal{F}(1) = (2\pi)^2\delta$$

On the other hand,

$$\begin{aligned}\mathcal{F}(zf) &= \mathcal{F}(xf) + i\mathcal{F}(yf)\\ &= i\frac{\partial}{\partial\xi}\hat{f} - \frac{\partial}{\partial\eta}\hat{f}\\ &= i\left(\frac{\partial}{\partial\xi} + i\frac{\partial}{\partial\eta}\right)\hat{f}\\ &= 2i\,\bar{\partial}\,\hat{f}\end{aligned}$$

Therefore $i\hat{f}/2\pi^2$ is a fundamental solution of the Cauchy–Riemann operator $\bar{\partial} = \frac{1}{2}[\partial/\partial\xi + i(\partial/\partial\eta)]$. By Example 4.12, $\hat{f}$ must satisfy

$$\frac{i}{2\pi}\hat{f}(\zeta) = \frac{1}{\zeta} + h(\zeta)$$

where h is an entire function.

Since f is homogeneous of degree -1 in $\mathbf{R}^2$, Theorem 4.12 implies that $\hat{f}$ is homogeneous of degree $-2 + 1 = -1$. If h is not identically 0, it must also have degree -1, and hence $h(t\zeta) = h(\zeta)/t$, $t > 0$, which becomes unbounded as $t \to 0$. Consequently, $h = 0$ and

$$\hat{f}(\zeta) = \mathcal{F}\left(\frac{1}{z}\right) = -\frac{2\pi i}{\zeta}$$

A linear transformation $\Lambda : \mathbf{R}^n \to \mathbf{R}^n$ is said to be *orthogonal* if $\Lambda^T = \Lambda^{-1}$. If Λ is orthogonal, then so is Λ^{-1} and $\det\Lambda = \pm 1$. Moreover, an orthogonal transformation Λ satisfies

$$|\Lambda x|^2 = \langle \Lambda x, \Lambda x \rangle = \langle x, \Lambda^T \Lambda x \rangle = \langle x, x \rangle = |x|^2$$

for all $x \in \mathbf{R}^n$. Hence $|\Lambda x| = |x|$. Conversely, if $|\Lambda x| = |x|$ for all $x \in \mathbf{R}^n$, then $\Lambda^T \Lambda$ = identity, which implies that Λ is orthogonal. In other words, the transformation Λ is orthogonal if and only if it is norm-preserving.

A distribution $T \in \mathcal{D}'$ is said to be *invariant* under the transformation $\Lambda : \mathbf{R}^n \to \mathbf{R}^n$ if $\Lambda^* T = T$. A function $f : \mathbf{R}^n \to \mathbf{C}$ is called *rotation-invariant*, or spherically symmetric, if there exists a function $g : [0, \infty) \to \mathbf{C}$ such that $f(x) = g(|x|)$ for all $x \in \mathbf{R}^n$. A function which is rotation-invariant is invariant under any orthogonal transformation Λ, since $\Lambda^* f(x) = f(\Lambda x) = g(|\Lambda x|) = g(|x|) = f(x)$. But a rotation in $\mathbf{R}^n$ is an orthogonal transformation, hence a function is rotation invariant if and only if it is invariant under orthogonal transformations.

Theorem 4.13 If $T \in \mathcal{S}'$ is invariant under orthogonal transformations, then $\hat{T}$ is also invariant under orthogonal transformations.

Proof. If Λ is an orthogonal transformation and T is any distribution in $\mathcal{S}'$ then, by (4.43), $(\Lambda^* T)^\wedge = \Lambda^* \hat{T}$. Consequently, $\Lambda^* T = T$ if and only if $\Lambda^* \hat{T} = \hat{T}$. □

When a distribution is represented by a rotation-invariant function, the distribution is also said to be rotation-invariant. The above theorem then implies that if $T \in \mathcal{S}'$ is rotation invariant and $\hat{T}$ is a function, then $\hat{T}$ is also rotation invariant.

Our final aim in this investigation is, of course, to study the form and properties of fundamental solutions in $\mathcal{S}'$ to linear differential operators of interest. These solutions turn out to be smooth functions when the differential operator L has constant coefficients, except for a singularity at the origin, whose strength depends on the order of L and the dimension n, and which accounts for the δ distribution in $LE = \delta$.

With this in mind, let f be a homogeneous function on $\mathbf{R}^n$-{0} of degree λ. Then we can write

$$\begin{aligned} f(x) &= |x|^\lambda f(x/|x|) \\ &= |x|^\lambda f(\omega) \qquad x \in \mathbf{R}^n - \{0\} \end{aligned} \tag{4.46}$$

where ω lies on the unit sphere $S_{n-1} = \{x \in \mathbf{R}^n : |x| = 1\}$. If $\lambda > -n$, the function f is integrable, and hence lies in $\mathcal{S}'$, otherwise it is not integrable. We shall now use the results of Section 2.8 to extend f as a distribution in $\mathcal{S}'$ beyond $\lambda > -n$. The representation obtained in Section 2.8 for x_+^λ, x_-^λ, $|x|^\lambda$ and $|x|^\lambda \operatorname{sgn} x$ as distributions in $\mathcal{D}'(\mathbf{R})$ remain valid in $\mathcal{S}'(\mathbf{R})$ because, as functions on $\mathbf{R} - \{0\}$, these powers of x have polynomial growth as $|x| \to \infty$. Thus for any $\phi \in \mathcal{S}(\mathbf{R}^n)$ we can write

$$\langle f,\phi\rangle = \int_{\mathbf{R}^n} |x|^\lambda f(\omega)\phi(x)dx \qquad \lambda > -n$$

$$= \int_0^\infty r^{\lambda+n-1} \int_{S_{n-1}} f(\omega)\phi(r\omega)d\omega\, dr$$

where $r = |x| = [x_1^2 + \cdots + x_n^2]^{1/2}$ and $d\omega$ is the element of area on S_{n-1}.

By defining the function

$$\sigma_\phi(r) = \int_{S_{n-1}} f(\omega)\phi(r\omega)d\omega \tag{4.47}$$

on $r \geqslant 0$, we can write

$$\langle f,\phi\rangle = \int_0^\infty r^{\lambda+n-1}\,\sigma_\phi(r)dr \tag{4.48}$$

Now, with $\phi \in C^\infty(\mathbf{R}^n)$, it is clear that $\sigma_\phi \in C^\infty(0,\infty)$.

The function ϕ has a Taylor series expansion about $x = 0$ of the form

$$\phi(x) = \sum_{|\alpha|\leq 2m} \frac{1}{\alpha!} x^\alpha \partial^\alpha \phi(0) + R_{2m+1}(x) \tag{4.49}$$

Writing $x^\alpha = r^{|\alpha|}\omega^\alpha$ and integrating on S_{n-1} according to (4.47),

$$\sigma_\phi(r) = a_0 + a_1 r + a_2 r^2 + \cdots + a_{2m} r^{2m} + M_{2m+1}(r) \tag{4.50}$$

where the coefficients a_k are constant for all $k \in \{0, \ldots, 2m\}$, and $M_{2m+1}(r)$ is of order r^{2m+1} in the neighborhood of $r = 0$. Thus the function σ_ϕ is differentiable up to order $2m$ at $r = 0$. But since m is arbitrary we conclude that σ_ϕ is infinitely differentiable at $r = 0$.

To extend σ_ϕ to $\mathbf{R}$ we note that (4.47) gives

$$\sigma_\phi(-r) = \int_{S_{n-1}} f(\omega)\,\phi(-r\omega)d\omega$$

$$= \int_{S_{n-1}} f(-\omega)\,\phi(r\omega)d\omega$$

so σ_ϕ may be extended as an even function to **R** provided $f(-\omega) = f(\omega)$, i.e. provided the homogeneous function f is invariant under reflection in the origin. This is always true when f is rotation-invariant. In that case $f(\omega)$ is a constant on S_{n-1} which we can assume, without loss of generality, to be 1. $f(x)$ is therefore reduced to $|x|^\lambda$. With this assumption the odd powers of r drop out of the Taylor series (4.50).

Now we shall determine the values of λ for which $|x|^\lambda$ is a tempered distribution in $\mathbf{R}^n$. For every pair of indices $j,k \in \mathbf{N}_0$, there is a pair $\alpha,\beta \in \mathbf{N}_0^n$ such that

$$\sup_{t\in\mathbf{R}} |t^j\partial^k\sigma_\phi(t)| \leqslant \sup_{x\in\mathbf{R}^n} |x^\alpha\partial^\beta\phi(x)| \int_{S_{n-1}} d\omega$$
$$= \omega_{n-1} \sup_{x\in\mathbf{R}^n} |x^\alpha\partial^\beta\phi(x)|$$

which means that σ_ϕ is in $\mathcal{S}(\mathbf{R})$ whenever ϕ is in $\mathcal{S}(\mathbf{R}^n)$. Thus the representation (4.48), which is valid on $\lambda > -n$, may now be expressed by

$$\langle|x|^\lambda, \phi\rangle = \langle t^\mu,\sigma_\phi\rangle \qquad \mu = \lambda + n - 1 > -1$$

and continued as a function of μ into $\mathbf{C} - \mathbf{Z}^-$ by using the results of Example 2.26. The residue of the function $\langle t_+^\mu,\sigma_\phi\rangle$ at $\mu = -m$ is given by $\sigma_\phi^{(m-1)}(0)/(m-1)!$. But since $\sigma_\phi^{(m-1)}(0)$ vanishes when $m - 1$ is odd, we see that t_+^μ has its poles at $\mu = -1,-3-5,\ldots$. This means that the poles of $|x|^\lambda$ are located at $\lambda = -n,-n-2,-n-4,\ldots$.

To investigate the Fourier transform of $|x|^\lambda$, we invoke Theorems 4.12 and 4.13. Since $|x|^\lambda$ is a homogeneous, rotation invariant distribution of degree λ for all $\lambda \in \mathbf{C} - \{-n - 2m : m \in \mathbf{N}_0\}$, its Fourier transform is homogeneous of degree $-n - \lambda$ and rotation invariant. On the interval $\lambda > -n$, the function $|x|^\lambda$ is locally integrable, so it is reasonable to expect that $\mathcal{F}(|x|^\lambda)$ is also a function in that range. But then, in view of the representation (4.46) of any homogeneous function on $\mathbf{R} - \{0\}$, $\mathcal{F}(|x|^\lambda)$ can only have the form

$$\mathcal{F}(|x^\lambda|) = c(n,\lambda)\,|\xi|^{-\lambda-n} \qquad \xi \in \mathbf{R}^n - \{0\} \tag{4.51}$$

which is locally integrable on $\lambda < 0$. To evaluate $c(n,\lambda)$ we shall use Parseval's relation

$$\langle\mathcal{F}(|x|^\lambda), \hat{\gamma}(\xi)\rangle = (2\pi)^n\langle|x|^\lambda, \gamma(x)\rangle \tag{4.52}$$

with $\gamma(x) = \exp(-(1/2)|x|^2)$ and $\hat{\gamma}(\xi) = (2\pi)^{n/2}\exp(-(1/2)|\xi|^2)$. Strictly speaking, Parseval's relation holds for L^2 functions, so we must further restrict λ to the interval $-(1/2)n < \lambda < 0$ and apply the relation to the truncated function which equals $|x|^\lambda$ on $B(0,R)$ and 0 on $\mathbf{R}^n - B(0,R)$. In the limit as $R \to \infty$ we obtain (4.52).

Equation (4.52) now gives

$$c(n,\lambda)(2\pi)^{(1/2)n} \int |\xi|^{-\lambda-n} \exp\left(-(1/2)|\xi|^2\right) d\xi = (2\pi)^n \int |x|^\lambda \exp\left(-(1/2)|x|^2\right) dx$$

In spherical coordinates this becomes

$$c(n,\lambda)\omega_{n-1} \int_0^\infty \rho^{-\lambda-1} e^{-(1/2)\rho^2}\, d\rho = (2\pi)^{(1/2)n} \omega_{n-1} \int_0^\infty r^{\lambda+n-1} e^{-(1/2)r^2}\, dr$$

Using the formula

$$\int_0^\infty r^\mu \exp\left(-\tfrac{1}{2}r^2\right) dr = \int_0^\infty e^{-y}(2y)^{(1/2)(\mu-1)}\, dy = 2^{(1/2)(\mu-1)}\Gamma\left(\tfrac{1}{2}(\mu - 1)\right)$$

we arrive at

$$c(n,\lambda) = 2^{\lambda+n}\, \pi^{(1/2)n}\Gamma\left(\tfrac{1}{2}(\lambda + n)\right)/\Gamma\left(-\tfrac{1}{2}\lambda\right)$$

and the representation

$$\mathcal{F}(|x|^\lambda) = 2^{\lambda+n}\pi^{(1/2)n} \frac{\Gamma(\frac{1}{2}(\lambda + n))}{\Gamma(-(1/2)\lambda)} |\xi|^{-\lambda-n} \tag{4.53}$$

which is valid in $-(1/2)\lambda < \operatorname{Re}\lambda < 0$. But we have just seen that $|x|^\lambda$ may be continued analytically into $\mathbf{C}$ except for the simple poles at $\lambda = -n, -n - 2, -n - 4, \ldots$, so the same is true of $\mathcal{F}(|x|^\lambda)$. Thus equation (4.53) is valid for all λ in $\mathbf{C} - \{-n - 2k : k \in \mathbf{N}_0\}$ Note that the poles of $|\xi|^{-\lambda-n}$ on the right-hand side of (4.53) cancel those of $\Gamma(-(1/2)\lambda)$, and only the poles of $\Gamma((1/2)(\lambda + n))$ at $\lambda = -n - 2k$, $k \in \mathbf{N}_0$, remain.

Example 4.16 Let log: $\mathbf{R} - \{0\} \to \mathbf{C}$ be the function defined by

$$\log x = \begin{cases} \log x & \text{if } x > 0 \\ \log|x| + i\pi & \text{if } x < 0 \end{cases}$$

Then we can write

$$\log x = \log|x| + i\pi H(-x)$$

From Example 2.12 we have

$$\frac{d}{dx}(\log x) = pv\,\frac{1}{x} - i\pi\delta$$

Taking the Fourier transform, and using Example 4.5,

$$i\xi\mathcal{F}(\log x) = -2\pi i H(\xi) + \pi i - i\pi$$

$$\mathcal{F}(\log x) = -\frac{2\pi}{\xi} H(\xi) \qquad x \in \mathbf{R} - \{0\}$$

To compute $\mathcal{F}(\log |x|)$, we note that

$$\frac{d}{dx}(\log |x|), = \frac{d}{dx}(\log x) + i\pi\delta$$

$$\begin{aligned} i\xi\mathcal{F}(\log |x|) &= i\xi\mathcal{F}(\log x) + i\pi \\ &= -2\pi i H(\xi) + i\pi \\ &= \pi i[H(-\xi) - H(\xi)] \end{aligned}$$

$$\begin{aligned} \mathcal{F}(\log |x|) &= \frac{\pi}{\xi}[H(-\xi) - H(\xi)] \\ &= -\pi/|\xi| \qquad \xi \in \mathbf{R} - \{0\} \end{aligned}$$

If we denote $\lim_{\varepsilon\to 0^\pm} 1/(x + \varepsilon)$ by $1/(x \pm 0)$, then we also deduce from this example that

$$\begin{aligned} \frac{1}{x + i0} = \lim_{\varepsilon\to 0^+} \frac{1}{x + i\varepsilon} &= \lim_{\varepsilon\to 0^+} \left[\frac{d}{dx}\log(x + i\varepsilon)\right] \\ &= \frac{d}{dx}(\log x) = pv\,\frac{1}{x} - i\pi\delta \end{aligned}$$

since the differential operator d/dx is continuous in $\mathcal{D}'$.

Similarly,

$$\begin{aligned} \frac{1}{x - i0} = \lim_{\varepsilon\to 0^-} \frac{1}{x + i\varepsilon} &= \lim_{\lambda\to 0^+} \frac{1}{x - i\lambda} \qquad \lambda = -\varepsilon > 0 \\ &= \lim_{\lambda\to 0^+} \left(-\frac{1}{y + i\lambda}\right) \qquad y = -x \in \mathbf{R} - \{0\} \\ &= -\left(pv\,\frac{1}{y} - i\pi\delta\right) \\ &= pv\,\frac{1}{x} + i\pi\delta \end{aligned}$$

Example 4.17 To calculate the Fourier transform of $\log|x|$ with $x \in \mathbf{R}^n - \{0\}$, we proceed as follows:

$$\partial_j |x|^\lambda = \lambda\, x_j |x|^{\lambda-2} \qquad j = 1, \ldots, n$$

by the chain rule. Applying the Fourier transformation and using (4.51), we obtain

$$i\xi_j\, c(n,\lambda)\, |\xi|^{-\lambda-n} = \lambda \mathcal{F}(x_j|x|^{\lambda-2}) \qquad \lambda \neq -n - 2k \qquad k \in \mathbf{N}_0$$

Hence, assuming $\lambda \neq 0$,

$$\mathcal{F}(x_j|x|^{\lambda-2}) = \frac{c(n,\lambda)}{\lambda}\, i\xi_j |\xi|^{-\lambda-n}$$

$$\begin{aligned}\frac{c(n,\lambda)}{\lambda} &= 2^{\lambda+n}\, \pi^{(1/2)n} \frac{\Gamma(\frac{1}{2}(\lambda + n))}{\lambda\Gamma(-\frac{1}{2}\lambda)} \\ &= 2^{\lambda+n}\, \pi^{(1/2)n} \frac{\Gamma(\frac{1}{2}(\lambda + n))}{-2\Gamma(\frac{1}{2}(-\lambda + 2))} \\ &= -2^{\lambda+n-1}\, \pi^{(1/2)n} \frac{\Gamma(\frac{1}{2}(\lambda + n))}{\Gamma(\frac{1}{2}(-\lambda+2))} \end{aligned} \tag{4.54}$$

Since the right-hand side of (4.54) is defined at $\lambda = 0$, the equality holds at $\lambda = 0$, and we therefore have

$$\mathcal{F}(x_j|x|^{\lambda-2}) = -2^{\lambda+n-1}\, \pi^{(1/2)n} \frac{\Gamma(\frac{1}{2}(\lambda + n))}{\Gamma(\frac{1}{2}(-\lambda + 2))}\, i\xi_j\, |\xi|^{-\lambda-n} \tag{4.55}$$

for $\lambda \neq -n - 2k$, $k \in \mathbf{N}_0$. And now, since

$$\partial_j \log|x| = x_j|x|^{-2}$$

$$i\xi_j \mathcal{F}(\log|x|) = \mathcal{F}(x_j|x|^{-2})$$

we obtain $\mathcal{F}(\log|x|)$ by substituting $\lambda = 0$ into equation (4.55),

$$\mathcal{F}(\log|x|) = -2^{n-1}\, \pi^{(1/2)n}\, \Gamma\left(\tfrac{1}{2}n\right) |\xi|^{-n} \tag{4.56}$$

Note that this expression reduces to $-\pi/|\xi|$, as we found in Example 4.16, when $n = 1$.

The general partial differential operator of order m with constant coefficients is represented by

$$P(D) = \sum_{|\alpha| \leq m} a_\alpha D^\alpha \qquad \alpha \in \mathbf{N}_0^n \tag{4.57}$$

Its homogeneous part of order m defines the homogeneous polynomial

$$P_m(\xi) = \sum_{|\alpha|=m} a_\alpha \xi^\alpha$$

which is called the *characteristic polynomial* of the operator $P(D)$. $P(D)$ is said to be *elliptic* if $P_m(\xi) \neq 0$ for all $\xi \in \mathbf{R}^n - \{0\}$. The Laplacian and the Cauchy–Riemann operators are both elliptic. Note that the characteristic polynomial of $P(\partial)$ is $P_m(i\xi) = i^m P_m(\xi)$.

From a formal point of view, the construction of the fundamental solution of the operator P is straightforward. Starting with $P(\partial)E = \delta$ we obtain $P(i\xi)\hat{E} = 1$ and then E is simply $\mathcal{F}^{-1}(1/P(i\xi))$. E will be in $\mathcal{S}'$ if and only if $\hat{E}$ is in $\mathcal{S}'$. The burden of constructing E is therefore thrown on the feasibility of dividing by $P(i\xi)$ in $\mathcal{S}'$, and then taking the inverse Fourier transform of the resulting distribution $1/P(i\xi)$.

Example 4.18 The iterated Laplacian $P(\partial) = \Delta^m$, where m is a positive integer, is elliptic and homogeneous of order $2m$.

$$\begin{aligned} P(\xi) &= |\xi|^{2m} \\ P(i\xi) &= (-1)^m|\xi|^{2m} \\ \hat{E}(\xi) &= (-1)^m|\xi|^{-2m} \\ &= \frac{(-1)^m}{c(n,-n+2m)} \mathcal{F}(|x|^{-n+2m}) \end{aligned}$$

by equation (4.51) with $\lambda = -n + 2m$. Hence

$$\begin{aligned} E(x) &= \frac{(-1)^m}{c(n,2m-n)} |x|^{2m-n} \\ &= \frac{(-1)^m \,\Gamma(\frac{1}{2}n - m)}{2^{2m}\pi^{n/2}\Gamma(m)} |x|^{2m-n} \end{aligned} \tag{4.58}$$

The representation (4.58) is valid whenever $2m - n \neq 2k$, $k \in \mathbf{N}_0$.
When

$$2m - n = 2k \qquad k \in \mathbf{N}_0 \tag{4.59}$$

the function $|x|^{2m-n} = r^{2k}$ is annihilated by the iterated Laplacian Δ^m because

$$\begin{aligned} \Delta\, r^{2k} &= \left(\frac{d^2}{dr^2} + \frac{n-1}{r}\frac{d}{dr}\right) r^{2k} \\ &= 2k(2k+n-2)\, r^{2k-2} \end{aligned}$$

$\Delta^k r^{2k} =$ constant, and $m = k + (1/2)n > k$.

On the other hand,

$$\Delta^k (r^{2k} \log r) = c_1 \log r + c_2 \qquad c_1 \neq 0 \tag{4.60}$$

To see that equation (4.60) is true, note that

$$\Delta(r^{2k} \log r) = (a_1 \log r + a_2)r^{2k-2}$$

$$\Delta(a_1 \log r + a_2)r^{2k-2} = (b_1 \log r + b_2)r^{2k-4}$$

with a_1, a_2, b_1, b_2 constants determined by k and n, and both a_1 and b_1 not zero.

Taking the Fourier transform of equation (4.60), we obtain

$$(-1)^k|\xi|^{2k}\, \mathcal{F}(r^{2k} \log r) = c_1\, \mathcal{F}(\log r) + (2\pi)^n c_2 \delta$$

Setting $2k = 2m - n$ and multiplying by $|\xi|^n$ gives

$$(-1)^k|\xi|^{2m}\, \mathcal{F}(r^{2m-n} \log r) = c_1|\xi|^n\, \mathcal{F}(\log r)$$

But, in view of (4.56), the right-hand side of this equation is a constant. Consequently $C_{nm}|x|^{2m-n} \log |x|$, for a suitable choices of the constant C_{nm}, is a solution of $\hat{E}(\xi) = (-1)^m|\xi|^{-2m}$. Thus

$$E(x) = \begin{cases} C_{nm}|x|^{2m-n} & \tfrac{1}{2}(2m - n) \notin \mathbf{N}_0 \\ C_{nm}|x|^{2m-n} \log |x| & \tfrac{1}{2}(2m - n) \in \mathbf{N}_0 \end{cases} \tag{4.61}$$

where C_{nm} is the coefficient of $|x|^{2m-n}$ in equation (4.58) when $(1/2)(2m - n) \notin \mathbf{N}_0$. Note that, since $m - (1/2)n \in \mathbf{N}_0$ if and only if n is even and at most $2m$, the logarithmic solution appears only when the dimension is even and does not exceed $2m$. The fundamental solution (4.40) for Δ may be obtained from (4.61) by setting $m = 1$. In that case the logarithmic solution appears only when $n = 2$.

4.9 THE DIVISION PROBLEM IN $\mathcal{S}'$

Let us return to the equation $P(D)u = f$, where $f \in \mathcal{S}'$ and $P(D)$ is a linear partial differential operator of order m with constant coefficients. The Fourier transformation yields

$$P(\xi)\hat{u} = \hat{f}$$

If the polynomial $P(\xi)$ has no zeros in $\mathbf{R}^n$ then $\hat{f}/P$ lies in $\mathcal{S}'$, and its inverse transform is the solution of the differential equation. Example (4.18) shows how we can deal with the situation when P has a zero at the origin. Obviously, the problem becomes more difficult when P has other zeros in $\mathbf{R}^n$. The question then is: Given $g \in \mathcal{S}'$ and P a polynomial in $\mathbf{R}^n$, is there a distribution $v \in$

$\mathcal{S}'$ such that $Pv = g$? This is known as the *division problem* in $\mathcal{S}'$. The affirmative answer, which had been conjectured by L. Schwartz, was proved in the late fifties by Hörmander and Łojasiewics. The references to this topic and an account of the more general division problem in $\mathcal{E}'$ and $\mathcal{D}'$ may be found in [8]. The closure of $\mathcal{S}'$ under division of polynomials means that every linear partial differential equation $P(D)u = f$ with constant coefficients, and with $f \in \mathcal{S}'$, has a solution in $\mathcal{S}'$.

On the question of uniqueness of the solution to $P(D)u = f$, we naturally turn to the homogeneous equation $P(D)v = 0$. When P has no zeros in $\mathbf{R}^n$, such as $|\xi|^2 + 1$ (corresponding to the operator $-\Delta + 1$), $P(D)v = 0$ has only the trivial solution, so that the solution of $P(D)u = f$ in $\mathcal{S}'$ is unique. When P has a zero at $\xi_0 \in \mathbf{R}^n$ then $c\delta_{\xi_0}$, where c is an arbitrary constant, satisfies $Pv = 0$. Hence the solution of $P(D)u = f$ is not unique.

Example 4.19 To construct the solution of $xv = g$ in $\mathcal{S}'(\mathbf{R})$, where g is a given tempered distribution, we note that the equality $\langle xv,\phi\rangle = \langle v,x\phi\rangle$ for all $\phi \in \mathcal{S}(\mathbf{R})$ implies we should consider the equation

$$\langle v,x\phi\rangle = \langle g,\phi\rangle \qquad \phi \in \mathcal{S}(\mathbf{R}) \tag{4.62}$$

By using Taylor's series, it can be shown that a function $\psi \in \mathcal{S}(\mathbf{R})$ has the form $\psi = x\phi$, with $\phi \in \mathcal{S}(\mathbf{R})$, if and only if $\psi(0) = 0$. The function ϕ is then determined by

$$\phi(x) = \int_0^1 \psi'(xt)\,dt = \begin{cases} \psi(x)/x & \text{when } x \neq 0 \\ \psi'(0) & \text{at } x = 0 \end{cases}$$

and equation (4.62) becomes

$$\langle v,\psi\rangle = \left\langle g,\frac{\psi}{x}\right\rangle$$

Any function $\phi \in \mathcal{S}(\mathbf{R})$ may be expressed as

$$\phi(x) = [\phi(x) - \phi(0)\,\chi(x)] + \phi(0)\,\chi(x)$$

where $\chi \in \mathcal{S}(\mathbf{R})$ satisfies $\chi(0) = 1$, and $\phi - \phi(0)\chi$ vanishes at $x = 0$. Now we define v on $\mathcal{S}(\mathbf{R})$ by

$$\begin{aligned} \langle v,\phi\rangle &= \langle v,\phi - \phi(0)\chi\rangle + c\phi(0) \\ &= \left\langle g, \frac{\phi - \phi(0)\chi}{x}\right\rangle + c\langle\delta,\phi\rangle \end{aligned} \tag{4.63}$$

where c is an arbitrary complex number.

Since $\phi - \phi(0)\chi$ vanishes at $x = 0$ the quotient $[\phi - \phi(0)\chi]/x$ is in $\mathcal{S}(\mathbf{R})$ and, with $g \in \mathcal{S}'(\mathbf{R})$, the right-hand side of equation (4.63) defines a linear functional on $\mathcal{S}(\mathbf{R})$. If $\phi \to 0$ in $\mathcal{S}$ it is not difficult to see that $\langle v,\phi\rangle \to 0$; hence v is continuous. Equation (4.63) also shows that v is unique up to a constant multiple of δ.

If this process of solving $xv = g$ is reminiscent of the existence proof for Theorem 3.10 of the primitive of a distribution in $\mathcal{D}'(\mathbf{R})$, it is because the equation $xv = g$ is equivalent to $(\hat{v})' = -i\hat{g}$, whose solution is the primitive in $\mathcal{S}(\mathbf{R})$ of $-i\hat{g}$. Such a solution is unique up to an additive constant, corresponding to the Fourier transform of $c\delta$.

The solution of the more general equation

$$x^m v = g \qquad g \in \mathcal{S}'(\mathbf{R}) \quad m \in \mathbf{N} \tag{4.64}$$

can also be constructed by following the procedure of Example 4.19. It is given by

$$\langle v,\phi\rangle = \left\langle g, x^{-m}\left[\phi - \sum_{k=0}^{m-1} \frac{1}{k!} x^k \phi^{(k)}(0)\chi(x)\right]\right\rangle + \sum_{k=0}^{m-1} c_k\langle \delta^{(k)},\phi\rangle \tag{4.65}$$

where χ is a function in $\mathcal{S}(\mathbf{R})$ which satisfies $\chi(0) = 1$ and $\chi^{(k)}(0) = 0$ for all $k \in \{1, \ldots, m - 1\}$, and c_k are arbitrary constants.

Note that the component $\sum_{k=0}^{m-1} c_k\delta^{(k)}$ of v is the solution of the homogeneous equation $x^m v = 0$, in agreement with the result of Example 3.2. It is a by-product of the correction made on ϕ, i.e., the subtraction of the first m terms in its Taylor series expansion about $x = 0$, in order to allow for division by x^m. This "correction" is really a process of discarding the divergent terms from the bilinear form $\langle g, x^{-m}\phi\rangle$, in much the same way that the distribution x_+^λ was defined in Section 2.8 by taking the finite part of the integral $\int x_+^\lambda \phi$. In fact, when $g = 1$, which is the case when v is the Fourier transform of the fundamental solution of the ordinary differential operator D^m, the problem of solving $x^m v = 1$ is precisely the problem of representing the distribution x^{-m}, which has already been done in Section 2.8.

The division of a distribution by a polynomial P is therefore basically a problem of extracting the finite part of $\langle g,\phi/P\rangle$. This problem obviously becomes much more complicated as we move from $\mathbf{R}$ to $\mathbf{R}^n$, and as the number of zeros of P increases. On this point a number of worked out examples can be found in [11].

EXERCISES

4.1 Prove that a function $\phi \in C^\infty$ satisfies $\sup\{|x^\alpha \partial^\beta \phi(x)| : x \in \mathbf{R}^n\} < \infty$ for all $\alpha,\beta \in \mathbf{N}_0^n$ if and only if $\sup\{(1 + |x|^2)^m |\partial^\beta \phi(x)| : x \in \mathbf{R}^n, |\beta| \leq m\} < \infty$ for all $m \in \mathbf{N}_0$.

4.2 Prove that a closed and bounded subset of $\mathcal{S}$ is compact in $\mathcal{S}$.

4.3 Give an example of a sequence in $\mathcal{S}'$ which converges to 0 in $\mathcal{D}'$ but not in $\mathcal{S}'$, and one of a sequence in $\mathcal{E}'$ which converges to 0 in $\mathcal{S}'$ but not in $\mathcal{E}'$.

4.4 Prove that the tensor product of two tempered distributions, $T_1 \in \mathcal{S}'(\mathbf{R}^{n_1})$ and $T_2 \in \mathcal{S}'(\mathbf{R}^{n_2})$, as defined in Theorem 3.4 with $\mathcal{D}(\Omega_i)$ replaced by $\mathcal{S}(\mathbf{R}^{n_i})$, is a tempered distribution on $\mathbf{R}^{n_1} \times \mathbf{R}^{n_2}$.

4.5 Prove that if ϕ and ψ are in $\mathcal{S}$ then the product $\phi\psi$ is also in $\mathcal{S}$.

4.6 Use the following direct method to compute the Fourier transform of the Heaviside function H on $\mathbf{R}$:

(i) Show that $T = i\hat{H}$ satisfies $xT = 1$
(ii) Use Examples 3.2 and 4.5 to conclude that the general solution of $xT = 1$ is $T = c\delta + pv\,(1/x)$
(iii) Evaluate the constant c by computing $\langle \hat{H}, e^{-(1/2)x^2}\rangle$.

4.7 Use the expression for $\hat{H}$ obtained in Exercise 4.6 and the equation $1 = H(x) + H(-x)$ to obtain the expression for $\mathcal{F}(pv\,(1/x))$.

4.8 Compute the Fourier transform of $e^{i\lambda x}$, $\lambda \in \mathbf{R}$, and hence obtain $\mathcal{F}(\sin \lambda x)$ and $\mathcal{F}(\cos \lambda x)$ in $\mathbf{R}$.

4.9 Determine the Fourier transform, where it exists, of each of the following functions on $\mathbf{R}$:
(a) $\mathrm{sgn} = H - \check{H}$
(b) Hx^{-1}
(c) The characteristic function of the interval $(0,1)$.
(d) The polynomial $a_n x^n + \cdots + a_1 x + a_0$
(e) The power series $\Sigma_0^\infty a_k x^k$
(f) $Hx^k e^{-ax}$, $k \in \mathbf{N}$ and a = constant
(g) $(ix)^k e^{-iax}$

4.10 Establish the following equalities
(a) $\hat{\hat{f}} = (2\pi)^n (\mathcal{F}^{-1} f)\check{}$
(b) $(\hat{f})\check{} = (\check{f})\hat{}$
(c) $\hat{\check{f}} = \mathcal{F}\check{f} = \mathcal{F}(\check{f})$
(d) $\hat{\hat{f}} = (\hat{f})\check{}$

4.11 For any $\phi \in \mathcal{S}(\mathbf{R})$ prove Poisson's summation formula $\Sigma_{-\infty}^{\infty} \phi(k) = \Sigma_{-\infty}^{\infty} \hat{\phi}(2\pi k)$.

4.12 Prove the following formulas:

(a) $\mathcal{F}(x^k H) = i^k \pi \delta^{(k)} + k!/(i\xi)^{k+1} \qquad k \in \mathbf{N}_0 \qquad x \in \mathbf{R}$

(b) $\mathcal{F}(x^k \operatorname{sgn}) = 2k!/(i\xi)^{k+1}$

(c) $\mathcal{F}(x^{-k}) = \dfrac{\pi(-i)^k}{(k-1)!}\, \xi^{n-1} \operatorname{sgn} \xi$

where the singular powers of x and ξ are to be considered as tempered distributions.

4.13 Show that the convolution product $T * \phi$ of a tempered distribution $T \in \mathcal{S}'$ and a rapidly decreasing function $\phi \in \mathcal{S}$ may be defined as a distribution in $\mathcal{D}'$ by

$$\begin{aligned}\langle T * \phi, \psi\rangle &= T_x\left(\int \phi(y)\, \psi(x+y)\, dy\right) \qquad \psi \in \mathcal{D} \\ &= T_x\left(\int_{\operatorname{supp}\psi} \phi(\xi - x)\psi(\xi)\, d\xi\right)\end{aligned}$$

4.14 Show that the distribution $T * \phi$, where $T \in \mathcal{S}'$ and $\phi \in \mathcal{S}$, as defined in Exercise 4.13 is a tempered distribution.

4.15 Extend the formula $\mathcal{F}(fg) = (2\pi)^{-n}\hat{f} * \hat{g}$ from $\mathcal{S} \times \mathcal{S}$ to $\mathcal{S} \times \mathcal{S}'$.

4.16 Show that the convolution of $f, g \in L^2$ is well defined by

$$(f * g)(x) = \int f(x-y)g(y)\, dy$$

and that $f * g \in L^2$.

4.17 Show that the distribution $x_+^{\lambda-1}$ on $\mathbf{R}$ which was defined in Section 2.8 for all complex numbers $\lambda \neq 0, -1, -2, \ldots$ is a tempered distribution and that its Fourier transform is

$$\Gamma(\lambda)\,(\xi - i0)^{-\lambda} \exp\left(-\tfrac{1}{2} i\pi\lambda\right)$$

4.18 Show that the heat operator $\partial_t - \Delta_x$, where $x \in \mathbf{R}^n$, is not elliptic in $\mathbf{R}^{n+1}$. Is the wave operator $\partial_t^2 - \Delta_x$ elliptic?

4.19 Show that the fundamental solution of the *Schrödinger operator* $(1/i)\partial_t - \partial_x^2$ is $(1/2\sqrt{\pi t})\, H(t)\, e^{i\pi/4} \exp(-x^2/4it)$

4.20 Use a direct method to show that the distributions $\delta, \delta', \ldots, \delta^{(m)}$ on $\mathbf{R}$ are linearly independent over $\mathbf{C}$.

4.21 If $T \in \mathcal{D}'(\mathbf{R}^n)$ is homogeneous of degree d, show that $\partial^\alpha T$ is homogeneous of degree $d - |\alpha|$ and that $\sum_{k=1}^n x_k \partial_k T = dT$.

4.22 Let $T \in \mathcal{S}'(\mathbf{R})$ satisfy $T^{(m)} = \delta$ for some positive integer m. Show that T cannot have compact support.

4.23 Suppose $\kappa \in C^\infty(\mathbf{R}^n - \{0\})$ is homogeneous of degree $-n$ and

$$\int_{S_{n-1}} \kappa(\omega)\, d\omega = 0$$

Let K be the distribution defined by the Cauchy principal value of κ,

$$\langle K,\phi\rangle = \lim_{\varepsilon\to 0} \int_{|x|>\varepsilon} \kappa(x)\phi(x)\, dx \qquad \phi \in \mathcal{D}$$

Prove that $K \in \mathcal{S}'$, $\hat{K} \in C^\infty(\mathbf{R}^n - \{0\})$ and

$$\int_{S_{n-1}} \hat{K}(\omega)\, d\omega = 0$$

4.24 Prove that a function ϕ in $\mathcal{S}(\mathbf{R})$ satisfies $\phi^{(k)}(0) = 0$ for all $k \in \{0, \ldots, m-1\}$ if and only if $\phi = x^m\psi$ for some $\psi \in \mathcal{S}(\mathbf{R})$.

4.25 In the notation of example (4.17), show that $\mathcal{F}(\log x) \in \mathcal{S}'(\mathbf{R})$ is defined by

$$\langle \mathcal{F}(\log x),\phi\rangle = \left\langle -2\pi H(\xi), \frac{\phi(\xi) - \phi(0)\chi}{\xi}\right\rangle + c\langle\delta,\phi\rangle \qquad \phi \in \mathcal{S}(\mathbf{R})$$

Hint: $\xi\mathcal{F}(\log x) = -2\pi H(\xi)$ from Example 4.14.

4.26 Solve the equation $|x|^2 v = 1$ in $\mathcal{S}'(\mathbf{R}^2)$.

4.27 Derive the general solution of the equation $xu' + u = \delta$, and hence find the fundamental solution of $x(d/dx) + 1$ with support in $[0,\infty)$.

4.28 Derive equation (4.64).

5

Distributions in Hilbert Space

5.1 HILBERT SPACE

A *Hilbert space* $\mathcal{H}$ is a Banach space whose norm is defined by an *inner product*. In a (complex) Hilbert space the inner product of any pair of vectors, $u,v \in \mathcal{H}$ is a complex number (u,v) with the following properties:

(i) $(au + bv,w) = a(u,w) + b(v,w)$ for all $u,v,w \in \mathcal{H}$ and $a,b \in \mathbf{C}$
(ii) $(u,v) = \overline{(v,u)}$
(iii) $(u,u) > 0$ whenever $u \neq 0$.

We clearly have $(u,av) = \bar{a}\,(u,v)$, and the inner product of any vector with the zero vector is zero. The norm of any $u \in \mathcal{H}$, denoted by $\|u\|_{\mathcal{H}}$ or simply $\|u\|$, is defined by

$$\|u\| = \sqrt{(u,u)}$$

and it is not difficult to verify that, with this definition, the properties for the norm as given in Section 1.2 are satisfied. Furthermore, we can deduce two additional properties which are peculiar to inner product spaces:

$$|(u,v)| \leq \|u\|\,\|v\| \tag{5.1}$$

$$\|u + v\|^2 + \|u - v\|^2 = 2(\|u\|^2 + \|v\|^2) \tag{5.2}$$

The first is called the *Schwarz inequality* and the second is the *parallelogram law*. The space $L^2(\Omega)$, where Ω is an open subset of $\mathbf{R}^n$, is an example of a Hilbert space in which the inner product of any two functions f and g is defined by

$$(f,g) = \int_\Omega f(x)\, \bar{g}(x)\, dx$$

Any two vectors $u,v \in \mathcal{H}$ are said to be *orthogonal* if $(u,v) = 0$. The notion of orthogonality admits a geometric structure in the Hilbert space that generalizes that of the (finite dimensional) Euclidean space $\mathbf{R}^n$, and it is this geometric structure which gives $\mathcal{H}$ its special features as a Banach space.

A linear functional T on the Hilbert space $\mathcal{H}$ is continuous if and only if

$$|T(\phi)| \leq M \, \|\phi\|_{\mathcal{H}} \qquad \phi \in \mathcal{H}$$

for some positive constant M. In the dual space $\mathcal{H}'$ we define the norm

$$\|T\|_{\mathcal{H}'} = \sup \{|T(\phi)| \, : \, \phi \in \mathcal{H}, \, \|\phi\|_{\mathcal{H}} = 1\}$$

This generates a topology on $\mathcal{H}'$ in which convergence of the sequence (T_i) to 0 is equivalent to the uniform convergence of $T_i(\phi)$ to 0 on every bounded subset of $\mathcal{H}$. In Section 2.4 this was called *strong* convergence in $\mathcal{H}'$.

For any vector ψ in $\mathcal{H}$, the map from $\mathcal{H}$ to $\mathbf{C}$ defined by

$$\phi \mapsto (\phi,\psi)$$

is obviously a linear functional on $\mathcal{H}$, and the Schwarz inequality

$$|(\phi,\psi)| \leq \|\psi\|_{\mathcal{H}} \, \|\phi\|_{\mathcal{H}}$$

shows that it is continuous. That every continuous linear functional on $\mathcal{H}$ is defined in this way is the content of the *Riesz representation theorem* for Hilbert space [6]. This states that to every continuous linear functional T on $\mathcal{H}$ there exists a unique vector $\psi \in \mathcal{H}$ such that

$$T(\phi) = (\phi,\psi) \qquad \phi \in \mathcal{H}$$

and

$$\|T\|_{\mathcal{H}'} = \|\psi\|_{\mathcal{H}}$$

Thus the dual space $\mathcal{H}'$ of continuous linear functionals on $\mathcal{H}$ is also a Hilbert space and the correspondence

$$\psi \leftrightarrow T_\psi$$

defines a norm-preserving bijection, or isometry, between $\mathcal{H}$ and $\mathcal{H}'$. $\mathcal{H}$ and $\mathcal{H}'$ may be identified as sets, but not as linear spaces, since the linear combination $a_1\psi_1 + a_2\psi_2$ in $\mathcal{H}$ corresponds to the *conjugate* linear combination $\bar{a}_1 T_{\psi_1} + \bar{a}_2 T_{\psi_2}$, unless of course $\mathcal{H}$ is a real Hilbert space.

However the *second dual* of $\mathcal{H}$,

$$\mathcal{H}'' = (\mathcal{H}')'$$

composed of the continuous linear functionals on $\mathcal{H}'$ may be identified with $\mathcal{H}$ since linearity in this case is restored to the correspondence between the elements of $\mathcal{H}$ and the elements of $\mathcal{H}''$. This is referred to as the *reflexive* property of the Hilbert space $\mathcal{H}$.

5.2 SOBOLEV SPACES

For any integer $m \in \mathbf{N}_0$ and any open $\Omega \subset \mathbf{R}^n$ we define the *Sobolev space* $H^m(\Omega)$ to be the set of all functions ϕ in $L^2(\Omega)$ whose distributional derivatives $\partial^\alpha\phi$ are also in $L^2(\Omega)$ for every $\alpha \in \mathbf{N}_0^n$ with $|\alpha| \leq m$, i.e.

$$H^m(\Omega) = \{\phi \in L^2(\Omega) : \partial^\alpha\phi \in L^2(\Omega), |\alpha| \leq m\} \tag{5.3}$$

Thus $H^m(\Omega)$ is the subspace of distributions $\phi \in \mathcal{D}'(\Omega)$ such that $\partial^\alpha\phi \in L^2(\Omega)$ for all $|\alpha| \leq m$, and we clearly have

$$\mathcal{D}'(\Omega) \supset L^2(\Omega) = H^0(\Omega) \supset H^1(\Omega) \supset H^2(\Omega) \supset \cdots$$

The inner product of two functions $\phi_1,\phi_2 \in H^m(\Omega)$ is defined by

$$(\phi_1,\phi_2)_m = \sum_{|\alpha|\leq m} \int_\Omega \partial^\alpha\phi_1(x)\, \partial^\alpha\bar{\phi}_2(x)\, dx \tag{5.4}$$

and the norm of $\phi \in H^m(\Omega)$ by

$$\|\phi\|_{m,2} = \sqrt{(\phi,\phi)_m} = \left[\sum_{|\alpha|\leq m} \int_\Omega |\partial^\alpha\phi(x)|^2\, dx\right]^{1/2} \tag{5.5}$$

Hence

$$\|\phi\|_{m,2}^2 = \sum_{|\alpha|\leq m} \|\partial^\alpha\phi\|_2^2 \qquad m \in \mathbf{N}_0 \tag{5.6}$$

In particular, when $m = 0$, $\|\phi\|_{0,2} = \|\phi\|_2$.

The subscript 2 in the symbol $\|\cdot\|_{m,2}$ indicates that this norm is defined through the L^2 norm. If L^2 is replaced by L^p in the definition (5.3), $1 \leq p < \infty$, then the resulting norm

$$\|\phi\|_{m,p} = \left[\sum_{|\alpha| \leq m} \|\partial^\alpha \phi\|_p^p \right]^{1/p}$$

generates the Banach spaces $H^{m,p}(\Omega)$ (see Exercises 5.1 and 5.2). $H^{m,p}(\Omega)$ is a Hilbert space only when $p = 2$, and we shall only be concerned with this case. Consequently $H^{m,2}(\Omega)$ has been abbreviated to $H^m(\Omega)$.

Similarly $\|\cdot\|_{m,2}$ will be abbreviated to $\|\cdot\|_m = \|\cdot\|_{H^m}$, which should not be confused with the L^p norm $\|\cdot\|_p = \|\cdot\|_{L^p}$. It is usually clear from the context which norm is being considered. In any case only L^2 will be relevant to the Hilbert space theory of distributions, and the L^2 norm $\|\cdot\|_2$ will henceforth be designated by $\|\cdot\|_0$. Equations (5.5) and (5.6) can then be rewritten as

$$\|\phi\|_m = \left[\sum_{|\alpha| \leq m} \|\partial^\alpha \phi\|_0^2 \right]^{1/2} = \left[\sum_{|\alpha| \leq m} \int_\Omega |\partial^\alpha \phi(x)|^2 dx \right]^{1/2} \tag{5.6}$$

Example 5.1 Let Ω be an open interval in $\mathbf{R}$ containing the closed interval $[a,b]$. If f is the characteristic function of $[a,b]$ then $f' = \delta_a - \delta_b$ and consequently $f \notin H^1(\Omega)$. But if Ω is bounded, f is continuous on Ω, and f' is bounded except at a finite number of points in Ω, then $f \in H^1(\Omega)$ (see Example 2.10).

Theorem 5.1 $H^m(\Omega)$ is a Hilbert space.

Proof Since $H^m(\Omega)$ is a normed linear space whose norm (5.5) is derived from the inner product (5.4), it suffices to show that $H^m(\Omega)$ is complete. Let (ϕ_k) be a Cauchy sequence in $H^m(\Omega)$. The convergence

$$\|\phi_k - \phi_j\|_m \to 0$$

implies, by equation (5.6), that

$$\|\partial^\alpha \phi_k - \partial^\alpha \phi_j\|_0 \to 0 \qquad |\alpha| \leq m$$

so that the sequence $(\partial^\alpha \phi_k)$ is a Cauchy sequence in $L^2(\Omega)$ for every α satisfying $|\alpha| \leq m$. Since $L^2(\Omega)$ is complete, the sequence $(\partial^\alpha \phi_k)$ converges in $L^2(\Omega)$ to some function $\phi_\alpha \in L^2(\Omega)$. By the Schwarz inequality

$$\left| \int_\Omega [\partial^\alpha \phi_k(x) - \phi_\alpha(x)]\, \psi(x)\, dx \right| \leq \|\partial^\alpha \phi_k - \phi_\alpha\|_0 \, \|\psi\|_0 \qquad \psi \in \mathcal{D}(\Omega)$$

we see that, as $k \to \infty$,

$$\int_\Omega \partial^\alpha \phi_k(x)\psi(x)\, dx \to \int_\Omega \phi_\alpha(x)\psi(x)dx \qquad \psi \in \mathcal{D}(\Omega)$$

On the other hand, since $f_k \to f$ in $L^2(\Omega)$ implies $f_k \to f$ in $\mathcal{D}'(\Omega)$, we have for all $\psi \in \mathcal{D}(\Omega)$

$$\int_\Omega \partial^\alpha \phi_k(x)\psi(x)\, dx = \langle \partial^\alpha \phi_k, \psi \rangle = (-1)^{|\alpha|} \langle \phi_k, \partial^\alpha \psi \rangle \to (-1)^{|\alpha|} \langle \phi, \partial^\alpha \psi \rangle$$

where $\phi = \lim \phi_k$ in $L^2(\Omega)$. But

$$(-1)^{|\alpha|}\langle \phi, \partial^\alpha \psi \rangle = \langle \partial^\alpha \phi, \psi \rangle = \int_\Omega \partial^\alpha \phi(x)\psi(x)\, dx$$

Hence $\phi_\alpha = \partial^\alpha \phi$, and ϕ, which is clearly in $H^m(\Omega)$, is the limit of ϕ_k in the H^m norm. □

In order to apply the Fourier transformation to $H^m(\Omega)$, we shall take $\Omega = \mathbf{R}^n$. A function f is in $H^m = H^m(\mathbf{R}^n)$ if and only if $\partial^\alpha f$ is in L^2 for all $|\alpha| \leq m$, hence $H^m \subset L^2 \subset \mathcal{S}'$. But, in view of Theorem 4.7 and equations (4.5) and (5.6), we have

$$\begin{aligned} \|f\|_m &= \left[\sum_{|\alpha| \leq m} \|\partial^\alpha f\|_0^2 \right]^{1/2} \\ &= (2\pi)^{(-1/2)n} \left[\sum_{|\alpha| \leq m} \|\xi^\alpha \hat{f}\|_0^2 \right]^{1/2} \\ &\leq c_1 \|(1 + |\xi|^2)^{(1/2)m} \hat{f}\|_0 \end{aligned}$$

where c_1 is a positive constant (which depends on m). Similarly, there is a positive constant c_2 such that

$$\|(1 + |\xi|^2)^{(1/2)m} \hat{f}\|_0 \leq c_2 (2\pi)^{-n/2} \left[\sum_{|\alpha| \leq m} \|\xi^\alpha \hat{f}\|_0^2 \right]^{1/2} = c_2 \|f\|_m$$

This means that a tempered distribution f is in H^m if and only if $(1 + |\xi|^2)^{(1/2)m} \hat{f} \in L^2$. We may therefore redefine the Sobolev space H^m, $m \in \mathbf{N}_0$, as the space of tempered distributions $f \in \mathcal{S}'$ such that

$$(1 + |\xi|^2)^{(1/2)m} \hat{f} \in L^2$$

with the scalar product

$$(f,g)_m^{\wedge} = \int (1 + |\xi|^2)^m \hat{f}(\xi)\bar{\hat{g}}(\xi)\, d\xi \tag{5.7}$$

and the norm

$$\|f\|_m^{\wedge} = \left[\int (1 + |\xi|^2)^m |\hat{f}(\xi)|^2\, d\xi\right]^{1/2} \tag{5.8}$$

Note that the norms $\|f\|_m$ and $\|f\|_m^{\wedge}$ as defined by (5.5) and (5.8), though equivalent, are not equal. In particular we note, from Plancherel's theorem, that $\|f\|_0^{\wedge} = (2\pi)^{(1/2)n} \|f\|_0$.

This definition of H^m is, of course, equivalent to the previous one when m is a nonnegative integer and $\Omega = \mathbf{R}^n$, but it allows us to extend the definition to any real number.

Definition For any $s \in \mathbf{R}$ we define $H^s(\mathbf{R}^n)$ to be the tempered distributions whose Fourier transforms are square-integrable with respect to the measure $(1 + |\xi|^2)^s\, d\xi$, i.e.

$$H^s(\mathbf{R}^n) = \{f \in \mathcal{S}'(\mathbf{R}^n) : (1 + |\xi|^2)^{(1/2)s}\hat{f}(\xi) \in L^2(\mathbf{R}^n)\}$$

With the inner product

$$(f,g)_s^{\wedge} = \int (1 + |\xi|^2)^s \hat{f}(\xi)\, \bar{\hat{g}}(\xi)\, d\xi \tag{5.9}$$

and the norm

$$\|f\|_s^{\wedge} = \sqrt{(f,f)_s^{\wedge}} = \left[\int (1 + |\xi|^2)^s |\hat{f}(\xi)|^2\, d\xi\right]^{1/2} \tag{5.10}$$

it is not difficult to see that H^s is a Hilbert space. For if (f_k) is a Cauchy sequence in H^s then $(1 + |\xi|^2)^{(1/2)s}\hat{f}_k$ is a Cauchy sequence in L^2. By the completeness of L^2, $(1 + |\xi|)^2)^{(1/2)s}\hat{f}_k$ converges to some $g \in L^2$, and therefore $f_k \to f = \mathcal{F}^{-1}\,[(1 + |\xi|^2)^{-s/2}\, g]$ in H^s for every $s \in \mathbf{R}$. But $(1 + |\xi|^2)^{(1/2)s}\hat{f} = g$ is in L^2, hence f is in H^s.

When $s \geq 0$, the inequality

$$\|f\|_0 = (2\pi)^{-(1/2)n}\, \|f\|_0^{\wedge} \leq (2\pi)^{-(1/2)n}\, \|f\|_s^{\wedge}$$

implies that $H^s \subset L^2$. But in general we have the following inclusion relations:

Theorem 5.2 For all real numbers s and t with $s > t$, we have

$$\mathcal{S} \subset H^s \subset H^t \subset \mathcal{S}'$$

and the identity mappings

$$\mathcal{S} \to H^s \to H^t \to \mathcal{S}'$$

are continuous. Furthermore, $\mathcal{S}$ is dense in H^s for all $s \in \mathbf{R}$.

Proof The inclusion relations between the spaces as sets is obvious. It is also clear that if $\phi_k \to 0$ in $\mathcal{S}$ then $\|\phi_k\|^\wedge_s \to 0$ for any $s \in \mathbf{R}$. Since $\|\phi_k\|^\wedge_t \leq \|\phi_k\|^\wedge_s$ whenever $t < s$, this implies that $\|\phi_k\|^\wedge_t \to 0$.

For any $\psi \in \mathcal{S}$, $\mathcal{F}^{-1}(\psi)$ is also in $\mathcal{S}$ and

$$\begin{aligned}\langle \phi_k,\psi\rangle &= \langle \hat{\phi}_k,\mathcal{F}^{-1}(\psi)\rangle \\ &= \langle (1 + |x|^2)^{(1/2)t}\hat{\phi}_k, (1 + |x|^2)^{-(1/2)t}\mathcal{F}^{-1}(\psi)\rangle \\ &\leq \|\phi_k\|^\wedge_t \, \|(1 + |x|^2)^{-(1/2)t}\mathcal{F}^{-1}(\psi)\|_0\end{aligned}$$

which means that $\phi_k \to 0$ in $\mathcal{S}'$ when $\phi_k \to 0$ in H^t.

Finally, if $f \in H^s$ then $(1 + |\xi|^2)^{(1/2)s}\hat{f}$ is in L^2 and, since $\mathcal{S}$ is dense in L^2, there is a sequence (ϕ_k) in $\mathcal{S}$ such that $\phi_k \to (1 + |\xi|^2)^{(1/2)s}\hat{f}$ in L^2. But $\hat{\psi}_k = (1 + |\xi|^2)^{-(1/2)s}\phi_k$ is in $\mathcal{S}$ for every $s \in \mathbf{R}$. Hence

$$\|(1 + |\xi|^2)^{(1/2)s}(\hat{f} - \hat{\psi}_k)\|_0 \to 0$$

or

$$\|f - \psi_k\|^\wedge_s \to 0$$

where (ψ_k) is clearly a sequence in $\mathcal{S}$. □

Since C_0^∞ is dense in $\mathcal{S}$, this theorem implies that it is also dense in H^s, and we can therefore conclude that

Corollary H^s is the completion of C_0^∞ under the norm $\|\cdot\|^\wedge_s$.

As a Hilbert space H^s has a dual space with respect to the inner product $(f,g) \mapsto (f,g)^\wedge_s$ which may be identified with H^s. But that is not the same as its dual in the bilinear form $(f,g) \mapsto \langle f,g\rangle = (f,\bar{g})_0$, except when $s = 0$ and the space is real. A function $f \in \mathcal{S}$ defines a continuous linear functional on $\mathcal{S}$ by

$$T_f(\phi) = \langle f,\phi\rangle = \int f(x)\,\phi(x)\,dx \qquad \phi \in \mathcal{S}$$
$$= (2\pi)^{-n}\int \hat{f}(\xi)\,\hat{\phi}(-\xi)\,d\xi$$

where the last equality follows from Parseval's relation. If we write

$$\hat{f}(\xi)\,\hat{\phi}(-\xi) = (1 + |\xi|^2)^{-(1/2)s}\hat{f}(\xi)\,(1 + |\xi|^2)^{(1/2)s}\hat{\phi}(-\xi)$$

and use Schwarz' inequality, we obtain

$$|\langle f,\phi\rangle| \leqslant (2\pi)^{-n}\|f\|^{\wedge}_{-s}\,\|\phi\|^{\wedge}_{s} \tag{5.11}$$

Since $\mathcal{S}$ is dense in H^s for all s, the bilinear form $\langle f,\phi\rangle$ can be extended from $\mathcal{S}\times\mathcal{S}$ to $H^{-s}\times H^s$ with the inequality (5.11) still valid; and since the dual of H^s is a subset of $\mathcal{S}'$, it follows that H^{-s} is a subset of $(H^s)'$.

To show that $(H^s)' \subset H^{-s}$, let $T \in (H^s)'$ be arbitrary. Then, by the Riesz representation theorem for the Hilbert space H^s, there is a function $f \in H^s$ such that

$$T(\phi) = (\phi,f)^{\wedge}_{s} = \int (1 + |\xi|^2)^s\hat{\phi}(\xi)\,\bar{\hat{f}}(\xi)\,d\xi \qquad \phi \in H^s$$
$$= \int \phi(x)\,\bar{h}(x)\,dx$$

where $h(x) = (2\pi)^n\mathcal{F}^{-1}((1 + |\xi|^2)^s\hat{f}(\xi))$. Now the function

$$(1 + |\xi|^2)^{-(1/2)s}\hat{h}(\xi) = (2\pi)^n\,(1 + |\xi|^2)^{(1/2)s}\hat{f}(\xi)$$

is in L^2, and this means that $\bar{h}$ is in H^{-s} and represents T in the sense that $T(\phi) = \langle\bar{h},\phi\rangle$ for all $\phi \in H$.

Thus we have proved

Theorem 5.3 H^{-s} represents the topological dual of H^s for all $s \in \mathbf{R}$, and

$$|\langle f,\phi\rangle| \leqslant (2\pi)^{-n}\,\|f\|^{\wedge}_{-s}\,\|\phi\|^{\wedge}_{s}$$

for all $\phi \in H^s$ and $f \in H^{-s}$.

When s is a nonnegative integer we have the following characterization of H^{-s}:

Theorem 5.4 $f \in H^{-m}$, where $m \in \mathbf{N}_0$, if and only if f is a finite sum of derivatives of order less than or equal to m of L^2 functions.

Proof Let $f \in H^{-m}$, so that the function $(1 + |\xi|^2)^{-(1/2)m}\hat{f}(\xi)$ is in L^2. Then

$$\begin{aligned}(1 + |\xi|^2)^{(1/2)m} &\leq (1 + |\xi|)^m \\ &= [1 + (\xi_1^2 + \cdots + \xi_n^2)^{1/2}]^m \\ &\leq (1 + |\xi_1| + \cdots + |\xi_n|)^m \\ &= 1 + \sum_{1 \leq |\alpha| \leq m} c_\alpha |\xi^\alpha| \end{aligned} \tag{5.12}$$

where c_α are nonnegative integers and α is a multi-index in $\mathbf{N}_0^n$. The inequality (5.12) implies that the function

$$\hat{g}(\xi) = \left(1 + \sum_{1 \leq |\alpha| \leq m} c_\alpha |\xi^\alpha|\right)^{-1} \hat{f}(\xi)$$

satisfies

$$|\hat{g}(\xi)| \leq (1 + |\xi|^2)^{-(1/2)m} |\hat{f}(\xi)|$$

Hence g is also in L^2.

Now we can write

$$\begin{aligned}\hat{f}(\xi) &= \left(1 + \sum_{1 \leq |\alpha| \leq m} c_\alpha |\xi^\alpha|\right) \hat{g}(\xi) \\ &= \sum_{|\alpha| \leq m} \xi^\alpha \hat{g}_\alpha(\xi) \end{aligned} \tag{5.13}$$

where $\hat{g}_\alpha(\xi) = \hat{g}(\xi)$ when $|\alpha| = 0$ and $\hat{g}_\alpha(\xi) = c_\alpha |\xi^\alpha| \xi^{-\alpha} \hat{g}(\xi)$ when $1 \leq |\alpha| \leq m$. Clearly $\hat{g}_\alpha$ is in L^2 whenever $\hat{g}$ is in L^2. Taking the inverse Fourier transform of (5.13), and using equation (4.5), gives

$$f(x) = \sum_{|\alpha| \leq m} D^\alpha g_\alpha(x)$$

with $g_\alpha \in L^2$ for all $|\alpha| \leq m$.

Conversely, if $f = \sum_{|\alpha| \leq m} \partial^\alpha g_\alpha$ with $g_\alpha \in L^2$, then $\partial^\alpha g_\alpha \in H^{-m}$ for all $|\alpha| \leq m$. Consequently $f \in H^{-m}$. □

Since every $g \in L^2$ is a distribution of order 0 by the inequality $|\langle f, \phi \rangle| \leq M|\phi|_0$, where $M = \|f\|_0$ [volume (supp ϕ)], we conclude that

Corollary Every element of H^m, $m \in \mathbf{Z}$, is a distribution of finite order.

This result, of course, also follows from the inclusion $H^s \subset \mathcal{S}'$ for all $s \in \mathbf{R}$.

Example 5.2 Since $\hat{\delta} = 1$ and $(1 + |\xi|^2)^{(1/2)s} \in L^2(\mathbf{R})$ provided $s < -(1/2)n$, we see that $\delta \in H^s$ for all $s < -(1/2)n$. When $n = 1$, the Dirac measure δ lies in $H^{-1}(\mathbf{R})$ and is consequently a sum of the form $f_1 + f_2'$ with $f_1, f_2 \in L^2(\mathbf{R})$. One possible choice for these functions is given by

$$f_1(x) = \tfrac{1}{2}e^{-|x|}$$

$$f_2(x) = \tfrac{1}{2}e^{-|x|}\,\mathrm{sgn}\,x$$

Here we use the fact that $(\mathrm{sgn}\,x)^2 = 1$ almost everywhere, and $(\mathrm{sgn}\,x)' = 2\delta$.

For $s \geqslant 0$, H^s is a subspace of L^2 and its functions would be expected to achieve higher degrees of smoothness with increasing values of s, as their derivatives of higher order have to lie in L^2. The *Sobolev imbedding theorem* gives the value of s above which H^s is made up of continuous functions.

Theorem 5.5 If $s > n/2$, then $H^s \subset C^0$ with continuous injection.

Proof The function $(1 + |\xi|^2)^{-s}$ is integrable if and only if $s > (1/2)n$. Therefore, when $s > (1/2)n$ and $f \in H^s$, we have

$$\int |\hat{f}(\xi)|d\xi = \int (1 + |\xi|^2)^{-(1/2)s}(1 + |\xi|^2)^{(1/2)s}\,|\hat{f}(\xi)|d\xi$$

$$\leqslant \|f\|_s \int (1 + |\xi|^2)^{-s}\,d\xi \tag{5.14}$$

by the Schwarz inequality. This implies that $\hat{f}$ is in $L^1 \subset \mathcal{S}'$, and the inverse Fourier transform f exists and satisfies the relation (4.15), i.e.,

$$f(x) = (2\pi)^{-n}\hat{\hat{f}}(-x) \tag{5.15}$$

Since $\hat{f}$ is in L^1, its Fourier transform

$$\hat{\hat{f}}(x) = \int e^{-i(x,\xi)}\,\hat{f}(\xi)\,d\xi$$

is continuous on $\mathbf{R}^n$, hence so is f.

To show that the topology of H^s, when $s > \frac{1}{2}n$, is stronger than that of C^0, let $\|f_k\|_s \to 0$. The inequality (5.14) implies that $\|\hat{f}_k\|_{L^1} \to 0$. Since the Fourier transformation is continuous from L^1 to C^0, we see that $f_k(x) = (2\pi)^{-n}\hat{\hat{f}}_k(-x) \to 0$ in C^0. □

It should be noted here that, as an element of H^s, f is really a class of functions which are equal almost everywhere, and by writing equations (5.15) we are choosing the continuous representative of that class. It is in that sense that the inclusion $H^s \subset C^0$ should be understood.

Another point which is worth nothing is that, since $\hat{f} \in L^1$, $\hat{\hat{f}} \to 0$ as $|x| \to \infty$ by the Riemann–Lebesgue lemma. Thus, when $s > (1/2)n$, H^s actually lies in the subspace C^0_∞ of C^0 which consists of all continuous functions on $\mathbf{R}^n$ that vanish at ∞.

Corollary If $s > (1/2)n + k$, where k is a nonnegative integer, then $H^s \subset C^k$ with continuous injection.

Proof If $f \in H^s$ then $\partial^\alpha f \in H^{s-|\alpha|}$. When $s > (1/2)n + k$ and $|\alpha| \leq k$, $s - |\alpha| \geq s - k > (1/2)n$ and Theorem 5.5 gives $\partial^\alpha f \in C^0$. Since the distributional derivative coincides with the ordinary derivative when it is continuous, we conclude that $f \in C^k$. □

Example 5.3 If $u(x) = \exp(-|x|)$, $x \in \mathbf{R}$, then $\hat{u}(\xi) = 2/(1 + \xi^2)$ and $u \in H^s$ if and only if $(1 + \xi^2)^{(1/2)(s-2)} \in L^2$, i.e., whenever $s < 3/2$. With $n = 1$, Sobolev's imbedding theorem then guarantees the continuity of u but not its differentiability. This is consistent with the fact that $\exp(-|x|)$ is continuous but not differentiable on $\mathbf{R}$.

Example 5.4 A fundamental solution in $\mathcal{S}'$ of the differential operator $1 - \Delta$, where Δ is the Laplacian in $\mathbf{R}^n$, is a tempered distribution E which satisfies $(1 - \Delta)E = \delta$. Its Fourier transform is therefore

$$\hat{E}(\xi) = \frac{1}{1 + |\xi|^2}$$

Since $(1 + |\xi|^2)^{(1/2)s}\hat{E} = (1 + |\xi|^2)^{(1/2)(s-2)}$ is in L^2 for all $s - 2 < -\frac{1}{2}n$, the operator has a fundamental solution in H^s for all $s < 2 - (1/2)n$. From the Sobolev imbedding Theorem 5.5, we see that such a fundamental solution is continuous if $(1/2)n < s < 2 - (1/2)n$. Since these inequalities can only be satisfied when $n = 1$, the operator $1 - \Delta$ has a continuous fundamental solution on $\mathbf{R}$. This is given by $\frac{1}{2}\exp(-|x|)$, as we have seen in Example 5.3.

Similarly $(1 - \Delta)^m$ has a fundamental solution E in H^s for all $s < 2m - (1/2)n$. In this case E is continuous provided $(1/2)n < s < 2m - (1/2)n$, i.e., whenever $m > (1/2)n$, and it attains higher degrees of smoothness for larger values of m.

By defining

$$H^\infty(\mathbf{R}^n) = \bigcap_{s \in \mathbf{R}} H^s(\mathbf{R}^n)$$

$$H^{-\infty}(\mathbf{R}^n) = \bigcup_{s \in \mathbf{R}} H^s(\mathbf{R}^n)$$

we see from the above corollary that H^∞ is a subspace of C^∞. A function ϕ in C^∞ lies in H^∞ if $\partial^\alpha \phi \in L^2$ for all $\alpha \in \mathbf{N}_0^n$. That means $\phi(x)$ and all its partial derivatives tend to 0 as $|x| \to \infty$. The topologies of H^∞ and $H^{-\infty}$ are defined so that the inclusion relations

$$H^\infty \subset H^s \subset H^{-\infty}$$

for any real number s become imbeddings. This is accomplished by defining the topology of H^∞ to be the weakest locally convex topology such that the identity mapping from H^∞ to H^s is continuous for every s. This is the projective limit topology of $\{H^s : s \in \mathbf{R}\}$ introduced in Section 2.7.

By Theorem 5.2, we have $\bigcap_{|s|\leq m} H^s = H^m$ and $\bigcup_{|s|\leq m} H^s = H^{-m}$. Since $(H^m)' = H^{-m}$, by Theorem 5.3, we conclude that

$$\left(\bigcap_{|s|\leq m} H^s\right)' = \bigcup_{|s|\leq m} H^s \qquad \text{for all } m \in \mathbf{N}$$

and hence $(H^\infty)' = H^{-\infty}$. This defines the topology of $H^{-\infty}$ as the inductive limit of the topologies on $\{H^s : s \in \mathbf{R}\}$, i.e. as the strongest locally convex topology such that the identity map from H^s to $H^{-\infty}$ is continuous for every s. In fact, as we have seen in Section 2.7 in connection with $\mathcal{D}_F$ and $\mathcal{D}'_F$, these two methods of defining a topology on a linear space, that is the projective limit and the inductive limit, generally produce dual topological vector spaces. The interested reader can refer to [2] on this point.

Now, since $\mathcal{D} \subset \mathcal{S} \subset H^\infty \subset \mathcal{E}$ and $\mathcal{D}$ is dense in $\mathcal{E}$, we clearly have the inclusions

$$\mathcal{E}' \subset H^{-\infty} \subset \mathcal{S}' \subset \mathcal{D}'$$

The relation $\mathcal{E}' \subset H^{-\infty}$ can also be obtained from the following example, which relates the order of a distribution in $H^s \cap \mathcal{E}'$ to s:

Example 5.5 Let $T \in \mathcal{E}'$. Since every distribution with compact support is of finite order, suppose that the order of T is m. We shall prove that $T \in H^s$ if $s \leq -\frac{1}{2}n - m$.

According to Theorem 4.8, $\hat{T}$ is a C^∞ function given by $\hat{T}(\xi) = \langle T_x, e^{-i\langle x,\xi\rangle}\rangle$. Since T is of order m, there exists a compact set $K \subset \mathbf{R}^n$ and a positive constant M such that

$$\begin{aligned}|\hat{T}(\xi)| &\leq M \sum_{|\alpha|\leq m} \sup\{|\partial^\alpha e^{-i\langle x,\xi\rangle}| : x \in K\} \\ &\leq M \sum_{|\alpha|\leq m} |\xi^\alpha|\end{aligned}$$

Therefore

$$
\begin{aligned}
|\hat{T}(\xi)|^2 &\leq M^2 \left(\sum_{|\alpha| \leq m} |\xi^\alpha| \right)^2 \\
&\leq M_1 \sum_{|\alpha| \leq m} |\xi^\alpha|^2 \\
&\leq M_2 (1 + |\xi|^2)^m
\end{aligned}
$$

for some positive constants M_1 and M_2. The last inequality follows from

$$
\begin{aligned}
|\xi^\alpha|^2 &= \xi_1^{2\alpha_1} \cdots \xi_n^{2\alpha_n} \\
&\leq (1 + \xi_1^2 + \cdots + \xi_n^2)^{\alpha_1} \cdots (1 + \xi_1^2 + \cdots + \xi_n^2)^{\alpha_n} \\
&= (1 + \xi_1^2 + \cdots + \xi_n^2)^{\alpha_1 + \cdots + \alpha_n} \\
&\leq (1 + |\xi|^2)^m
\end{aligned}
$$

since $|\alpha| \leq m$. Hence $(1 + |\xi|^2)^s |\hat{T}|^2 \leq M_2(1 + |\xi|^2)^{m+s}$, and $T \in H^s$ when $(1 + |\xi|^2)^{m+s} \in L^1$, i.e., when $m + s < -\frac{1}{2}n$.

Thus all distributions with compact support and zero order are contained in H^s for $s < -(1/2)n$. Since δ is included in this set, and in view of Example 5.2, this estimate cannot be made any sharper.

5.3 SOME PROPERTIES OF H^S SPACES

Sobolev spaces provide a useful way of measuring the differentiability properties of functions on $\mathbf{R}^n$. From the definition of the Sobolev space $H^s = H^s(\mathbf{R}^n)$, the differential operator ∂_k, where $k \in \{1, \ldots, n\}$, is a continuous linear operator from H^s to H^{s-1}, since

$$
\begin{aligned}
\|\partial_k f\|_{s-1} &= \|\xi_k (1 + |\xi|^2)^{(1/2)(s-1)} \hat{f}\|_0 \\
&\leq \|(1 + |\xi|^2)^{(1/2)s} \hat{f}\|_0 \\
&= \|f\|_s \qquad \text{for all } f \in H^s.
\end{aligned}
$$

Consequently, if P is a polynomial on $\mathbf{R}^n$ with constant coefficients and degree $\leq m$, then $P(D)$ is a continuous linear operator from H^s to H^{s-m}.

When the polynomial has no zeros in $\mathbf{R}^n$, the mapping $P(D) : H^s \to H^{s-m}$ is also bijective, as the next example shows.

Example 5.6 Here we shall prove that the operator $k^2 - \Delta$, $k \neq 0$, is a homeomorphism from H^{s+2} onto H^s.

(i) The continuity of $(k^2 - \Delta) : H^{s+2} \to H^s$ is obvious since this operator has constant coefficients.

(ii) To show that $(k^2 - \Delta)$ is bijective, let $(k^2 - \Delta)u = 0$ for some $u \in H^{s+2}$. Then $(k^2 + |\xi|^2)\,\hat{u} = 0$ and, since $k \neq 0$, $\hat{u} = 0$. Hence $u = 0$ and $(k^2 - \Delta)$ is injective.

If $v \in H^s$ then $u = \hat{v}/(k^2 + |\xi|^2) \in \mathcal{S}'$ and $(k^2 - \Delta)\mathcal{F}^{-1}(u) = v$. Thus $k^2 - \Delta$ is surjective if $\mathcal{F}^{-1}(u) \in H^{s+2}$. But this follows from the inequality

$$\begin{aligned}(1 + |\xi|^2)^{(1/2)(s+2)}\,|\mathcal{F}(\mathcal{F}^{-1}(u))| &= (1 + |\xi|^2)^{(1/2)s+1}|u| \\ &= (1 + |\xi|^2)^{(1/2)s+1}\frac{|\hat{v}|}{k^2 + |\xi|^2} \\ &\leq c(1 + |\xi|^2)^{(1/2)s+1}\frac{|\hat{v}|}{1 + |\xi|^2} \\ &= c(1 + |\xi|^2)^{(1/2)s}|\hat{v}|\end{aligned}$$

and the fact that $(1 + |\xi|^2)^{(1/2)s}\hat{v} \in L^2$, since $v \in H^s$.

(iii) Having shown that $k^2 - \Delta$ is a continuous bijection from H^{s+2} to H^s, we now observe that, H^{s+2} and H^s being Banach spaces, the open mapping theorem [4] implies that $(k^2 - \Delta)^{-1} : H^s \to H^{s+2}$ is continuous. Therefore $k^2 - \Delta$ is a homeomorphism.

Before we can allow the coefficients in the differential operator $P(D) : H^s \to H^{s-m}$ to be functions, we have to investigate the feasibility of multiplying the elements of H^s by such functions.

Theorem 5.6 The mapping from $\mathcal{S} \times H^s$ into H^s defined by $(\phi,u) \mapsto \phi u$ is bilinear and continuous on $\mathcal{S}$ and H^s separately.

Proof It suffices to consider the case when $s \geq 0$, since $\langle \phi v,u\rangle = \langle v,\phi u\rangle$ for all $u \in H^s$, $v \in H^{-s}$ and $\phi \in \mathcal{S}$.

If we assume that ϕ and u are in $\mathcal{S}$, then their Fourier transforms $\hat{\phi}$ and $\hat{u}$ are also in $\mathcal{S}$ and, with $\mathcal{F}(\phi u) = (2\pi)^{-n}\hat{\phi} * \hat{u}$, we have

$$(1 + |\xi|^2)^{(1/2)s}\,|\mathcal{F}(\phi u)(\xi)| \leq (2\pi)^{-n}\int (1 + |\xi|^2)^{(1/2)s}\,|\hat{\phi}(\eta)\hat{u}(\xi - \eta)|\,d\eta \tag{5.16}$$

But since

$$\begin{aligned}1 + |\xi|^2 &= 1 + |\xi - \eta + \eta|^2 \\ &\leq 1 + |\xi - \eta|^2 + 2|\xi - \eta|\,|\eta| + |\eta|^2 \\ &\leq 1 + |\xi - \eta|^2 + 2\,|\eta|\,(1 + |\xi - \eta|^2) + |\eta|^2 \\ &\leq (1 + |\xi - \eta|^2)\,(1 + |\eta|)^2\end{aligned}$$

we can use this last inequality in (5.16) and integrate with respect to ξ to obtain

$$\|\phi u\|_s^\wedge \leqslant (2\pi)^{-n} \|u\|_s^\wedge \int (1 + |\eta|)^s |\hat{\phi}(\eta)| \, d\eta \tag{5.17}$$

$\mathcal{S}$ being dense in H^s, the inequality (5.17) may be extended by continuity to all u in H^s. Thus the product ϕu is in H^s and depends continuously on $\phi \in \mathcal{S}$ and $u \in H^s$. □

Corollary If P is a polynomial on $\mathbf{R}^n$ with coefficients in $\mathcal{S}$ and degree m, then $P(D)$ is a continuous linear differential operator from H^s to H^{s-m}.

Given any real number t, a linear operator L defined on $H^{-\infty}$ is said to have *order t* if it maps H^s into H^{s-t} for every $s \in \mathbf{R}$. Thus the differential operator ∂^α has order $|\alpha|$ and $P(D)$, where P is a polynomial of degree m, with coefficients in $\mathcal{S}$, has order m. If the function f defined on $\mathbf{R}^n$ is bounded (almost everywhere) then the mapping $u \mapsto v$ defined by $\hat{v} = f\hat{u}$ is clearly an operator of order 0. On the other hand, if $\hat{v} = (1 + |\xi|^2)^{(1/2)t} \hat{u}$ then the mapping $u \mapsto v$ is an operator of order t whose inverse has order $-t$. Theorem 5.6 implies that the mapping $u \mapsto fu$, with $f \in \mathcal{S}$, is an operator of order 0.

When $u \in H^s \subset \mathcal{S}'$ and $v \in \mathcal{E}'$ the convolution product $u * v$ is well-defined and, from Theorem 4.9, we have $\mathcal{F}(u * v) = \hat{v}\,\hat{u}$ with $\hat{v}$ in C^∞. In general $\hat{v}$ is not bounded on $\mathbf{R}^n$, so neither is $(1 + |\xi|^2)^{(1/2)t}\, \hat{v}(\xi)$ for any $t \in \mathbf{R}$. But if we restrict $v \in \mathcal{E}'$ so that

$$\|v\|_{t,\infty}^\wedge = \sup_{\xi \in \mathbf{R}^n} (1 + |\xi|^2)^{(1/2)t} |\hat{v}(\xi)| < \infty \tag{5.18}$$

then the set

$$\{v \in \mathcal{E}' : \|v\|_{t,\infty}^\wedge < \infty\}$$

is a linear subspace of $\mathcal{E}'$ on which $\|\cdot\|_{t,\infty}^\wedge$ defines a norm. The closure of this subspace in $\mathcal{S}'$ under the norm $\|\cdot\|_{t,\infty}^\wedge$ is a normed linear subspace of $\mathcal{S}'$, which we shall denote by $H^{t,\infty}$,

$$H^{t,\infty} = \{v \in \mathcal{S}' : \|v\|_{t,\infty}^\wedge < \infty\}$$

The notation used here is suggested by that of the Banach space L^∞ of measurable function on $\mathbf{R}^n$ which are bounded almost everywhere. Recall that the norm of $f \in L^\infty$ is defined as the essential supremum of $|f|$ on $\mathbf{R}^n$, from which we conclude that the inequality $|f(x)| \leqslant \|f\|_{L^\infty} = \|f\|_\infty$ holds almost everywhere in $\mathbf{R}^n$. Equation (5.18) clearly implies that

$$\|u\|_{s,\infty}^\wedge = \|(1 + |\xi|^2)^{(1/2)s}\hat{u}\|_\infty \qquad u \in H^{s,\infty} \tag{5.19}$$

which suggests that $H^{s,\infty}$ is related to L^{∞} in the same way that H^s is related to L^2.

Theorem 5.7 The convolution$(u,v) \mapsto u * v$ is a bilinear mapping of $H^s \times H^{t,\infty}$ into H^{s+t} which is continuous on H^s and $H^{t,\infty}$ separately.

Proof Let $u \in H^s$ and $v \in H^{t,\infty}$, then

$$(1 + |\xi|^2)^{(1/2)(s+t)} \mathcal{F}(u * v)(\xi) = (1 + |\xi|^2)^{(1/2)s} \hat{u}(\xi)\,(1 + |\xi|^2)^{(1/2)t} \hat{v}(\xi)$$

By integrating the absolute value of each side of this equation, and using (5.19), we arrive at the inequality

$$\|u * v\|_{s \hat{+} t} \leqslant \|u\|_{\hat{s}} \|v\|_{\hat{t},\infty} \tag{5.20}$$

which is all we need. □

Corollary When $u \in H^s$ and $v \in \mathcal{S}$ then $u * v \in H^{\infty}$.

Because the distributions in H^s, for real values of s, are defined through their Fourier transforms, they are necessarily distributions in $\mathbf{R}^n$. In the next section we shall return to Sobolev spaces over open subsets of $\mathbf{R}^n$. But we can also consider distributions which are, so to speak, *locally* in H^s.

Definition Let Ω be an open subset of $\mathbf{R}^n$. A distribution $u \in \mathcal{D}'(\Omega)$ is said to be in $H^s_{\text{loc}}(\Omega)$ if, for every bounded open set ω in Ω with $\overline{\omega} \subset \Omega$, there is a distribution $v \in H^s$ such that $u = v$ on ω.

Thus the distributions in $H^s_{\text{loc}}(\Omega)$ enjoy the smoothness properties of H^s on Ω without being subjected to its global integrability condition. Moreover, any distribution in $H^s_{\text{loc}}(\Omega)$ with compact support is necessarily in $H^s(\Omega)$.

The next theorem gives a useful characterization of $H^s_{\text{loc}}(\Omega)$.

Theorem 5.8 $u \in H^s_{\text{loc}}(\Omega)$ if and only if $\phi u \in H^s$ for every $\phi \in C_0^{\infty}(\Omega)$.

Proof If $u \in H^s_{\text{loc}}(\Omega)$ and $\phi \in C_0^{\infty}(\Omega)$ then there is a $v \in H^s$ such that $u = v$ on supp ϕ. Since ϕv lies in H^s by Theorem 5.6, so does ϕu.

If, on the other hand, $\phi u \in H^s$ for all $\phi \in C_0^{\infty}(\Omega)$ and ω is any bounded open set in Ω, whose closure also lies in Ω, then we can choose $\phi \in C_0^{\infty}(\Omega)$ with $\phi = 1$ on $\overline{\omega}$ and $u = \phi u \in H^s$ on ω. □

When $u \in H^s$ and ϕ is any function in $C_0^{\infty}(\Omega)$, Theorem 5.6 implies that $\phi u \in H^s$. From Theorem 5.8 we conclude that $u \in H^s_{\text{loc}}(\Omega)$. Thus

Corollary $H^s \subset H^s_{loc}(\Omega)$ for every $\Omega \subset \mathbf{R}^n$.

Example 5.7 For any open set $\Omega \subset \mathbf{R}^n$ we can show that $\bigcap_{s\in R} H^s_{loc}(\Omega) = C^\infty(\Omega)$.

(i) The first step is to show that if $s > \frac{1}{2}n + k$ then $H^s_{loc}(\Omega) \subset C^k(\Omega)$: Suppose u $\in H^s_{loc}(\Omega)$. For any $x \in \Omega$ let U be a bounded neighborhood of x, and $\phi \in C_0^\infty(\Omega)$ be such that $\phi = 1$ on U. Then $\phi u \in H^s(\mathbf{R}^n) \subset C^k(\mathbf{R}^n)$ by Theorem 5.8 and the corollary to Theorem 5.5. Since $\phi = 1$ on U, this implies that $u \in C^k(U)$ for every U. Therefore $u \in C^k(\Omega)$. Thus $\bigcap_{s\in R} H^s_{loc}(\Omega) \subset C^\infty(\Omega)$.

(ii) To show the inclusion in the other direction, let $u \in C^\infty(\Omega)$. For any $\phi \in C_0^\infty(\Omega)$ the product ϕu is in $C_0^\infty \subset \mathcal{S} \subset H^s$ for all s. Thus $u \in H^s_{loc}(\Omega)$ for every s, and therefore $u \in \bigcap H^s_{loc}(\Omega)$.

The union of $H^s_{loc}(\Omega)$ spaces, on the other hand, is not quite $\mathcal{D}'(\Omega)$, as we shall see in the next example. Although the qualification "loc" removes all restrictions on the growth behavior of the elements of $H^s_{loc}(\Omega)$, the limitation comes from the index s, since every distribution in $H^s_{loc}(\Omega)$ is of finite order.

Example 5.8 Here we shall show that $\bigcup H^s_{loc}(\Omega) = \mathcal{D}'_F(\Omega)$, where $\mathcal{D}'_F(\Omega)$ is the space of distributions in $\mathcal{D}'(\Omega)$ of finite order.

(i) To show that $\bigcup H^s_{loc}(\Omega) \subset \mathcal{D}'_F(\Omega)$, let $u \in H^s_{loc}(\Omega)$ be arbitrary. Take $\phi \in \mathcal{D}(\Omega)$ with compact support K. By the definition of $H^s_{loc}(\Omega)$, there exists a distribution $v \in H^s$ such that $u = v$ in a neighborhood of K.

If we now define u on $\mathcal{D}(\Omega)$ by $\langle u,\phi\rangle = \langle v,\phi\rangle$ it is straightforward to verify that u is a distribution in Ω. By Theorem 5.4, v can be expressed as a finite sum of derivatives of order $\leq |s| + 1$ of L^2 functions. Consequently, u has finite order.

(ii) Now suppose that $u \in \mathcal{D}'_F(\Omega)$. By Theorem 3.9, u is a derivative of a continuous function in Ω. Since any continuous function is locally square integrable, u is locally a derivative of finite order, say m, of an L^2 function. For any $\phi \in C_0^\infty(\Omega)$, ϕu is also a finite sum of derivatives of order $\leq m$ of L^2 functions, and therefore lies in H^{-m}. Thus $u \in H^{-m}_{loc}$.

We have seen in the corollary to Theorem 5.6 that a linear differential operator L of order m, with coefficients in $\mathcal{S}$, maps H^s into H^{s-m}, but we do not know whether $Lu \in H^s$ implies that $u \in H^{s+m}$, i.e., whether L^{-1} is an operator of order $-m$. In fact this is not true in general, since the equation $\partial_x\partial_y u = 0$ on $\mathbf{R}^2$ is satisfied by the sum $u(x,y) = f(x) + g(y)$ of any pair of differentiable functions on $\mathbf{R}$. Thus, although $\partial_x\partial_y u \in H^\infty$, the function u is not necessarily in H^∞.

However, when L is elliptic, we have the following important result, also known as the *local regularity theorem*. Recall that the linear differential operator $L = \sum_{|\alpha|\leq m} c_\alpha \partial^\alpha$ is elliptic if $\sum_{|\alpha|=m} c_\alpha \xi^\alpha \neq 0$ whenever $\xi \neq 0$.

Theorem 5.9 Let $L = \sum_{|\alpha|\leq m} c_\alpha \partial^\alpha$ be a linear elliptic differential operator in Ω of order m with coefficients $c_\alpha \in C^\infty(\Omega)$. If $Lu = f \in H^s_{\text{loc}}(\Omega)$ for some $s \in \mathbf{R}$, then $u \in H^{s+m}_{\text{loc}}(\Omega)$.

Denoting $\cap H^s_{\text{loc}}(\Omega)$ by $H^\infty_{\text{loc}}(\Omega)$ and $\cup H^s_{\text{loc}}(\Omega)$ by $H^{-\infty}_{\text{loc}}(\Omega)$, we have seen in Examples 5.7 and 5.8 that $H^\infty_{\text{loc}}(\Omega) = C^\infty(\Omega)$ and $H^{-\infty}_{\text{loc}}(\Omega) = \mathfrak{D}'_F(\Omega)$. This theorem, therefore, yields two significant results for the elliptic operator L:

Corollary 1 If $Lu \in C^\infty(\Omega)$ then $u \in C^\infty(\Omega)$. Hence any solution of the homogeneous equation $Lu = 0$ is in $C^\infty(\Omega)$. In particular, every harmonic distribution in $\mathfrak{D}'(\Omega)$ is a C^∞ harmonic function in Ω.

Corollary 2 Any fundamental solution of L, i.e. a solution of $LE = \delta$ on $\mathbf{R}^n$, is infinitely differentiable on $\mathbf{R}^n - \{0\}$.

At the origin the regularity of the fundamental solution E will naturally depend on the dimension n and on the order m of L. Since $\delta \in H^s$ for all $s < -(1/2)n$, the fundamental solution of the Laplacian operator Δ, for example, is locally in H^{s+2} on a neighborhood of 0 for all $s < -(1/2)n$. Thus E is in $H^1 \subset C_0$ when $n = 1$, H^0 when $n = 2$ or 3, and so on. This is consistent with our previous findings that, at the origin, E is continuous when $n = 1$ and has a square integrable singularity when $n = 2$ or 3.

In order to simplify the proof of Theorem 5.9, we shall assume that the leading coefficients in L, i.e. those c_α in which $|\alpha| = m$, are constants.

Proof Let $\phi \in C_0^\infty(\Omega)$ be arbitrary and $t \leq s + m - 1$. We shall first prove that if $\psi u \in H^t$ for some $\psi \in C_0^\infty(\Omega)$ which is 1 on an open set containing supp ϕ, then $\phi u \in H^{t+1}$.

The distribution $v = L(\phi u) - \phi Lu = L(\phi u) - \phi f$ has its support in supp ϕ, so u may be replaced by ψu in this equation to give

$$\begin{aligned} v &= L(\phi\psi u) - \phi L(\psi u) \\ &= \sum_{|\alpha|\leq m} c_\alpha[\partial^\alpha(\phi\psi u) - \phi\partial^\alpha(\psi u)] \end{aligned}$$

Since the derivatives of ψu of order m cancel out, this sum is a linear combination of derivatives of ψu of orders $\leq m - 1$ with coefficients in $C_0^\infty(\mathbf{R}^n)$. With $\psi u \in H^t$ we conclude from the corollary to Theorem 5.6 that $v \in H^{t-m+1}$.

Since $\phi f \in H^s$ and $t - m + 1 \leq s$ we also have $\phi f \in H^{t-m+1}$. Thus

$$L(\phi u) \in H^{t-m+1} \tag{5.21}$$

and from this we wish to conclude that $\phi u \in H^{t+1}$.

Let $L = P(\partial) + Q(\partial)$, where P is the characteristic polynomial of L defined by

$$P(y) = \sum_{|\alpha|=m} c_\alpha y^\alpha \qquad y \in \mathbf{R}^n$$

and Q is a polynomial of degree $\leq m - 1$. Since P, by assumption, has constant coefficients, we have for any $w \in H^s$

$$\begin{aligned}\mathcal{F}(P(\partial)w) &= P(i\xi)\,\hat{w}\\ &= [|\xi|^{-m}(1 + |\xi|^m) - |\xi|^{-m}]P(i\xi)\,\hat{w}\\ &= \mathcal{F}([P_2(\partial) - P_1(\partial)]\,w)\end{aligned}$$

where $P_1(\partial)$ and $P_2(\partial)$ are operators on $H^{-\infty}(\Omega)$ defined on $\mathbf{R}^n - \{0\}$ by

$$\mathcal{F}(P_1(\partial)w) = |\xi|^{-m}P(i\xi)\,\hat{w}$$

$$\mathcal{F}(P_2(\partial)w)) = (1 + |\xi|^m)|\xi|^{-m}P(i\xi)\,\hat{w}$$

Since P is a homogeneous polynomial of degree m whose only zero is $\xi = 0$, both $P(i\xi)/|\xi|^m$ and $|\xi|^m/P(i\xi)$ are bounded functions on $\mathbf{R}^n - \{0\}$. Hence $P_1(\partial)$ and $P_1^{-1}(\partial)$ are operators of order 0. On the other hand, we have

$$1 + |\xi|^m = (1 + |\xi|^2)^{(1/2)m}\,\frac{1 + |\xi|^m}{(1 + |\xi|^2)^{(1/2)m}}$$

where $(1 + |\xi|^m)\,(1 + |\xi|^2)^{-(1/2)m}$ and its reciprocal are bounded on $\mathbf{R}^n$. Since the mapping $w \mapsto z$ defined by $\hat{z} = (1 + |\xi|^2)^{(1/2)m}\hat{w}$ is of order m, the same is true of the mapping defined by $\hat{z} = (1 + |\xi|^2)^{(1/2)m}g(\xi)\,\hat{w}$, where g and its inverse are bounded in $\mathbf{R}^n$. Therefore $P_2(\partial)$ is an operator of order m whose inverse has order $-m$.

From equation (5.21) we have

$$[P_2(\partial) - P_1(\partial) + Q(\partial)]\,(\phi u) = L(\phi u) \in H^{t-m+1}$$

But since $\phi u = \phi\psi u$ and $\psi u \in H^t$, we conclude that $\phi u \in H^t$, by theorem (5.6). And since $Q(\partial)$ has order $m - 1$ and $P_1(\partial)$ has order 0,

$$[Q(\partial) - P_1\partial)] (\phi u) \in H^{t-m+1}$$

Therefore $P_2(\partial)(\phi u) \in H^{t-m+1}$ and, since $P_2^{-1}(\partial)$ has order $-m$, $\phi u \in H^{t+1}$. It remains to show that $\phi u \in H^{s+m}$.

Choose $\phi_0 \in C_0^\infty(\Omega)$ such that $\phi_0 = 1$ on a neighborhood of supp ϕ, which we shall call U_0. Since $\phi_0 u$ has compact support it lies in H^t for some t, which we may take to be $s + m - k$ for some positive integer k. Now we choose the open sets $U_1, \ldots, U_k$ such that U_j properly contains $\overline{U}_{j+1}$, $0 \leq j \leq k - 1$, and $\overline{U}_k$ = supp ϕ. And we finally choose the C_0^∞ functions $\phi_1, \ldots, \phi_k$ such that $\phi_j = 1$ on U_j and supp $\phi_j = \overline{U}_{j-1}$, where $1 \leq j \leq k - 1$, and $\phi_k = \phi$. From the preceding argument, we conclude that $\phi_1 u \in H^{t+1}$, $\phi_2 u \in H^{t+2}, \ldots, \phi_k u = \phi u \in H^{t+k} = H^{s+m}$. □

5.4 MORE ON THE SPACE $H^m(\Omega)$

Because of their Hilbert space structure, Sobolev spaces provide a convenient framework for applying the theory of distributions to boundary value problems. It would naturally be desirable to extend the definition of $H^s(\mathbf{R}^n)$ to $H^s(\Omega)$ for any open set Ω in $\mathbf{R}^n$. This can be done in more than one way (see [12], for example), but, for the purposes which we have in mind, it suffices to fall back on the Definition 5.3 of $H^m(\Omega)$, where m is a nonnegative integer.

Thus $H^m(\Omega)$ is the Hilbert space of functions on Ω such that $\partial^\alpha u \in L^2(\Omega)$ for all $|\alpha| \leq m$. It is equipped with the norm

$$\|u\|_m = \left[\sum_{|\alpha|\leq m} \|\partial^\alpha u\|_0^2\right]^{1/2} \qquad m \in \mathbf{N}_0$$

Theorem 5.5 implies that $H^1(\mathbf{R}) \subset C^0(\mathbf{R})$. The same inclusion can be shown to hold when $\mathbf{R}$ is replaced by any open interval in $\mathbf{R}$.

Example 5.9 Let $f \in H^1(a,b)$ and define the function g on (a,b) by $g(x) = \int_a^x f'(t)\,dt$. Then g is continuous and $g' = f'$ in the sense of distributions, since

$$\begin{aligned}\langle g',\phi\rangle &= -\langle g,\phi'\rangle \\ &= -\int_a^b\left[\int_a^x f'(t)\,dt\right]\phi'(x)\,dx \\ &= -\int_a^b\int_a^b H(x - t)f'(t)\phi'(x)\,dt\,dx\end{aligned}$$

$$= -\int_a^b \left[\int_t^b \phi'(x)dx\right] f'(t)\, dt$$

$$= \int_a^b \phi(t) f'(t)\, dt$$

$$= \langle f', \phi \rangle \qquad \phi \in \mathcal{D}(a,b)$$

Therefore $g = f +$ constant by Example 2.21, and f is continuous a.e. in (a,b). Thus $H^1(a,b) \subset C^0(a,b)$.

That this is not true when $n \geqslant 2$ may be seen by referring to Exercise 5.4.

Although $C_0^\infty(\mathbf{R}^n)$ is dense in $H^m(\mathbf{R}^n)$, as we have observed in the corollary to Theorem 5.2, this is not true of Ω in general. It is not even true that $C^\infty(\overline{\Omega}) \cap H^m(\Omega)$ is dense in $H^m(\Omega)$, where $C^\infty(\overline{\Omega})$ denotes the restriction to $\overline{\Omega}$ of the functions in $C^\infty(\mathbf{R}^n)$, unless $\partial\Omega$ is smooth enough. This last point will be addressed in Section 6.1. The advantage that $\mathbf{R}^n$ enjoys, it turns out, is that it has no boundary.

We now define $H_0^m(\Omega)$ to be the closure of $C_0^\infty(\Omega)$ in $H^m(\Omega)$. Consequently $H_0^m(\mathbf{R}^n) = H^m(\mathbf{R}^n)$, but in general $H_0^m(\Omega)$ is a proper closed subspace of $H^m(\Omega)$, and therefore a Hilbert space in the induced structure. In fact, the next example shows that $H_0^1(\Omega) \neq H^1(\Omega)$ if Ω is a bounded set in $\mathbf{R}^n$.

Example 5.10 Let $u \in H_0^1(\Omega)$ and define u_0 in $\mathbf{R}^n$ by

$$u_0(x) = \begin{cases} u(x) & \text{if } x \in \Omega \\ 0 & \text{if } x \in \mathbf{R}^n - \Omega \end{cases}$$

To see that $u_0 \in H^1(\mathbf{R}^n)$, let ϕ be any function in $C_0^\infty(\Omega)$. Then $\phi_0 \in C_0^\infty(\mathbf{R}^n)$ and $\|\phi_0\|_1 = \|\phi\|_1$. Consequently the map

$$\lambda_1 : C_0^\infty(\Omega) \to H^1(\mathbf{R}^n)$$

defined by $\lambda_1(\phi) = \phi_0$ is a linear isometry which extends by continuity to a continuous linear map from $H_0^1(\Omega)$ to $H^1(\mathbf{R}^n)$.

Since $u \in H_0^1(\Omega)$, there is a sequence (u_k) in $C_0^\infty(\Omega)$ which converges to u in $H^1(\Omega)$. By the continuity of λ_1, $\lambda_1(u_k) \to \lambda_1(u)$ in $H^1(\mathbf{R}^n)$ and therefore in $L^2(\mathbf{R}^n)$. Consequently there is a subsequence $(u_{k'})$ of (u_k) such that $\lambda_1(u_{k'}) \to \lambda_1(u)$ a.e. in $\mathbf{R}^n$. Hence $u_0 = \lambda_1(u)$ lies in $H^1(\mathbf{R}^n)$.

Now if Ω is a bounded open set in $\mathbf{R}^n$, and $u = 1$ on Ω, then $u \in H^1(\Omega)$. But, by Example 5.1, $u_0 \notin H^1(\mathbf{R}^n)$, and hence $u \notin H_0^1(\Omega)$.

Corresponding to the Sobolev imbedding theorem for H^m, we have the following result for $H_0^m(\Omega)$:

Example 5.11 If Ω is an open set in $\mathbf{R}^n$ and $u \in H_0^m(\Omega)$, then $\lambda_m(u) = u_0 \in H^m(\mathbf{R}^n)$ and $\lambda_m(u) \in C^k(\mathbf{R}^n)$ if $m > \frac{1}{2}n + k$ by the corollary to Theorem 5.5. Since $\lambda_m(u) = u$ on Ω, we conclude that $H_0^m(\Omega) \subset C^k(\overline{\Omega})$ when $m > (1/2)n + k$.

In the special case when $m = 0$, $H_0^0(\Omega)$ coincides with $H^0(\Omega)$ since $C_0^\infty(\Omega)$ is dense in $L^2(\Omega)$. Any function $v \in H_0^0(\Omega) = L^2(\Omega)$ defines a continuous linear functional T_v on $H_0^m(\Omega)$ by

$$\begin{aligned} T_v(u) &= \langle v,u \rangle \\ &= (v,\bar{u})_0 \qquad u \in H_0^m(\Omega) \end{aligned} \tag{5.22}$$

Since

$$|\langle v,u \rangle| \leq \|v\|_0 \, \|u\|_0 \leq \|v\|_0 \, \|u\|_m$$

by the Schwarz inequality, we see that T_v is bounded by $\|v\|_0$. We define the "negative norm" of $v \in L^2(\Omega)$ by

$$\|v\|_{-m} = \sup_{u \in H_0^m(\Omega)} \frac{|\langle v,u \rangle|}{\|u\|_m}$$

so that

$$|\langle v,u \rangle| \leq \|v\|_{-m} \, \|u\|_m$$

Since $\|u\|_0 \leq \|u\|_m$, we have

$$\|v\|_{-m} \leq \sup_{u \in H_0^m(\Omega)} \frac{|\langle v,u \rangle|}{\|u\|_0} = \|v\|_0$$

and it is straightforward to verify that $\|\cdot\|_{-m}$ satisfies the properties of a norm. Now we define $H^{-m}(\Omega)$ to be the completion of $L^2(\Omega)$ in the norm $\|\cdot\|_{-m}$.

Theorem 5.10 The dual space$(H_0^m)'(\Omega)$ of the space $H_0^m(\Omega)$ may be identified with $H^{-m}(\Omega)$ for all $m \geq 0$.

Proof Let F be the set of continuous linear functionals T_v on $H_0^m(\Omega)$ defined by equation (5.22). This is clearly a subspace of the Hilbert space $(H_0^m)'(\Omega)$, and we now show that it is a dense subspace.

If F is not dense in $(H_0^m)'(\Omega)$, then there is a nonzero element $S \in (H_0^m)''(\Omega)$ such that $S(T_v) = 0$ for all $T_v \in F$. By the reflexive property of the Hilbert space $H_0^m(\Omega)$, there is an element $w \in H_0^m(\Omega)$ such that $S(T) = T(w)$ for all $T \in (H_0^m)'(\Omega)$. Therefore $S(T_v) = T_v(w) = \langle v,w\rangle = 0$ for all $v \in L^2(\Omega)$. Since $H_0^m(\Omega) \subset L^2(\Omega)$ we can choose $v = w$ and conclude that $w = 0$. This contradicts the assumption that $S \neq 0$.

Thus $\bar{F} = (H_0^m)'(\Omega)$. Now $(H_0^m)'(\Omega)$ can be identified with $H^{-m}(\Omega)$ by the correspondence $T_v \leftrightarrow v$ and $\|T_v\| = \|v\|_{-m}$. □

In the next example the characterization of $H^{-m}(\mathbf{R}^n)$, as given by Theorem 5.4, is carried over to $H^{-m}(\Omega)$.

Example 5.12 We shall now prove that $v \in H^{-m}(\Omega)$ if and only if $v = \sum_{|\alpha|\le m} \partial^\alpha v_\alpha$, where $v_\alpha \in L^2(\Omega)$. By Theorem 5.10 this is equivalent to proving that a distribution T belongs to $(H_0^m)'(\Omega)$ if and only if T is of the form T_v, where $v = \sum_{|\alpha|\le m} \partial^\alpha v_\alpha$, $v_\alpha \in L^2(\Omega)$.

(i) Let T be a distribution of the form T_v with $v = \sum_{|\alpha|\le m} \partial^\alpha v_\alpha$, and $u \in C_0^\infty(\Omega)$. Then

$$\begin{aligned} T_v(u) &= \left\langle \sum_{|\alpha|\le m} \partial^\alpha v_\alpha, u \right\rangle \\ &= \sum_{|\alpha|\le m} (-1)^{|\alpha|}\langle v_\alpha, \partial^\alpha u\rangle \end{aligned}$$

Since $C_0^\infty(\Omega)$ is dense in $H_0^\infty(\Omega)$, this equality holds even when $u \in H_0^\infty(\Omega)$. Hence

$$\begin{aligned} |T_v(u)| &= \left| \sum_{|\alpha|\le m} (-1)^{|\alpha|}(v_\alpha, \partial^\alpha \bar{u})_0 \right| \\ &\le \sum_{|\alpha|\le m} \|v_\alpha\|_0 \, \|u\|_m \end{aligned}$$

which clearly shows that T_v lies in $(H_0^m)'(\Omega)$.

(ii) On the other hand, if $T \in (H_0^m)'(\Omega)$, then there exists a function $g \in H_0^m(\Omega)$ such that, for all $f \in H_0^m(\Omega)$,

$$\begin{aligned} T(f) &= (f,g)_m \qquad \text{by the Riesz representation theorem} \\ &= \sum_{|\alpha|\le m} (\partial^\alpha f, \partial^\alpha g)_0 \end{aligned}$$

In particular, if $\phi \in C_0^\infty(\Omega) \subset H_0^m(\Omega)$,

$$T(\phi) = \sum_{|\alpha|\le m} (\partial^\alpha \phi, \partial^\alpha g)_0$$

$$= \sum_{|\alpha| \leq m} (-1)^{|\alpha|} \langle \phi, \partial^{2\alpha} \overline{g} \rangle$$

$$= \sum_{|\alpha| \leq m} \langle \phi, \partial^{\alpha} g_{\alpha} \rangle$$

where $g_\alpha = (-1)^{|\alpha|}\partial^\alpha \overline{g} \in L^2(\Omega)$. Since $C_0^\infty(\Omega)$ is dense in $H_0^m(\Omega)$, T has the form T_v with $v = \sum_{|\alpha| \leq m} \partial^\alpha g_\alpha$, $g_\alpha \in L^2(\Omega)$.

Example 5.13 Consider now the differential operator $(1 - \Delta)^m : H^m(\Omega) \to \mathcal{D}'(\Omega)$, where Δ is the Laplacian operator in $\mathbf{R}^n$ and $m \in \mathbf{N}_0$. If $u \in H^m(\Omega)$ and $\phi \in C_0^\infty(\Omega)$,

$$\begin{aligned} \langle (1 - \Delta)^m u, \phi \rangle &= \sum_{|\alpha| \leq m} c_\alpha (-1)^{|\alpha|} \langle \partial^{2\alpha} u, \phi \rangle \\ &= \sum_{|\alpha| \leq m} c_\alpha \langle \partial^\alpha u, \partial^\alpha \phi \rangle \\ &= \sum_{|\alpha| \leq m} c_\alpha (\partial^\alpha u, \partial^\alpha \overline{\phi})_0 \end{aligned}$$

where c_α are the binomial coefficients of $(1 - \sum_{k=1}^n \partial_k^2)^m$. Since $1 \leq c_\alpha \leq c_m$ for some integer c_m, we have

$$\begin{aligned} |\langle (1 - \Delta)^m u, \phi \rangle| &\leq c_m \sum_{|\alpha| \leq m} \|\partial^\alpha u\|_0 \, \|\partial^\alpha \phi\|_0 \\ &\leq c_m \|u\|_m \, \|\phi\|_m \end{aligned}$$

This means that the mapping

$$\phi \mapsto \langle (1 - \Delta)^m u, \phi \rangle = ((1 - \Delta)^m u, \overline{\phi})_0$$

is a continuous linear functional on $C_0^\infty(\Omega)$, bounded in the H^m norm, which may therefore be extended by continuity to $H_0^m(\Omega)$. Thus $(1 - \Delta)^m u \in (H_0^m)'(\Omega) = H^{-m}(\Omega)$, and we conclude that the linear differential operator $(1 - \Delta)^m$ maps $H^m(\Omega)$ continuously into $H^{-m}(\Omega)$.

5.5 FOURIER SERIES AND PERIODIC DISTRIBUTIONS

The classical theory of Fourier series lends itself to an elegant development within the context of Sobolev spaces of periodic functions, where the integral which defines the inner product is taken over the period of the function instead of $\mathbf{R}^n$. This gives rise to periodic distributions which come up naturally in the study of certain differential equations, such as the heat or the wave equations, in a finite domain.

Let

$$u(x) = \sum_{|\alpha| \leq k} a_\alpha e^{i\langle \alpha, x\rangle} \tag{5.23}$$

be a finite sum of exponential functions with $x \in \mathbf{R}^n$, $\alpha = (\alpha_1, \ldots, \alpha_n) \in \mathbf{Z}^n$, $|\alpha| = \sum_{j=1}^{n} |\alpha_j|$ and $\langle \alpha, x\rangle = \alpha_1 x_1 + \cdots + \alpha_n x_n$. The coefficients a_α are complex numbers which satisfy $a_{-\alpha} = \bar{a}_\alpha$ when u is a real function. For any integer m we define the inner product of u with

$$v(x) = \sum_{|\alpha| \leq k} b_\alpha e^{i\langle \alpha, x\rangle}$$

by

$$(u,v)_m = (2\pi)^n \sum_{|\alpha| \leq k} (1 + |\alpha|^2)^m a_\alpha \bar{b}_\alpha \tag{5.24}$$

from which we conclude that

$$(u,v)_0 = (2\pi)^n \sum_{|\alpha| \leq k} a_\alpha \bar{b}_\alpha$$

$$= \int u(x)\, \bar{v}(x)\, dx \tag{5.25}$$

where the integral is taken over the cube $[-\pi, \pi]^n$, and that

$$a_\alpha = (2\pi)^{-n}(u, e^{i\langle \alpha, x\rangle})_0$$

The norm generated by the inner product (5.24) is

$$\|u\|_m = \sqrt{(u,u)_m}$$

$$= (2\pi)^{n/2} \left[\sum_{|\alpha| \leq k} (1 + |\alpha|^2)^m |a_\alpha|^2 \right]^{1/2} \tag{5.26}$$

and the Schwarz inequality

$$|(u,v)_m| \leq \|u\|_m \|v\|_m$$

may be generalized to

$$|(u,v)_m| \leq \|u\|_{m+l} \|v\|_{m-l} \tag{5.27}$$

for any integer l.

If we complete the linear space of trigonometric polynomials of the form (5.23) under the norm $\|\cdot\|_m$, the resulting space, denoted by $\tilde{H}^m$, is clearly a Hilbert space. The elements of $\tilde{H}^m$ are represented by infinite sums of the form $\sum a_\alpha e^{i\langle \alpha, x\rangle}$ such that the norm

$$(2\pi)^{n/2}\left[\sum (1 + |\alpha|^2)^m |a_\alpha|^2\right]^{1/2}$$

is finite.

When m is a nonnegative integer, this implies the convergence of the series in the $L^2([-\pi,\pi]^n)$ norm. $\Sigma a_\alpha e^{i\langle a,x\rangle}$ is then a Fourier series expansion of a periodic function in $\mathbf{R}^n$ whose Fourier coefficients, in the classical sense, are a_α. Though we have chosen the period here to be $(2\pi)^n$, there is no loss of generality in this choice, since the period can be changed by an appropriate change of scale of x.

For every multi-index $\alpha \in \mathbf{N}_0^n$ we have

$$\partial^\alpha \sum a_\beta e^{i\langle\beta,x\rangle} = \sum (i\beta)^\alpha a_\beta e^{i\langle\beta,x\rangle}$$

where

$$(i\beta)^\alpha = (i\beta_1)^{\alpha_1} (i\beta_2)^{\alpha_2} \cdots (i\beta_n)^{\alpha_n}$$

Hence

$$\begin{aligned}\left\|\partial^\alpha \sum a_\beta e^{i\langle\beta,x\rangle}\right\|_0^2 &= (2\pi)^n \sum |\beta^\alpha a_\beta|^2 \\ &\leq (2\pi)^n \sum (1 + |\beta|^2)^{|\alpha|} |a_\beta|^2 \\ &= \left\|\sum a_\beta e^{i\langle\beta,x\rangle}\right\|_{|\alpha|}^2\end{aligned}$$

or $\|\partial^\alpha u\|_0 \leq \|u\|_{|\alpha|}$ for any $u \in \tilde{H}^m$, $|\alpha| \leq m$. More generally, we have

$$\|\partial^\alpha u\|_l \leq \|u\|_{l+|\alpha|} \qquad u \in \tilde{H}^m \tag{5.28}$$

for all l and α such that $l + |\alpha| \leq m$. This implies, in particular, that ∂^α is a bounded linear operator from $\tilde{H}^m$ to $\tilde{H}^{m-|\alpha|}$, $|\alpha| \leq m$.

If $u \in \tilde{H}^m$ and $l < m$ then $\|u\|_l \leq \|u\|_m$, so that $\tilde{H}^m \subset \tilde{H}^l$. $\tilde{H}^0$ is therefore the space of periodic functions which are square integrable over $[-\pi,\pi]^n$. It obviously includes $\tilde{C}^0$, the continuous periodic functions in $\mathbf{R}^n$.

When $m > 0$, $\tilde{H}^m$ is the space of periodic functions whose (distributional) derivatives up to order m are square integrable. In order to characterize $\tilde{H}^m$ when $m < 0$, we first consider the following examples:

Example 5.14 Let $\sum a_\alpha e^{i\langle a,x\rangle} \in \tilde{H}^m$, $m \geq 0$. Then there exists a $u \in \tilde{H}^m$ such that, for any $\varepsilon > 0$, $\|u(x) - \Sigma_{|\alpha|\leq k}\, a_\alpha e^{i\langle\alpha,x\rangle}\|_m < \varepsilon$ when k is large enough. Consequently

$$\left\| \partial^\beta u - \sum_{|\alpha| \le k} (i\alpha)^\beta a_\alpha e^{i\langle\alpha,x\rangle} \right\|_0 = \left\| \partial^\beta \left[u - \sum_{|\alpha| \le k} a_\alpha e^{i\langle\alpha,x\rangle} \right] \right\|_0$$

$$\le \left\| u - \sum_{|\alpha| \le k} a_\alpha e^{i\langle\alpha,x\rangle} \right\|_{|\beta|} \qquad \text{by (5.28)}$$

$$\le \left\| u - \sum_{|\alpha| \le k} a_\alpha e^{i\langle\alpha,x\rangle} \right\|_m \qquad \text{for } |\beta| \le m$$

$$< \varepsilon$$

Therefore $\sum_{|\alpha| \le k} (i\alpha)^\beta a_\alpha e^{i\langle\alpha,x\rangle} \to \partial^\beta u$ in $\tilde{H}^0 = \tilde{L}^2$ for all $|\beta| \le m$.

Example 5.15 Since

$$(1 - \Delta) \sum a_\alpha e^{i\langle\alpha,x\rangle} = \sum (1 + |\alpha|^2) a_\alpha e^{i\langle\alpha,x\rangle}$$

it follows that

$$(1 - \Delta)^{-1} \sum a_\alpha e^{i\langle\alpha,x\rangle} = \sum (1 + |\alpha|^2)^{-1} a_\alpha e^{i\langle\alpha,x\rangle}$$

and therefore

$$(1 - \Delta)^l \sum a_\alpha e^{i\langle\alpha,x\rangle} = \sum (1 + |\alpha|^2)^l a_\alpha e^{i\langle\alpha,x\rangle} \qquad \text{for all } l \in \mathbf{Z}$$

Thus, for any pair of trigonometric polynomials $u_k = \sum_{|\alpha| \le k} a_\alpha e^{i\langle\alpha,x\rangle}$ and $v_k = \sum_{|\alpha| \le k} b_\alpha e^{i\langle\alpha,x\rangle}$, we have

$$((1 - \Delta)^l u_k, v_k)_m = (u_k, (1 - \Delta)^l v_k)_m = (u_k, v_k)_{m+l}$$

Taking $v_k = (1 - \Delta)^l u_k$,

$$\|(1 - \Delta)^l u_k\|_m^2 = (u_k, (1 - \Delta)^{2l} u_k)_m = (u_k, u_k)_{m+2l}$$

or

$$\|(1 - \Delta)^l u_k\|_m = \|u_k\|_{m+2l}$$

Since the trigonometric polynomials are dense in $\tilde{H}^m$, this equation can be extended by continuity to $\tilde{H}^m$. Hence, for any $u \in \tilde{H}^m$,

$$\|(1 - \Delta)^l u\|_m = \|u\|_{m+2l} \qquad l, m \in \mathbf{Z}$$

In particular, if m is replaced by $-m$ and $l = m$, then

$$\|(1 - \Delta)^m u\|_{-m} = \|u\|_m \qquad m \in \mathbf{Z} \tag{5.29}$$

Equation (5.29) implies that the linear mapping

$$(1 - \Delta)^m : \tilde{H}^m \to \tilde{H}^{-m} \qquad m \in \mathbf{Z}$$

is bijective and norm preserving. Since $\tilde{H}^m$ is a Hilbert space, the Riesz representation theorem provides another norm-preserving isomorphism between $\tilde{H}^m$ and $(\tilde{H}^m)'$. Consequently there is a norm-preserving isomorphism between $(\tilde{H}^m)'$ and $\tilde{H}^{-m}$ which allows us to identify these two spaces for all integral values of m.

When $\tilde{H}^m$ is real, i.e., where the expansion coefficients satisfy $a_{-\alpha} = \bar{a}_\alpha$, and T is a continuous linear functional on $\tilde{H}^m$ in the (weak) topology defined by $(\cdot,\cdot)_0$, we have for any $u \in \tilde{H}^m$

$$|T(u)| \leqslant M \, \|u\|_0 \leqslant M \, \|u\|_m$$

provided $m \geqslant 0$. Hence T is also continuous in the (strong) topology of $\tilde{H}^m$ defined by $(\cdot,\cdot)_m$. By the Riesz representation theorem, there is a unique $v \in (\tilde{H}^m)' = \tilde{H}^{-m}$ such that $T(u) = (v,u)_0$ and

$$\|T\| = \sup \{(v,u)_0 : \|u\|_m = 1\} = \|v\|_{-m}$$

Thus, corresponding to Theorem 5.10, we have proved the following representation theorem:

Theorem 5.11 For all $m \geqslant 0$, $\tilde{H}^{-m}$ is the dual space of the real Hilbert space $\tilde{H}^m$ with respect to the inner product $(\cdot,\cdot)_0$, in the sense that T is a continuous linear functional on $\tilde{H}^m$ if and only if there is a unique $v \in \tilde{H}^{-m}$ such that

$$T(u) = \langle v,u \rangle = (v,u)_0 \qquad u \in \tilde{H}^m$$

Furthermore,

$$\|T\| = \|v\|_{-m}$$

Example 5.17 The trigonometric polynomial

$$f_p(x) = \sum_{k=-p}^{p} e^{ikx} \qquad x \in \mathbf{R}$$

converges in $\tilde{H}^m$ whenever $m \leqslant -1$. According to Theorem 5.11, the limit to which f_p converges in $\tilde{H}^{-1}$ may be determined by considering the (weak) limit

$$\lim_{p\to\infty} \sum_{-p}^{p} (e^{ikx},\phi(x))_0 = \lim_{p\to\infty} \sum_{-p}^{p} (e^{-ikx},\phi(x))_0$$

where ϕ is an arbitrary function in $\tilde{H}^1$. But since $(e^{-ikx},\phi(x))_0$ is the expansion coefficient a_k of ϕ, this limit is simply $\phi(0)$. Thus $\lim f_p = \tilde{\delta}$, where $\tilde{\delta}$ denotes a periodic version of the Dirac distribution. Its mth derivative is $\tilde{\delta}^{(m)} = \Sigma(ik)^m e^{ikx}$.

Corresponding to Theorem 5.5 we have a Sobolev imbedding theorem for $\tilde{H}^m$, which states that, when $m > \frac{1}{2}n$, $\tilde{H}^m$ may be identified with a subspace of $\tilde{C}^0$. This follows from the observation that, if $m > (1/2)n$ and if $u = \Sigma a_\alpha e^{i(\alpha,x)} \in \tilde{H}^m$, then

$$\begin{aligned}
\left(\sum|a_\alpha|\right)^2 &= \left[\sum(1 + |\alpha|^2)^{-(1/2)m}(1 + |\alpha|^2)^{(1/2)m}|a_\alpha|\right]^2 \\
&\leq \left[\sum(1 + |\alpha|^2)^{-m}\right]\left[\sum(1 + |\alpha|^2)^m|a_\alpha|^2\right] \\
&= \sum(1 + |\alpha|^2)^{-m}\,(2\pi)^{-n}\|u\|_m^2 \\
&< \infty
\end{aligned}$$

Thus the Fourier series of u converges uniformly and, since $e^{i(\alpha,x)}$ is continuous, the sum $\Sigma a_\alpha e^{i(\alpha,x)}$ is continuous.

This result is easily generalized to

Theorem 5.12 If $m > (1/2)n + k$, where k is a nonnegative integer, then $\tilde{H}^m$ is a subspace of $\tilde{C}^k$.

Setting $\tilde{H}^\infty = \cap \tilde{H}^m$ we therefore conclude that $\tilde{H}^\infty = \tilde{C}^\infty$. $\tilde{H}^\infty$ is a locally convex topological vector space in the projective limit topology defined by $\{\tilde{H}^m : m \in \mathbf{N}_0\}$.

Similarly, we define $\tilde{H}^{-\infty} = \cup \tilde{H}^m$. That $\tilde{H}^{-\infty}$ represents the dual of $\tilde{H}^\infty$ in the inner product $(\cdot,\cdot)_0$ may be seen by following the same argument as was used for H^s: Since $\tilde{H}^m \subset \tilde{H}^l$ when $l < m$, we have for any positive integer m

$$\bigcup_{|k|\leq m} \tilde{H}^k = \tilde{H}^{-m} \qquad \bigcap_{|k|\leq m} \tilde{H}^k = \tilde{H}^m$$

Now Theorem 5.11 implies that

$$\bigcup_{|k|\leq m} \tilde{H}^k = \left(\bigcap_{|k|\leq m} \tilde{H}^k\right)'$$

from which we conclude, m being arbitrary, that $\tilde{H}^{-\infty} = (\tilde{H}^\infty)'$.

$\tilde{H}^{-\infty}$ is the space of *periodic distributions*, whose elements are continuous linear functionals on $\tilde{C}^\infty$ in the (weak) topology defined by $\langle\cdot,\cdot\rangle = (\cdot,\cdot)_0$. That means u$\in \tilde{H}^{-\infty}$ if $|(u,\phi_k)_0| \leq \|u\|_{-m}\ \|\phi_k\|_m \to 0$ for every sequence ϕ_k in $\tilde{C}^\infty$ such that $\|\phi_k\|_m \to 0$ for all $m \geq 0$; or, equivalently, whenever $\partial^\alpha\phi_k \to 0$ uniformly for all $\alpha \in \mathbf{N}_0^n$.

When m is a positive integer we have already shown that $\tilde{H}^{-m}$ represents the continuous linear functionals on $\tilde{H}^m$. But since $\tilde{H}^\infty = \tilde{C}^\infty$ is dense in $\tilde{H}^m$ (Exercise 5.19), $\tilde{H}^{-m}$ may also be identified with the subspace of $\tilde{H}^{-\infty}$ consisting of the distributions u for which $(u,\phi_k)_0 \to 0$ whenever ϕ_k is a sequence in $\tilde{C}^\infty$ which converges to 0 in $\tilde{H}^m$.

To justify calling the distributions in $\tilde{H}^{-\infty}$ "periodic," we note that the translation by $(2\pi, \ldots, 2\pi)$, denoted by $\tau_{2\pi}$, of $u \in \tilde{H}^{-\infty}$ satisfies $\langle\tau_{2\pi}u,\phi\rangle = \langle u,\tau_{-2\pi}\phi\rangle = \langle u,\phi\rangle$ for all $\phi \in \tilde{C}^\infty$. The completeness of the trigonometric polynomials in $\tilde{H}^{-\infty}$, in the sense that every $u \in \tilde{H}^{-\infty}$ is the limit of a sum

$$\sum_{|\alpha|\leq k} b_\alpha e^{i(\alpha,x)}$$

as $k \to \infty$, follows from the completeness of these polynomials in $\tilde{H}^m$, for every m, and the equality $\tilde{H}^{-\infty} = \cup \tilde{H}^m$.

Note that any $\phi \in \tilde{C}^\infty$ is represented by a series $\sum a_\alpha e^{i(\alpha,x)}$ where $|\alpha|^m a_\alpha \to 0$ as $|\alpha| \to \infty$ for all $m \geq 0$. Hence

$$\begin{aligned}\langle u,\phi\rangle &= \sum (u,a_\alpha e^{i(\alpha,x)})_0 \\ &= \sum \bar{a}_\alpha\ (u,e^{i(\alpha,x)})_0\end{aligned}$$

where $a_\alpha = (2\pi)^{-n}(\phi(x),e^{i(\alpha,x)})_0$ are the Fourier coefficients of ϕ. Denoting $(u,e^{i(\alpha,x)})_0$ by $(2\pi)^n b_\alpha$, we therefore have

$$\begin{aligned}\langle u,\phi\rangle &= (2\pi)^n\sum b_\alpha\bar{a}_\alpha \\ &= (\sum b_\alpha e^{i(\alpha,x)},\sum a_\alpha e^{i(\alpha,x)})_0 \\ &= (\sum b_\alpha e^{i(\alpha,x)},\phi)_0\end{aligned}$$

Thus u is represented by the series

$$u(x) = \sum b_\alpha e^{i(\alpha,x)}$$

Example 5.18 Let $n = 1$. If $f(x) = \Sigma a_k e^{ikx} \in \tilde{H}^m$, $m \geq 1$, then

$$|f(x) - \sum_{|k|\leq l} a_k e^{ikx}| \leq \sum_{|k|>l} |a_k|$$

$$\leq \left[\sum_{|k|>l} (1 + k^2)^m |a_k|^2\right]^{1/2} \left[\sum_{|k|>l} (1 + k^2)^{-m}\right]^{1/2}$$

$$\leq (2\pi)^{-(1/2)} \|f\|_m \left[\sum_{|k|>l} (1 + k^2)^{-m}\right]^{1/2}$$

Since

$$\sum_{|k|>l} (1 + k^2)^{-m} \leq 2 \int_l^\infty \frac{dt}{(1 + t^2)^m} \leq 2 \int_l^\infty t^{-2m}\, dt = \frac{2}{2m - 1} l^{-2m+1},$$

we obtain the estimate

$$\sup_{x \in \mathbb{R}} \left| f(x) - \sum_{|k| \leq l} a_k e^{ikx} \right| \leq c l^{-m+1/2}$$

where c is a positive constant which depends on f and m.

Thus $l^{-m+1/2}$ indicates the rate of convergence of the Fourier series for f as l increases. The greater the (positive) integer m, the smoother the function f, and the faster the convergence of the series.

The inclusion relations

$$\tilde{C}^m \subset \tilde{H}^m \subset \tilde{C}^{m-1} \tag{5.30}$$

can also be shown to hold when $n = 1$ and $m \geq 1$:

Let $f(x) = \Sigma a_k e^{ikx} \in \tilde{C}^m$. Then $f^{(m)}(x) = \Sigma (ik)^m a_k e^{ikx} \in \tilde{C}^0 \subset \tilde{H}^0 = \tilde{L}^2$. Hence $\Sigma k^{2m} |a_k|^2 < \infty$ and consequently $\Sigma (1 + k^2)^m |a_k|^2 < \infty$, which means that $f \in \tilde{H}^m$. The inclusion $\tilde{H}^m \subset \tilde{C}^{m-1}$ follows directly from Theorem 5.12.

Example 5.19 If $\tilde{L}^1$ is the space of periodic functions which are locally integrable, we can show that $\tilde{L}^1 \subset \tilde{H}^{-1}$ when $n = 1$.

To see this, let $u = \Sigma a_k e^{ikx} \in \tilde{L}^1$. If $\phi(x) = \Sigma b_k e^{ikx}$ is any function in $\tilde{H}^1$, then

$$\langle u, \phi \rangle = (u, \phi)_0 = 2\pi \sum a_k \bar{b}_k.$$

By the Riemann–Lebesgue lemma, $a_k \to 0$ as $|k| \to \infty$, so there is a positive integer l such that $|a_k| \leq 1$ for all $|k| \geq l$. Given $\varepsilon > 0$, we have

$$\sum_{|k|\geq l} |a_k b_k| \leq \sum_{|k|\geq l} |b_k|$$

$$\leq \left[\sum_{|k|\geq l} (1 + k^2)^{-1}\right]^{1/2} \left[\sum_{|k|\geq l} (1 + k^2)|b_k|^2\right]^{1/2}$$

$$< \varepsilon$$

provided l is large enough, since the series $\Sigma(1 + k^2)^{-1}$ and $\Sigma(1 + k^2)|b_k|^2$ both converge, the latter due to the fact that $\phi \in \tilde{H}^1$.

Hence $|\langle u,\phi\rangle| \leq 2\pi \, \Sigma|a_k b_k| \leq c\|\phi\|_1$ for some constant c, and u defines a continuous linear functional on $\tilde{H}^1$, i.e., $u \in \tilde{H}^{-1}$

Since $L^2_{\text{loc}} \subset L^1_{\text{loc}}$, we also have $\tilde{L}^2 \subset \tilde{L}^1$. Combining this with (5.30), we arrive at

$$\tilde{H}^1 \subset \tilde{C}^0 \subset \tilde{L}^2 \subset \tilde{L}^1 \subset \tilde{H}^{-1} \tag{5.31}$$

Corresponding to Theorem 5.6 and its corollary, we have

Theorem 5.13 If $\phi \in \tilde{C}^\infty$ and $u \in \tilde{H}^m$ then $\phi u \in \tilde{H}^m$ and $\|\phi u\|_m \leq c\|u\|_m$, where c is a constant which depends on ϕ and m.

Proof If $m \geq 0$ then, by Leibnitz formula, $\|\phi u\|_m$ is bounded by a constant multiple of $\|u\|_m$.

If $m < 0$ then

$$\begin{aligned} \|\phi u\|_m &= \sup \{(\phi u, v)_0/\|v\|_{|m|} : v \in \tilde{H}^{|m|}\} \\ &\leq \sup_{v \neq 0} \|u\|_m \|\phi v\|_{|m|}/\|v\|_{|m|} \end{aligned}$$

But $\|\phi v\|_{|m|} \leq c\|v\|_{|m|}$, hence $\|\phi u\|_m \leq c\|u\|_m$. □

Since ∂^α maps $\tilde{H}^m$ continuously into $\tilde{H}^{m-|\alpha|}$, this theorem implies

Corollary The linear differential operator of order l,

$$L = \sum_{|\alpha| \leq l} a_\alpha \partial^\alpha$$

with coefficients in $\tilde{C}^\infty$ maps $\tilde{H}^m$ continuously into $\tilde{H}^{m-l}$ with

$$\|Lu\|_{m-l} \leq c\|u\|_m \quad u \in \tilde{H}^m$$

The converse of this result, i.e., that $Lu \in \tilde{H}^m$ implies that $u \in \tilde{H}^{m+l}$, is true provided L is elliptic, and then it may be deduced from the local regularity Theorem 5.9.

If it is beginning to look as though we are repeating results obtained earlier in this chapter, it is because of the striking analogy between Fourier series and Fourier transforms, where the weight function $(1 + |\xi|^2)^s$ in the integral which defines the inner product in H^s corresponds to $(1 + |\alpha|^2)^m$ in the summation formula (5.24). This analogy, so apparent in the classical theory, comes through even more clearly in the theory of distributions.

EXERCISES

5.1 For $1 \leq p < \infty$, $L^p(\Omega)$ denotes the linear space of complex Lebesgue measurable functions f on $\Omega \subset \mathbf{R}^n$ such that $|f|^p \in L^1(\Omega)$. $L^p(\Omega)$ is a Banach space in the norm

$$\|f\|_p = \left[\int_\Omega |f(x)|^p \, dx\right]^{1/p}$$

According to the Riesz representation theorem for $L^p(\Omega)$, if $p > 1$ and $q = p/(p - 1)$, then $L^q(\Omega)$ is the dual of $L^p(\Omega)$ in the bilinear form

$$\langle f,g\rangle = \int_\Omega f(x)g(x)dx \qquad f \in L^p(\Omega),\ g \in L^q(\Omega)$$

and we have Hölder's inequality

$$|\langle f,g\rangle| \leq \|f\|_p \, \|g\|_q$$

Prove that a function f in $L^p(\Omega)$, $p \geq 1$, defines a distribution in $\mathfrak{D}'(\Omega)$ by

$$\langle f,\phi\rangle = \int_\Omega f(x)\, \phi(x)\, dx \qquad \phi \in \mathfrak{D}(\Omega)$$

5.2 For any nonnegative integer m, let

$$H^{m,p}(\Omega) = \{f \in L^p(\Omega) : \partial^\alpha f \in L^p(\Omega),\ |\alpha| \leq m\}$$

where $\partial^\alpha f$ is the distributional derivative of f as an element of $\mathfrak{D}'(\Omega)$. Show that $H^{m,p}(\Omega)$ is a Banach space in the norm

$$\|f\|_{m,p} = \left[\sum_{|\alpha|\leq m} \|\partial^\alpha f\|^p\right]^{1/p}$$

When $p = 2$, $H^{m,2}(\Omega)$ is a Hilbert space which coincides with the Sobolev space $H^m(\Omega)$. Prove that $H^{m,p}(\Omega) \subset H^{l,p}(\Omega)$ whenever $m > l$.

5.3 Given $f \in H^m$, determine an upper bound for each of the constants c_1 and c_2 in

$$\|f\|_m \leq c_1 \,\|(1 + |\xi|^2)^{(1/2)m} \hat{f}\|_0 = c_1\|f\|\hat{_0}$$
$$\|f\|\hat{_0} = \|(1 + |\xi|^2)^{(1/2)m}\hat{f}\|_0 \leq c_2\|f\|_m$$

5.4 Show that the function $f(x) = (\log(1/|x|))^k$, where $x \in \Omega = B(0,\rho) - \{0\} \subset \mathbf{R}^2$ and $\rho < 1$, is in $H^1(\Omega)$ if $0 < k < 1/2$. Use this to show that $H^1(\Omega) \not\subset C^0(\Omega)$ when $B(0,\rho) \subset \mathbf{R}^n$ and $n \geq 2$.

5.5 Let m be a positive integer, $f \in H_0^m(\Omega)$ with Ω open in $\mathbf{R}^n$, and

$$g = \begin{cases} f & \text{on } \Omega \\ 0 & \text{on } \mathbf{R}^n - \Omega \end{cases}$$

(a) Show that $g \in H^m(\mathbf{R}^n)$.

(b) Why is this conclusion false when f is merely in $H^m(\Omega)$?

5.6 If $\Omega_1 \subset \Omega_2$ show that $H_0^m(\Omega_1)$ may be identified with a subspace of $H_0^m(\Omega_2)$. When can the same be said of $H^m(\Omega_1)$ and $H^m(\Omega_2)$?

5.7 If $u \in H^m(\Omega)$ and $\phi \in C_0^l(\Omega)$, when is the product ϕu in $H_0^m(\Omega)$?

5.8 Show that the orthogonal complement of $H_0^1(\Omega)$ in the Hilbert space $H^1(\Omega)$ is given by the null space of the operator $\Delta - 1$ in $H^1(\Omega)$. Find a basis for this null space in the special case when Ω is the finite open interval (a,b) in $\mathbf{R}$.

5.9 Show that $(u,v) = \sum_{|\alpha|=1} \int_\Omega \partial^\alpha u \partial^\alpha \bar{v}$ defines an inner product on $H_0^1(\Omega)$ and that the topology generated by the norm $\|u\| = \sqrt{(u,u)}$ is equivalent to the standard topology of $H_0^1(\Omega)$.

5.10 Let K be the distribution defined in Exercise 4.23. Show that $u \mapsto K * u$ is a continuous mapping from H^s into H^s.

5.11 For $1 \leq p < \infty$ we use l^p to denote the Banach space of all complex sequences $(c_k : k \in \mathbf{N})$ equipped with the norm $[\Sigma|c_k|^p]^{1/p}$. Discuss the relation between l^p and H^m.

5.12 Define an appropriate inner product on the linear space of trigonometric polynomials of period λ^n,

$$u(x) = \sum_{|\alpha| \leq k} a_\alpha \exp(i2\pi\langle\alpha,x\rangle/\lambda)$$

where λ is a positive constant, and hence define the Hilbert space $\tilde{H}_\lambda^m$ of periodic distributions with period λ. Find an expression for the periodic Dirac distribution $\tilde{\delta} \in \tilde{H}_\lambda^{-1}(\mathbf{R})$ and show that it is related to $\delta \in \mathcal{D}'(\mathbf{R})$ by $\tilde{\delta} = \Sigma_{k=-\infty}^{\infty} \delta_{k\lambda}$.

5.13 Determine the Fourier coefficients of a periodic distribution u on $\mathbf{R}$ whose period is λ. What are the Fourier coefficients of its mth derivative? Determine the Fourier coefficients of $\tilde{\delta}$ in Exercise 5.12.

5.14 Show that the periodic distribution $u = \Sigma a_\alpha e^{i\langle\alpha,x\rangle}$ is zero if and only if $a_\alpha = 0$ for all $\alpha \in \mathbf{Z}^n$.

5.15 Indicate which of the following sums represent periodic distributions on $\mathbf{R}$:

(a) $\Sigma(k^2 - k + 1)e^{ikx}$
(b) $\Sigma|k|^k e^{ikx}$
(c) $\Sigma k^2!\, e^{ikx}$

5.16 Prove that every periodic distribution is a tempered distribution. Hence determine the Fourier transform of $u(x) = \Sigma a_\alpha e^{i(\alpha, x)}$ and express a_α in terms of $\hat{u}$.

5.17 Show that the function defined on $\mathbf{R} - \{2k\pi : k \in \mathbf{Z}\}$ by $1/\sin x$ is not a distribution. Determine the periodic distribution which is defined by its Cauchy principal value.

5.18 Repeat Example 5.2 with $n = 2$.

5.19 Use the density of $\tilde{C}^\infty$ in $\tilde{H}^0$ and equation (5.29) to prove that $\tilde{C}^\infty$ is dense in $\tilde{H}^m$ for all integers m.

5.20 Prove the inequality (5.27).

5.21 Use the Sobolev imbedding theorem to prove that Δ has a continuous periodic fundamental solution in $\mathbf{R}^n$ provided $n = 1$. What is the analogous result for the iterated Laplacian Δ^m?

5.22 Investigate where the spaces $\tilde{L}^p$ and $\tilde{L}^q$, $1/p + 1/q = 1$, fit in the inclusion relations (5.31).

6

Applications to Boundary Value Problems

6.1 THE TRACE OPERATOR IN $H^m(\Omega)$

As we have seen in some of the examples of Section 3.6, the treatment of boundary value problems on an open domain in $\mathbf{R}^n$ will necessarily involve a treatment of the behavior of distributions on the boundary of such a domain. At this point, though, it is not clear what it means for a distribution on Ω to assume "values" on $\partial\Omega$. For a sufficiently well-behaved function which is continuous, for example, on Ω this may be interpreted to mean that the function is first extended by continuity to $\overline{\Omega}$ and then restricted to $\partial\Omega$. As we shall now see, this is precisely how we define the "boundary values," or *trace*, of a distribution in $H^m(\Omega)$. But, as we should also expect, the possibility of defining the extension depends on the smoothness properties of $\partial\Omega$.

The simplest boundary that we can think of in $\mathbf{R}^n$ is the hyperplane $\{x = (x_1, \ldots, x_n) \in \mathbf{R}^n : x_n = 0\}$, so we shall start with the open set

$$\mathbf{R}^n_+ = \mathbf{R}^{n-1} \times (0,\infty) = \{x = (\tilde{x},x_n) : \tilde{x} \in \mathbf{R}^{n-1}, x_n > 0\}$$

In accordance with our notation convention, $C_0^m(\overline{\Omega})$ will denote the linear space of functions in $C_0^m(\mathbf{R}^n)$ restricted to $\overline{\Omega}$. In contrast with functions in $C_0^m(\Omega)$, a function in $C_0^m(\overline{\Omega})$ does not necessarily vanish, together with its derivatives, on $\partial\Omega$; but every function in $C_0^m(\Omega)$, once extended by continuity to $\overline{\Omega}$, is in $C_0^m(\overline{\Omega})$.

Let $\phi \in C_0^1(\overline{\mathbf{R}}^n_+)$ and denote $\phi(\bar{x},0)$ by $\tilde{\phi}(\bar{x})$, where $\tilde{\phi}$ is now a function in $C_0^1(\partial\mathbf{R}^n_+) = C_0^1(\mathbf{R}^{n-1})$. Then

$$|\tilde{\phi}(\bar{x})|^2 = -\int_0^\infty \partial_n\, |\phi(\bar{x},x_n)|^2\, dx_n$$

and, after integrating on $\mathbf{R}^{n-1}$, we obtain

$$\begin{aligned}\|\tilde{\phi}\|_0^2 &= \left|\int_{\mathbf{R}^n_+} [(\partial_n\phi)\,\overline{\phi} + \phi(\partial_n\overline{\phi})]\, dx\right| \\ &\leq 2\, \|\partial_n\phi\|_0\, \|\phi\|_0 \\ &\leq \|\phi\|_0^2 + \|\partial_n\phi\|_0^2\end{aligned}$$

Thus

$$\|\tilde{\phi}\|_0 \leq \|\phi\|_1 \tag{6.1}$$

Now, since $C_0^m(\mathbf{R}^n)$ is dense in H^m, $C_0^m(\overline{\mathbf{R}}^n_+)$ is clearly dense in $H^m(\mathbf{R}^n_+)$, and we have just shown that the linear mapping from $C_0^1(\overline{\mathbf{R}}^n_+)$ to $C_0^0(\partial\mathbf{R}^n_+)$ defined by $\phi \mapsto \tilde{\phi}$ is continuous in the induced topologies of $H^1(\mathbf{R}^n_+)$ and $H^0(\partial\mathbf{R}^n_+)$, respectively. This mapping, therefore, has a unique continuous linear extension to $H^1(\mathbf{R}^n_+)$, which we shall denote by γ. To show that γ has a dense range in $H^0(\partial\mathbf{R}^n_+)$, let $\tilde{\phi}$ be an arbitrary function in $C_0^1(\mathbf{R}^{n-1})$, and define the function ϕ in $C_0^1(\overline{\mathbf{R}}^n_+)$ by $\phi(x) = \tilde{\phi}(\bar{x})\phi_0(x_n)$, where ϕ_0 is a function in $C_0^1(\overline{\mathbf{R}}_+)$ with $\phi_0(0) = 1$. It then follows that $\gamma(\phi) = \tilde{\phi}$, and, since $C_0^1(\mathbf{R}^{n-1})$ is dense in $L^2(\mathbf{R}^{n-1}) = H^0(\partial\mathbf{R}^n_+)$, we see that the linear mapping

$$\gamma : H^1(\mathbf{R}^n_+) \to H^0(\partial\mathbf{R}^n_+)$$

called the *trace operator*, is continuous on $H^1(\mathbf{R}^n_+)$ and its range $\gamma(H^1(\mathbf{R}^n_+))$ is dense in $H^0(\partial\mathbf{R}^n_+)$.

We can also prove that the kernel of γ in $H^1(\mathbf{R}^n_+)$ is $H_0^1(\mathbf{R}^n_+)$, i.e. a function $u \in H^1(\mathbf{R}^n_+)$ lies in $H_0^1(\mathbf{R}^n_+)$ if and only if $\gamma(u) = 0$. Since any $u \in H_0^1(\mathbf{R}^n_+)$ is the limit in the norm $\|\cdot\|_1$ of a sequence (u_k) in $C_0^\infty(\mathbf{R}^n_+)$, and since $\gamma(u_k) = 0$ for every k, $\gamma(u) = \lim \gamma(u_k) = 0$ by the continuity of γ. This proves the "only if" part.

Now let $u \in H^1(\mathbf{R}^n_+)$, with $\gamma(u) = 0$, and define

$$u_0(x) = H(x_n)u(x) = \begin{cases} u(x) & x \in \mathbf{R}^n_+ \\ 0 & x \notin \mathbf{R}^n_+ \end{cases}$$

where H is the Heaviside function. Then $u_0 \in H^0$ and $\partial_j u_0 \in H^0$ for all $j =$

$1, \ldots, n-1$. But $\partial_n u_0 = H(x_n)\partial_n u + v$ belongs to H^{-1}, by Theorem 5.4, and $H(x_n)\partial_n u \in H^0$. Thus

$$v = \partial_n u_0 - H(x_n)\partial_n u \in H^{-1} = (H^1)'$$

For any $\phi \in C_0^m$ we have

$$\begin{aligned}
\langle v,\phi\rangle &= -\langle u_0,\partial_n\phi\rangle - \langle H(x_n)\partial_n u,\phi\rangle \\
&= -\int_{\mathbf{R}^n_+} (u\partial_n\phi + \phi\partial_n u)\, dx \\
&= -\int_{\mathbf{R}^{n-1}}\int_0^\infty \partial_n(\phi u)\, dx_n\, d\bar{x} \\
&= \int_{\mathbf{R}^{n-1}} (\phi u)(\bar{x},0)\, d\bar{x} \\
&= \int_{\mathbf{R}^{n-1}} \gamma(\phi u)(\bar{x})\, d\bar{x} \\
&= \int_{\mathbf{R}^{n-1}} \gamma(\phi)\,\gamma(u)\, d\bar{x} \\
&= 0
\end{aligned}$$

because $\gamma(u) = 0$. Thus $v = 0$, $\partial_n u_0 = H(x_n)\partial_n u \in H^0$, and therefore $u_0 \in H^1$.

To show that $u \in H_0^1(\mathbf{R}^n_+)$ we shall assume, without loss of generality, that $u \in H^1(\mathbf{R}^n_+)$ has compact support in $\overline{\mathbf{R}}^n_+$. For any positive number h, let $\tau_h u_0$ be the translation of u_0 along the nth coordinate x_n defined by $\tau_h u_0(\bar{x},x_n) = u_0(\bar{x},x_n - h)$. Then $\tau_h u_0 \in H^1$ and $\tau_h u_0 \to u_0$ in H^1 as $h \to 0$. Taking γ_k to be a regularizing sequence of δ in $C_0^\infty(\mathbf{R}^n)$, the sequence $\gamma_k * \tau_h u_0$ has compact support in $\mathbf{R}^n_+$, when k is large enough, and converges to $\tau_h u_0$ as $k \to \infty$. Consequently, there is a sequence in $C_0^\infty(\mathbf{R}^n_+)$ which converges to u in $H^1(\mathbf{R}^n_+)$, so u must be in $H_0^1(\mathbf{R}^n_+)$.

Thus we have proved

Theorem 6.1 The trace operator $\gamma : H^1(\mathbf{R}^n_+) \to H^0(\mathbf{R}^{n-1})$ which is defined on $C_0^1(\overline{\mathbf{R}}^n_+)$ by

$$\gamma(u)(\bar{x}) = u(\bar{x},0)$$

is a continuous linear operator whose range is dense in $H^0(\mathbf{R}^{n-1})$ and whose null space in $H^1(\mathbf{R}^n_+)$ is $H_0^1(\mathbf{R}^n_+)$.

Example 6.1 Green's formulas for any pair of functions $u, v \in H^1(\mathbf{R}^n_+)$ take the form

$$\text{(i)} \quad \int_{\mathbf{R}^n_+} (u\,\partial_i v + v\,\partial_i u)\,dx = 0 \qquad \text{when } 1 \leq i \leq n-1$$

$$\text{(ii)} \quad \int_{\mathbf{R}^n_+} (u\,\partial_n v + v\,\partial_n u)\,dx = -\int_{\mathbf{R}^{n-1}} u(\bar{x},0)v(\bar{x},0)\,d\bar{x}$$

To prove this, first suppose that $u, v \in C_0^\infty(\overline{\mathbf{R}}^n_+)$. Then formula (i) is evident, and (ii) follows from

$$\int_{\mathbf{R}^n_+} (u\,\partial_n v + v\,\partial_n u)\,dx = \int_{\mathbf{R}^{n-1}} \int_0^\infty \partial_n(uv)\,dx_n\,d\bar{x}$$

$$= -\int_{\mathbf{R}^{n-1}} u(\bar{x},0)v(\bar{x},0)\,d\bar{x}$$

Since $C_0^\infty(\mathbf{R}^n_+)$ is dense in $H^1(\mathbf{R}^n_+)$ and the trace operator is continuous from $H^1(\mathbf{R}^n_+)$ into $H^0(\mathbf{R}^{n-1}) = L^2(\mathbf{R}^{n-1})$ by Theorem 6.1, the general result easily follows.

Note that the minus sign on the right-hand side of formula (ii) is consistent with the fact that the differential operator $\partial_n = \partial/\partial x_n$ is in the direction of the *inward* normal to $\partial\mathbf{R}^n_+ = \mathbf{R}^{n-1}$.

Example 6.2 The linear map $P : C_0^\infty(\overline{\mathbf{R}}^n_+) \to H^1(\mathbf{R}^n)$ defined by

$$Pu(\bar{x},x_n) = \begin{cases} u(\bar{x},x_n) & \text{if } x_n > 0 \\ u(\bar{x},-x_n) & \text{if } x_n < 0 \end{cases}$$

is clearly continuous and $\|Pu\|_1 = 2\|u\|_1$.

Since $C_0^\infty(\mathbf{R}^n_+)$ is dense in $H^1(\mathbf{R}^n_+)$, P can be extended by continuity to a linear map P: $H^1(\mathbf{R}^n_+) \to H^1(\mathbf{R}^n)$ such that $Pu = u$ a.e. in $\mathbf{R}^n_+$, known as the "1-extension by reflection."

If $m > 1$ we can still define the reflection map P, but the definition becomes more complicated. For $n = 2$, P: $H^2(\mathbf{R}^n_+) \to H^2(\mathbf{R}^n)$ is defined by

$$Pu(\bar{x},x_n) = \begin{cases} u(\bar{x},x_n) & \text{if } x_n > 0 \\ 3u(\bar{x},-x_n) - 2u(\bar{x},-2x_n) & \text{if } x_n < 0 \end{cases}$$

Example 6.3 When $n \leq 3$ we have $H^2(\mathbf{R}^n) \subset C^0(\mathbf{R}^n)$ by Theorem 5.5. Using the reflection map P: $H^2(\mathbf{R}^n_+) \to H^2(\mathbf{R}^n)$ defined in Example 6.2, we obtain the inclusion $H^2(\overline{\mathbf{R}}^n_+) \subset C^0(\overline{\mathbf{R}}^n_+)$. Hence every $u \in H^2(\mathbf{R}^n_+)$ is continuous on $\overline{\mathbf{R}}^n_+$ whenever $n \leq 3$.

In order that Theorem 6.1 apply to a more general open set $\Omega \subset \mathbf{R}^n$, its boundary should be smooth enough for defining the trace of any function in $H^1(\Omega)$ in the way that we have defined it when $\partial\Omega = \mathbf{R}^{n-1}$. This basically means that $\overline{\Omega}$ should be locally homeomorphic to $\overline{\mathbf{R}}^n_+$, the homeomorphism satisfying certain compatibility and differentiability conditions that we shall now specify.

If U and V are open sets in $\mathbf{R}^n$ and $\phi: U \to V$ is a homeomorphism, then ϕ is called a *diffeomorphism* of class C^m, for some positive integer m, or a C^m diffeomorphism, if each component of ϕ is in $C^m(U)$ and each component of ϕ^{-1} is in $C^m(V)$. If such a ϕ exists then U is said to be C^m-diffeomorphic to V. The boundary $\partial\Omega$ of $\Omega \subset \mathbf{R}^n$ is called a *hypersurface* in $\mathbf{R}^n$ of class C^m, or a C^m hypersurface, if

1. Every point $x \in \partial\Omega$ has an open neighborhood in $\mathbf{R}^n$, call it U_x, and a corresponding C^m diffeomorphism ϕ_x of U_x onto $\mathbf{R}^n$, which maps $U_x \cap \overline{\Omega}$ onto $\overline{\mathbf{R}}^n_+$.
2. If $\tilde{\phi}_x$ is the restriction of ϕ_x to $U_x \cap \partial\Omega = \tilde{U}_x$, then $\tilde{U}_x$ is homeomorphic to $\mathbf{R}^{n-1}$; and when $\tilde{U}_x \cap \tilde{U}_y = \tilde{U}_{xy} \neq \emptyset$ for any pair of points $x,y \in \partial\Omega$, then $\tilde{\phi}_y \circ \tilde{\phi}_x^{-1}$ is a C^m diffeomorphism of $\phi_x(\tilde{U}_{xy})$ onto $\phi_y(\tilde{U}_{xy})$.

Thus a hypersurface in $\mathbf{R}^n$ is locally diffeomorphic to the hyperplane $\mathbf{R}^{n-1}$, i.e., it is a differentiable manifold of dimension $n - 1$, the class of the diffeomorphism determining the class of the hypersurface. This allows us to define what it means for a function f defined on a C^m hypersurface $\partial\Omega$ to be of class C^k, where $k \leq m$. It simply means that, for every $x \in \partial\Omega$, the function $f \circ \tilde{\phi}_x^{-1}$ on $\tilde{\phi}_x(\tilde{U}_x) = \mathbf{R}^{n-1}$ is of class C^k.

The above definition of the hypersurface property of $\partial\Omega$ implies that Ω is locally on one side of $\partial\Omega$. For simplicity, we shall also assume that Ω is connected and lies globally on one side of $\partial\Omega$. We call Ω a *regular domain* in $\mathbf{R}^n$ if it satisfies all these properties. When Ω is bounded, its closure $\overline{\Omega}$ is compact and $C^\infty(\overline{\Omega})$ coincides with $C_0^\infty(\overline{\Omega})$.

Theorem 6.2 Let Ω be a bounded regular domain in $\mathbf{R}^n$ whose boundary is a C^1 hypersurface. Then there is a unique, continuous, linear mapping γ from $H^1(\Omega)$ to $H^0(\partial\Omega)$ such that, for every $u \in C^1(\overline{\Omega})$, $\gamma(u)$ is the restriction to $\partial\Omega$ of u. The kernel of γ is $H_0^1(\Omega)$ and its range is dense in $H^0(\partial\Omega)$.

Proof. Since Ω is bounded, both $\overline{\Omega}$ and $\partial\Omega$ are compact, and the collection $\{U_x : x \in \partial\Omega\}$ of open neighborhoods in $\mathbf{R}^n$ which covers $\partial\Omega$ contains a finite open subcover $\{U_1, \ldots, U_k\}$ of $\partial\Omega$. Let ϕ_i be a C^1 diffeomorphism of U_i onto $\mathbf{R}^n$ which maps $U_i \cap \overline{\Omega}$ onto $\overline{\mathbf{R}}^n_+$, $i = 1, \ldots, k$. By setting $U_0 = \Omega$ we obtain the finite open cover $\{U_0, U_1, \ldots, U_k\}$ of $\overline{\Omega}$, and we can always define a C^∞ diffeomorphism ϕ_0 from U_0 to $\mathbf{R}^n_+$ since both sets are open in $\mathbf{R}^n$.

Let $\{\psi_0, \ldots, \psi_k\}$ be a $C^\infty(\mathbf{R}^n)$ partition of unity on $\overline{\Omega}$ subordinate to $\{U_0, \ldots, U_k\}$, i.e., ψ_i is a $C^\infty(\mathbf{R}^n)$ function such that $0 \leq \psi_i \leq 1$, supp $\psi_i \subset U_i$ for all i, and $\Sigma_0^k \psi_i = 1$ on $\overline{\Omega}$.

Given any $u \in H^1(\Omega)$ we can then write $u = \Sigma_0^k \psi_i u = \Sigma_0^k u_i$ with $u_i = \psi_i u \in H^1(\Omega)$ and supp $u_i \subset U_i \cap \overline{\Omega}$. Since ϕ_i maps $U_i \cap \overline{\Omega}$ onto $\overline{\mathbf{R}}^n_+$, the composition $u_i \circ \phi_i^{-1}$ lies in $H^1(\mathbf{R}^n_+)$ for every $i \in \{0, \ldots, k\}$, so its trace $\gamma(u_i \circ \phi_i^{-1})$ is well-defined on $\mathbf{R}^{n-1}$. We now define the trace of $u \in H^1(\Omega)$ to be the restriction of the function $\Sigma_0^k \gamma(u_i \circ \phi_i^{-1}) \circ \phi_i$ to $\partial\Omega$, i.e.,

$$\gamma(u) = \sum_{i=1}^{k} \gamma(u_i \circ \phi_i^{-1}) \circ \tilde{\phi}_i$$

To check that this gives the desired result, let $x \in \overline{\Omega}$ be the point corresponding to any $\tilde{x} \in \partial\Omega$. x then lies in the intersection of a subcollection of $\{U_1, \ldots, U_k\}$, say $\{U_1, \ldots, U_j\}$ (after relabeling if necessary). Let $\phi_i(x) = y_i = (\tilde{y}_i, 0)$ be its image in $\mathbf{R}^n$ under ϕ_i, $1 \leq i \leq j$. Then $\tilde{\phi}_i(\tilde{x}) = \tilde{y}_i$ and

$$\gamma(u)(\tilde{x}) = \sum_{i=1}^{j} \gamma(u_i \circ \phi_i^{-1})(\tilde{y}_i)$$

If u is a C^1 function then so is $u_i \circ \phi_i^{-1}$ and, by Theorem 6.1, we can write

$$\gamma(u)(\tilde{x}) = \sum_{i=1}^{j} (u_i \circ \phi_i^{-1})\,(\tilde{y}_i, 0) = \sum_{i=1}^{j} u_i(x) = u(x)$$

Thus $\gamma(u)$ is the restriction to $\partial\Omega$ of u. Since $C^1(\overline{\Omega})$ is dense in $H^1(\Omega)$, γ has a unique continuous extension to $H^1(\Omega)$. Using Theorem 6.1, it is straightforward to check the other properties of γ stated in the theorem. □

Example 6.4 A generalized form of Green's first formula for a bounded regular domain Ω can now be given: For any pair of functions $u \in H^1(\Omega)$ and $v \in H^2(\Omega)$,

$$\int_{\Omega} (u\, \Delta v + \nabla u \cdot \nabla v)\, dx = \int_{\partial\Omega} u \partial_n v \, d\sigma \tag{6.2}$$

where $\nabla u \cdot \nabla v = \Sigma_{i=1}^{n} (\partial_i u)(\partial_i v)$, and η is the outward normal on $\partial\Omega$. Here, of course, the product $u\ \partial_\eta v$ in the integral over $\partial\Omega$ is to be interpreted as the trace $\gamma(u\partial_\eta v) = \gamma(u)\gamma(\partial_\eta v)$, which is well defined in $H^0(\partial\Omega)$ because u and $\partial_\eta v$ are in $H^1(\Omega)$.

To prove (6.2), we start with the formula for integration by parts,

$$\int_\Omega f\,\partial_i g\,dx + \int_\Omega g\,\partial_i f\,dx = \int_{\partial\Omega} fg\eta_i\,d\sigma$$

where η_i is the ith direction cosine of η. This formula is valid for any $f,g \in C_0^\infty(\overline{\Omega})$. Since $C_0^\infty(\overline{\Omega})$ is dense in $H^1(\Omega)$, it is also valid for any $f,g \in H^1(\Omega)$.

Now take any $u \in H^1(\Omega)$ and $v \in H^2(\Omega)$. By Theorem 6.2 the trace $\gamma(u)$ is in $H^0(\partial\Omega)$ and, since $\partial_i v \in H^1(\Omega)$, we also have $\gamma(\partial_i v) \in H^0(\partial\Omega)$. Therefore the trace of the normal derivative $\partial_\eta v = \Sigma(\partial_i v)\eta_i$ is well defined in $H^0(\partial\Omega)$. Replacing f by u and g by $\partial_i v$ and summing up over i from 1 to n, we obtain (6.2).

If u and v are both in $H^2(\Omega)$, we immediately obtain, by subtraction, Green's second identity

$$\int_\Omega (u\,\Delta v - v\,\Delta u)\,dx = \int_{\partial\Omega} (u\,\partial_\eta v - v\,\partial_\eta u)\,d\sigma$$

Theorem (6.2) may be generalized to treat boundary conditions where the derivatives of $u \in H^m(\Omega)$, $m \geqslant 1$, are L^2 functions on $\partial\Omega$. Note that the trace of $u \in H^0(\Omega)$ is meaningless since $\partial\Omega$ has measure zero in $\mathbf{R}^n$. When the boundary conditions involve the derivatives of $u \in H^m(\Omega)$, they are generally of order $\leqslant m - 1$. In that case we need to define the trace of $\partial^\alpha u$ for $|\alpha| \leqslant m - 1$. With $\partial^\alpha u \in H^1(\Omega)$ we can use Theorem 6.2 to define $\gamma(\partial^\alpha u)$ provided the diffeomorphisms ϕ_i are of class C^m. That means we should require $\partial\Omega$ to be a C^m hypersurface for the argument to go through.

In practice, the derivatives of u which are specified on $\partial\Omega$ are the *normal* derivatives $\partial_\eta^j u$, $0 \leqslant j \leqslant m - 1$, rather than $\partial^\alpha u$, $|\alpha| \leqslant m - 1$. When $\partial\Omega$ is a C^m hypersurface, $m \geqslant 1$, the (outward) unit normal vector η to $\partial\Omega$ is well defined, since it is well defined on the hypersurface $\partial\mathbf{R}_+^n$. Thus Theorem 6.2 may be restated in the more general form:

Theorem 6.3 Let Ω be a bounded regular domain in $\mathbf{R}^n$ whose boundary $\partial\Omega$ is a C^m hypersurface. Then

(i) For every $j \in \{0,1, \ldots ,m - 1\}$ there is a unique, continuous, linear mapping γ_j from $H^m(\Omega)$ to $H^{m-1-j}(\partial\Omega)$ such that $\gamma_j(u)$ is the restriction to $\partial\Omega$ of $\partial_\eta^j u$ when $u \in C^m(\overline{\Omega})$.

(ii) $u \in H^m(\Omega)$ lies in $H_0^m(\Omega)$ if and only if $\gamma_j(u) = 0$ for all $j = 0, \ldots, m - 1$.

(iii) $\gamma_j(H^m(\Omega))$ is dense in $H^{m-1-j}(\partial\Omega)$ for all $j = 0, \ldots, m - 1$.

Part (ii) merely states that, as a subspace of $H^m(\Omega)$, $H_0^m(\Omega)$ is completely characterized by the property that any $u \in H_0^m(\Omega)$ vanishes on $\partial\Omega$ together with all its derivatives up to order $m - 1$. This is really based on the inequality $\|\gamma(u)\|_{m-1} \leq \|u\|_m$, which generalizes (6.1), and the fact that any $u \in H_0^m(\Omega)$ is the limit of a sequence in $C_0^\infty(\Omega)$ which converges to u in the $\|\cdot\|_m$ norm. Since $\gamma_j(u_k) = 0$ for all k, $0 \leq j \leq m - 1$, we have

$$\|\gamma(u)\|_{m-1} = \|\gamma(u - u_k)\|_{m-1} \leq \|u - u_k\|_m \to 0 \text{ as } k \to \infty.$$

With $C_0^\infty(\overline{\mathbf{R}}_+^n)$ dense in $H^m(\mathbf{R}_+^n)$, $m \geq 0$, the proof of Theorem 6.2 shows that $u_i = \psi_i u \in H^1(U_i \cap \Omega)$ is the limit in $H^1(\Omega)$ of a sequence in $C_0^\infty(\overline{\mathbf{R}}_+^n)$, and consequently the same is true of $u = \Sigma u_i$. We can extend this to higher values of m and conclude that, under the hypothesis of Theorem 6.2, $C^\infty(\overline{\Omega})$ is dense in $H^m(\Omega)$ for all $m \geq 0$. In other words, the completion of $C^\infty(\overline{\Omega})$ in the $\|\cdot\|_m$ norm coincides with $H^m(\Omega)$. While this is always true when $\Omega = \mathbf{R}^n$, it is only true for a general open set if its boundary is smooth enough. From the definition of $H_{\text{loc}}^m(\Omega)$ we see that $H^m(\Omega) \subset H_{\text{loc}}^m(\Omega)$, ∂^α is a continuous mapping from $H^m(\Omega)$ to $H^{m-|\alpha|}(\Omega)$, and that multiplication by functions in $C^\infty(\overline{\Omega})$ is continuous on $H^m(\Omega)$. Thus a partial differential operator of order l, with coefficients in $C^\infty(\overline{\Omega})$, maps $H^m(\Omega)$ continuously into $H^{m-l}(\Omega)$.

Here we have only considered $H^m(\Omega)$ for integer values of m. For the treatment of $H^s(\Omega)$, with $s \in \mathbf{R}$, we can resort to Fourier transforms and use the order relation between these spaces to interpolate between the integer values of s. With this extension it is possible to sharpen the inequality $\|\tilde{\phi}\|_0 \leq \|\phi\|_1$, on which our definition of the trace is based, to $\|\tilde{\phi}\|_0 \leq \|\phi\|_{1/2}$, and thereby conclude that the trace operator is a continuous linear mapping of $H^m(\Omega)$ *onto* $H^{m-1/2}(\partial\Omega)$. The interested reader may wish to consult [12] or [13] on this point.

6.2 BOUNDARY VALUE PROBLEMS FOR ELLIPTIC EQUATIONS

Our aim in this section will be to illustrate how Sobolev spaces may be used to formulate boundary value problems for differential equations and to shed some light, through examples, on the form and properties of their solutions. Because of their close connection with Hilbert space methods, elliptic differential equations provide a natural setting for drawing such examples. However, no attempt will be made to give an account of the theory of elliptic equations, nor will the question of existence and uniqueness of their solutions be addressed, except as it comes up naturally in some instances.

Let Ω be a bounded regular domain in $\mathbf{R}^n$, $n \geqslant 2$. The appropriate boundary condition for the second order linear differential equation

$$\sum_{|\alpha|\leqslant 2} c_\alpha \partial^\alpha u = f \qquad \text{in } \Omega \quad c_\alpha \in C^\infty$$

is a system of first order differential equations which generally depends on the type of operator

$$L = \sum_{|\alpha|\leqslant 2} c_\alpha \partial^\alpha. \tag{6.3}$$

When L is elliptic, i.e. when $\Sigma_{|\alpha|=2}\, c_\alpha \xi^\alpha \neq 0$ for $\xi \neq 0$, it is possible by a suitable coordinate transformation to express L as

$$L = -\Delta + \sum_{|\alpha|=1} c_\alpha\, \partial^\alpha + c \tag{6.4}$$

where Δ is the Laplacian operator in $\mathbf{R}^n$ and the coefficients c_α and c are assumed to be real C^∞ functions on $\mathbf{R}^n$. The restriction to real functions is made at this point in order to avoid some unnecessary technical details. The boundary condition for $Lu = f$ then has the form

$$b_0 u + b_1\, \partial_\eta u = g \qquad \text{on } \partial\Omega \tag{6.5}$$

where η is the unit outward normal on $\partial\Omega$, which is well-defined at every point of the hypersurface $\partial\Omega$, and b_0 and b_1 are differentiable functions on $\partial\Omega$ (to the order permissible by the differentiable structure on $\partial\Omega$). f and g are given as part of the problem data and u is sought in some appropriate space $H^m(\Omega)$, the degree of smoothness of u in general being determined by that of f and g. The boundary condition (6.5) on u and $\partial_\eta u$ is, of course, to be interpreted in terms of their traces.

When $b_0 = 1$ and $b_1 = 0$ in (6.5) the resulting equation is called a *Dirichlet boundary condition*, and the system

$$\begin{aligned} Lu &= f \qquad \text{in } \Omega \\ u &= g \qquad \text{on } \partial\Omega \end{aligned} \tag{6.6}$$

is known as a *Dirichlet boundary value problem* for the operator L. The choice $b_0 = 0$ and $b_1 = 1$, on the other hand, yields the *Neumann boundary value problem*

$$\begin{aligned} Lu &= f \qquad \text{in } \Omega \\ \partial_\eta u &= g \qquad \text{on } \partial\Omega \end{aligned} \tag{6.7}$$

These two types of boundary conditions cover a wide class of physical problems, a typical example of which is to determine the temperature distribution in Ω when the temperature on the boundary $\partial\Omega$ is given (Dirichlet condition), or when the heat flow through $\partial\Omega$ is given (Neumann condition).

We shall start with the Dirichlet problem. The nonhomogeneous boundary condition in (6.6) may be made homogeneous by an appropriate transformation of u. If g is in $H^1(\partial\Omega)$ then it may be extended to a function in $H^1(\overline{\Omega})$, which we shall denote by G. This is accomplished by first noting that an $H^1(\mathbf{R}^{n-1})$ function can be extended into $\overline{\mathbf{R}}^n_+$ through multiplication by a function $\phi \in H^1([0,\infty))$ which satisfies $\phi(0) = 1$. Thus g may be extended locally from $U_i \cap \partial\Omega$ to $U_i \cap \overline{\Omega}$, and hence globally by the summation technique used in the proof of Theorem 6.2. If we define $U = u - G$ and take u in $H^1(\Omega)$, then $Lu = f \in H^{-1}(\Omega)$, and $LU = f - LG = F \in H^{-1}(\Omega)$. Dirichlet's problem then becomes: Given $F \in H^{-1}(\Omega)$, find $U \in H^1(\Omega)$ such that

$$\begin{aligned} LU &= F \qquad \text{on } \Omega \\ U &= 0 \qquad \text{on } \partial\Omega \end{aligned} \tag{6.8}$$

which is equivalent to seeking the solution of $LU = F$ in $H_0^1(\Omega)$.

The (formal) adjoint of $L = -\Delta + \Sigma_{|\alpha|=1} c_\alpha \partial^\alpha + c$ with respect to the bilinear form $\langle \cdot,\cdot \rangle$ is

$$L^* = -\Delta - \sum_{|\alpha|=1} \partial^\alpha c_\alpha + c$$

For L to be (formally) self-adjoint, we shall take $c_\alpha = 0$, so that

$$L = -\Delta + c \tag{6.9}$$

This simplifies the procedure without obscuring its fundamental features.

With the assumptions made on Ω, Green's first formula

$$\int_\Omega (v\,\Delta w + \nabla v \cdot \nabla w) = \int_{\partial\Omega} v\,\partial_\eta w \tag{6.10}$$

where $\nabla v \cdot \nabla w = \Sigma_{i=1}^n (\partial_i v)(\partial_i w)$, applies to any pair of functions $v \in H^1(\Omega)$, $w \in H^2(\Omega)$, as we have observed in Example 6.4, and the right-hand side of (6.10) is the integral over $\partial\Omega$ of the trace product $\gamma(v)\gamma(\partial_\eta w)$. If we choose $\gamma(v) = 0$ and $w = U$ in equation (6.10), then

$$(LU,v)_0 = (cU,v)_0 - (\Delta U,v)_0$$
$$= (cU,v)_0 + \sum_{i=1}^{n} (\partial_i U,\partial_i v)_0$$
$$= D(U,v)$$

where

$$D(U,v) = (cU,v)_0 + \sum_{i=1}^{n} (\partial_i U,\partial_i v)_0$$

is called the *Dirichlet integral* and D the *Dirichlet form* for L. The equation $LU = F$ therefore implies

$$(F,v)_0 = D(U,v) \tag{6.11}$$

and the Dirichlet problem (6.8) may be restated in yet another form: Given $F \in H^{-1}(\Omega)$, find $U \in H_0^1(\Omega)$ such that

$$D(U,v) = (F,v)_0 \tag{6.12}$$

for every $v \in H_0^1(\Omega)$.

If U is the difference between any two solutions to the Dirichlet problem, then $U \in H_0^1(\Omega)$, $LU = 0$ and

$$D(U,v) = (cU,v)_0 + \sum_{i=1}^{n} (\partial_i U,\partial_i v)_0 = 0$$

for every $v \in H_0^1(\Omega)$. If, furthermore, $c(x) \geqslant 0$ for all $x \in \Omega$ then, by choosing $v = U$, we see that $\partial_i U = 0$ on Ω for all $i = 1, \ldots ,n$. Therefore U, as an element of $H^1(\Omega)$, is constant on Ω. Since it vanishes on $\partial\Omega$, it is identically zero on $\overline{\Omega}$. Thus the solution of Dirichlet's problem for the operator $L = -\Delta + c$ is unique provided $c \geqslant 0$. In particular, it is unique for the Laplacian operator.

The Neumann problem (6.7) can be treated similarly to give

$$LU = F \qquad \text{in } \Omega$$
$$\partial_\eta U = 0 \qquad \text{on } \partial\Omega$$

with $U \in H^1(\Omega)$ and $F \in H^{-1}(\Omega)$. This leads to

$$D(U,v) = (F,v)_0 \qquad v \in H^1(\Omega)$$

Note here that U and v are both in $H^1(\Omega)$ rather than $H_0^1(\Omega)$, because the surface integral in Green's formula (6.10) vanishes due to the Neumann condition on U, and no boundary assumption on v is needed. Now the identity

$$(cU,U)_0 + \sum_{i=1}^{n} (\partial_i U, \partial_i U)_0 = 0$$

implies that $\partial_i U = 0$ on Ω provided $c \geqslant 0$ on Ω. If $c \neq 0$ then U also vanishes on Ω and we have uniqueness. But if $c = 0$ then U is a constant on Ω.

Thus the solution of Neumann's problem for Laplace's equation can only be determined up to an additive constant. This problem is also peculiar in that f and g cannot be specified arbitrarily because of the compatibility condition

$$\int_{\Omega} f + \int_{\partial\Omega} g = 0$$

which results from setting $v = 1$ in equation (6.10) and taking w as a solution of $-\Delta w = f$, $\gamma(\partial_\eta w) = g$.

The formulation of Dirichlet's problem in terms of equation (6.12) involves one order of differentiation less on U than in the original system (6.8), and we can take advantage of that to impose a higher degree of smoothness on F. Taking $v \in C_0^\infty(\Omega)$, we can write

$$\begin{aligned}(LU, v)_0 &= -\int_{\Omega} v\, \Delta U + \int_{\Omega} cUv \\ &= -\int_{\Omega} U\, \Delta v + \int_{\Omega} Ucv \\ &= (U, Lv)_0.\end{aligned}$$

Thus

$$D(U,v) = (U,Lv)_0 = (F,v)_0$$

for all $v \in C_0^\infty(\Omega)$, and hence for all $v \in H_0^1(\Omega)$, as $C_0^\infty(\Omega)$ is dense in $H_0^1(\Omega)$. Therefore we can take $F \in H^0(\Omega)$ in the formulation of Dirichlet's problem. In that case the differential equation $LU = F$ implies, by the local regularity theorem (5.9), that U is actually in $H_0^1(\Omega) \cap H_{loc}^2(\Omega)$.

Now we consider elliptic operators of higher order. It turns out, conveniently, that when the coefficients of the elliptic operator

$$L = \sum_{|\alpha| \leq k} c_\alpha \, \partial^\alpha$$

are real, its order k must be even. This follows from the observation that, if the polynomial

$$p(\xi) = \sum_{|\alpha| = k} c_\alpha \xi^\alpha$$

does not vanish on the unit sphere $|\xi| = 1$, then it cannot change sign, and hence $p(-\xi) = (-1)^k p(\xi)$ must have the same sign as $p(\xi)$, so that k must be even. We shall assume, therefore, that L is elliptic of order $2m$. With $u,v \in C_0^\infty(\Omega)$ we can use integration by parts to obtain

$$\begin{aligned} \int_\Omega (Lu)v &= \int_\Omega \Big(\sum_{|\alpha| \leq 2m} c_\alpha \, \partial^\alpha u \Big) v \\ &= \sum_{|\alpha|,|\beta| \leq m} \int_\Omega c_{\alpha\beta} \, (\partial^\alpha u)(\partial^\beta v), \end{aligned} \tag{6.13}$$

where $c_{\alpha\beta}$ are C^∞ functions obtained from the derivatives of c_α by Leibnitz' formula. The right-hand side of (6.13) is also called a Dirichlet integral (of order m) for the operator L, and denoted by $D(u,v)$. The Dirichlet form D, which is bilinear and bounded on the product space $C_0^\infty(\Omega) \times C_0^\infty(\Omega)$, may be extended by completion to $H_0^m(\Omega) \times H_0^m(\Omega)$ as

$$D(u,v) = \sum_{|\alpha|,|\beta| \leq m} (c_{\alpha\beta} \, \partial^\alpha u, \partial^\beta v)_0 \qquad u,v \in H_0^m(\Omega)$$

Now the Dirichlet problem for the elliptic operator L of order $2m$ is given by the system of equations

$$Lu = f \qquad \text{in } \Omega \tag{6.14}$$

$$\partial_\eta^j u = g_j \qquad \text{on } \partial\Omega, \; j = 0, \ldots, m-1 \tag{6.15}$$

Note that u has to satisfy m boundary conditions, which is half the order of L, hence the convenience of having the order of the operator L even.

We shall assume that there is a function $g \in H^m(\Omega)$ such that $\gamma(\partial_\eta^j g) = g_j \in H^{m-1-j}(\partial\Omega)$ for all $0 \leq j \leq m-1$. When $m = 2$ we have already seen how this function is constructed by extending g from $\partial\Omega$ to $\overline{\Omega}$. Now, if u is replaced by $u - g$ in (6.14) and (6.15), the boundary conditions become homogeneous, so there is no loss of generality in taking $g_j = 0$. Thus we seek a function u in $H_0^m(\Omega)$ such that

$$(u,L^*v)_0 = (f,v)_0 \qquad \text{for all } v \in C_0^\infty(\Omega)$$

where L^* is the adjoint of L.

Since $(u,L^*v)_0 = D(u,v)$, the function u must satisfy

$$D(u,v) = (f,v)_0$$

for every $v \in C_0^\infty(\Omega)$ and, by extension, for every $v \in H_0^m$. Hence Dirichlet's problem can finally be stated as: Given $f \in H^0(\Omega)$, find a function $u \in H_0^m(\Omega)$ such that $D(u,v) = (f,v)_0$ for every $v \in H_0^m(\Omega)$. But, once again, the local regularity theorem and the equation $Lu = f$ ensure that u will be in $H_0^m(\Omega) \cap H_{loc}^{2m}(\Omega)$.

If the equality $D(u,v) = (f,v)_0$ is to hold for every v in $C_0^\infty(\Omega)$ then the solution $u \in H_0^m$ of such an equation is a distributional, or *weak,* solution of the Dirichlet problem. A *strong* solution has continuous derivatives up to order $2m$ on Ω, and up to order m on $\overline{\Omega}$. If u is a weak solution which has continuous derivatives in Ω of order $2m$, then $(u,L^*v)_0 = (f,v)_0$, and integration by parts gives $(Lu,v)_0 = (f,v)_0$ for all $v \in C_0^\infty(\Omega)$. Since Lu is continuous, u is a strong solution of $Lu = f$. Furthermore, if u has continuous derivatives up to order m in $\overline{\Omega}$, then all its derivatives up to order $m - 1$ must vanish on $\partial\Omega$, by Theorem 6.3, since $u \in H_0^m(\Omega)$. Thus u satisfies the boundary equations in the strong sense. Every weak solution is therefore a strong solution if it has the required smoothness properties.

To construct the solution of the Dirichlet problem for Laplace's equation in the bounded regular domain Ω,

$$\Delta u = f \qquad \text{in } \Omega \tag{6.16}$$

$$u = g \qquad \text{on } \partial\Omega \tag{6.17}$$

we recall that any fundamental solution E on the Laplacian operator satisfies $\Delta E = \delta$ in $\mathbf{R}^n$, and that the convolution $f * E$ satisfies the differential equation (6.16), but not necessarily the boundary condition (6.17). It is reasonable, therefore, to expect that the (unique) solution of this pair of equations will result from a particular choice of the fundamental solution.

Since E is a C^∞ function in $\mathbf{R}^n - \{0\}$ which is locally integrable in $\mathbf{R}^n$, and $f \in L^2(\Omega)$ by assumption,

$$f * E(x) = \lim_{\varepsilon \to 0} \int_{\Omega_\varepsilon} f(\xi)\, E(\xi - x)\, d\xi$$

$$= \int_{\Omega} f(\xi)\, E(\xi - x)\, d\xi$$

$$= \int_{\Omega} E(\xi - x)\, \Delta u(\xi)\, d\xi,$$

where $\Omega_\varepsilon = \Omega - \bar{B}(x,\varepsilon) = \{\xi \in \Omega : |\xi - x| > \varepsilon\}$. Assuming that $u \in C^2(\Omega) \cap C^1(\bar{\Omega})$, we can apply Green's second formula to the pair of functions $u(\xi)$ and $E(\xi - x)$ in Ω_ε

$$\int_{\Omega_\varepsilon} [u(\xi)\, \Delta E(\xi - x) - E(\xi - x)\, \Delta u(\xi)] d\xi$$

$$= \int_{\partial\Omega_\varepsilon} [u(\xi)\, \partial_\eta E(\xi - x) - E(\xi - x)\partial_\eta u(\xi)]\, d\sigma$$

where $d\sigma$ is the Euclidean measure on the hypersurface $\partial\Omega_\varepsilon$. We have $\Delta E(\xi - x) = 0$ in Ω_ε and, since Ω is open, we can choose ε small enough so that $\bar{B}(x,\varepsilon) \subset \Omega$. Thus

$$\int_{\Omega_\varepsilon} E(\xi - x) f(\xi)\, d\xi = \int_{\partial\Omega} [E(\xi - x)\partial_\eta u(\xi) - u(\xi)\partial_\eta E(\xi - x)]\, d\sigma$$

$$- \int_{|\xi - x| = \varepsilon} [E(\xi - x)\, \partial_r u(\xi) - u(\xi)\, \partial_r E(\xi - x)]\, \varepsilon^{n-1}\, d\omega \tag{6.18}$$

where $r = |\xi - x|$ and $d\omega$ is the element of area on the unit sphere S_{n-1}.

Since the difference between any two fundamental solutions of Δ is a harmonic function in $\mathbf{R}^n$, the behavior of $E(\xi - x)$ for small values of $|\xi - x| = \varepsilon$ is determined by the expression in equation (4.40), which gives

$$\lim_{\varepsilon \to 0} \int_{|\xi - x| = \varepsilon} E(\xi - x)\, \partial_r u(\xi)\, \varepsilon^{n-1}\, d\omega = 0$$

$$\lim_{\varepsilon \to 0} \int_{|\xi - x| = \varepsilon} u(\xi)\, \partial_r E(\xi - x)\, \varepsilon^{n-1}\, d\omega = u(x)$$

Thus, in the limit as $\varepsilon \to 0$, equation (6.18) becomes

$$u(x) = \int_{\Omega} E(\xi - x) f(\xi)\, d\xi + \int_{\partial\Omega} [u(\xi)\partial_\eta E(\xi - x) - E(\xi - x)\partial_\eta u(\xi)]\, d\sigma. \tag{6.19}$$

In the Dirichlet problem we are not given $\partial_\eta u$ on the boundary and, in order to eliminate the last integral in (6.19), we can choose a fundamental solution which vanishes on $\partial\Omega$. Such a fundamental solution is necessarily unique and is called the *Green's function* $G(\xi,x)$ for the Dirichlet problem under consideration. The representation (6.19) then takes the form

$$u(x) = \int_{\Omega} G(\xi,x)f(\xi)\, d\xi + \int_{\partial\Omega} \partial_{\eta}G(\xi,x)g(\xi)\, d\sigma \tag{6.20}$$

Since $C^2(\Omega) \cap C^1(\overline{\Omega})$ is dense in $H^1(\Omega)$, equation (6.20) holds for any u in $H^1(\Omega)$.

The appropriate Green's function for the Neumann problem, on the other hand, should have a vanishing normal derivative on $\partial\Omega$ if u is to be expressed by the right-hand side of equation (6.19) in terms of the given data. Such a function is also referred to as *Neumann's function*.

While the fundamental solution of the Laplacian in $\Omega \subset \mathbf{R}^n$ is determined solely by the dimension n, the Green's function for Δ is determined by the shape of the domain Ω as well, since it has to vanish on $\partial\Omega$. In general it is not possible to construct G explicitly, even for the simple boundary value problem (6.16), (6.17), except when the geometry of the domain Ω is simple enough.

Perhaps the simplest domain we can consider is the half-space $\mathbf{R}^n_+$. When $n = 2$, we recall from (4.40) that the fundamental solution of Δ is $(1/2\pi)\log|x|$, hence Green's function has the form

$$G(\xi,x) = \frac{1}{2\pi}\log|\xi - x| + h(\xi,x)$$

where h is a harmonic function of both ξ and x in $\mathbf{R}^2_+ \times \mathbf{R}^2_+$ and $G(\xi,x) = 0$ on the line $\xi_2 = 0$. By choosing $h(\xi,x) = -(1/2\pi)\log|\bar{\xi} - x|$, where $\bar{\xi} = (\xi_1, -\xi_2)$ is the reflection of the point ξ in the line $\xi_2 = 0$, we meet both conditions on h (Figure 6.1). The required Green's function is therefore

$$G(\xi,x) = \frac{1}{2\pi}\log|\xi - x| - \frac{1}{2\pi}\log|\bar{\xi} - x|$$

The basic idea in this construction is that, starting with the fundamental solution $E(\xi - x)$, if we subtract its image $E(\bar{\xi} - x)$ in the boundary $\xi_2 = 0$, then the resulting function $G(\xi,x)$ is still a fundamental solution in the upper half-plane because $E(\bar{\xi} - x)$, as a function of ξ, is harmonic there for all $x \in \mathbf{R}^2_+$; and G vanishes on $\xi_2 = 0$ because it is an odd function of ξ_2. This method of constructing Green's function, known as the *method of images*, is used in the next two examples.

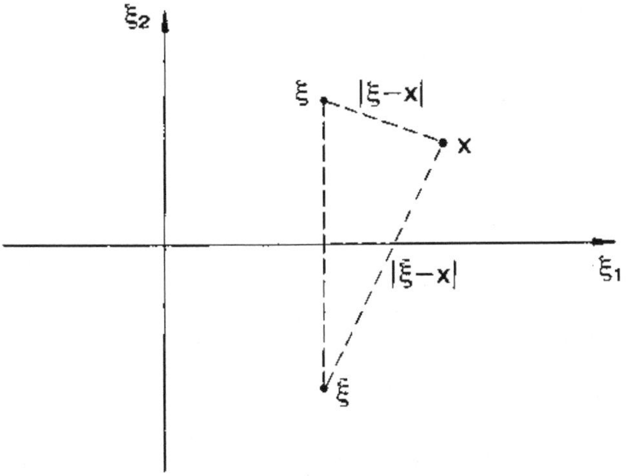

Figure 6.1

Example 6.5 When Ω is the unit ball $B = \{x \in \mathbf{R}^n : |x| < 1\}$ we can determine Green's function explicitly by subtracting from the fundamental solution $E(\xi - x)$, as expressed by (4.40), its reflection in the sphere $S_{n-1} = \partial B$. With its singularity outside $\bar{B}$, the reflected function $E(\xi/|\xi|^2 - x)$ is harmonic in B, and the two fundamental solutions coincide, and therefore cancel out, on ∂B. Thus, for $n = 2$, we have

$$G(\xi,x) = \frac{1}{2\pi}\left(\log|\xi - x| - \log\left|\frac{\xi}{|\xi|^2} - x\right|\right)$$
$$= \frac{1}{2\pi}\log\frac{r}{r'},$$

where $r = |\xi - x|$ and $r' = |\xi/|\xi|^2 - x|$ (Fig. 6.2).

Using polar coordinates, with $x = |x|e^{i\theta}$ and $\xi = |\xi|e^{i\phi}$, we find by straightforward computation that the normal derivative of G on the circle $|\xi| = 1$ is given by

$$\partial_\eta G(e^{i\phi}, |x|e^{i\theta}) = \frac{\partial}{\partial|\xi|}G(|\xi|e^{i\phi}, |x|e^{i\theta})\Bigg|_{|\xi|=1}$$
$$= \frac{1}{2\pi}\frac{1 - |x|^2}{1 - 2|x|\cos(\phi - \theta) + |x|^2}$$

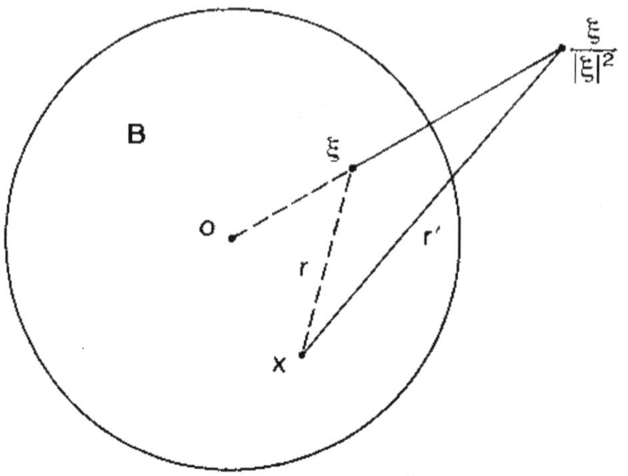

Figure 6.2

When $f = 0$, equation (6.20) gives the *Poisson Integral Representation* for u,

$$u(x) = \frac{1}{2\pi}\int_0^{2\pi} \frac{1 - |x|^2}{1 - 2|x|\cos(\phi - \theta) + |x|^2}\, g(e^{i\phi})\, d\phi. \tag{6.21}$$

This formula gives the value of the harmonic function u inside the unit disc in terms of its values on the boundary, and yields the well-known mean-value property

$$u(0) = \frac{1}{2\pi}\int_0^{2\pi} g(e^{i\phi})\, d\phi$$

$$= \frac{1}{2\pi}\int_0^{2\pi} u(e^{i\phi})\, d\phi$$

from which we can deduce the *maximum principle* for harmonic functions. This states that, if u is harmonic in B and continuous on $\overline{B}$, the maximum value of u cannot be attained in the interior of the disc B, unless of course u is a constant.

Example 6.6 To construct Green's function for the Dirichlet problem in the ball $B = \{x \in \mathbf{R}^n : |x| < R\}$ with n $\geq$ 3, note that the inverse image of $\xi \in B$ with respect to the sphere ∂B is $R^2\xi/|\xi|^2$, which lies outside $\overline{B}$ on the line through 0 and ξ. With $r = |\xi - x|$ and $r' = |R^2\xi/|\xi|^2 - x|$, the expression

$$G(\xi,x) = -\frac{n-2}{\omega_{n-1}}\left[\frac{1}{r^{n-2}} - \frac{1}{r'^{n-2}}\right]$$

is a fundamental solution of Δ in B which vanishes on the sphere $|\xi| = R$. It is therefore the required Green's function. By computing $\partial_\eta G$ on ∂B we can represent harmonic functions in B by a Poisson integral analogous to (6.21) when $f = 0$.

If we let $R \to \infty$, then $1/r'^{n-2} \to 0$ in the above expression for $G(\xi,x)$, and hence

$$G(\xi,x) = E(|\xi - x|) = -\frac{n-2}{\omega_{n-1}|\xi - x|^{n-2}}$$

Assuming, moreover, that $u(x) \to 0$ as $|x| \to \infty$, i.e., taking $g = 0$ in equation (6.20), we arrive at

$$u(x) = -\frac{n-2}{\omega_{n-1}}\int_{\mathbf{R}^n}\frac{f(\xi)}{|\xi - x|^{n-2}}\,d\xi \tag{6.22}$$

provided this integral, which is in fact the convolution $f * E$, exists. This result generalizes equation (3.19) for the representation of a Newtonian potential function.

Note that equation (6.20) provides an additional term to the expression (6.22) when the potential u is not 0 at ∞. Take, for example, the case where f is a bounded function with compact support and $u(x) \to u_0$ = constan as $|x| \to \infty$. Equation (6.20) gives

$$u(x) = -\frac{n-2}{\omega_{n-1}}\int\frac{f(\xi)}{|\xi - x|^{n-2}}\,d\xi + u_0,$$

which is to be expected since the integral tends to 0 as $|x| \to \infty$.

Example 6.7 Let $\mu \geq 0$ be a Radon measure in $\mathbf{R}^n$, $n \geq 3$, with compact support K. With $E(x) = -(n-2)/\omega_{n-1}|x|^{n-2}$, the convolution

$$u(x) = \mu * E(x) = \int E(|\xi - x|)\,d\mu(\xi)$$

is a well-defined function in $\mathbf{R}^n$ which represents the (generalized) Newtonian potential associated with the measure μ. Clearly $\Delta u = 0$ outside K, and u is therefore harmonic in $\mathbf{R}^n - K$.

If $\mu \geqslant 0$ does not necessarily have compact support then we define μ_k to be the restriction of μ to the ball $|x| < k$. Each $\mu_k * E$ is then well defined and, since E is negative, $\mu_k * E$ is a decreasing sequence of potentials. If its limit $u = \mu * E$ exists, then

$$\Delta u = \mu * \Delta E = \mu * \delta = \mu \geqslant 0$$

Conversely, suppose that u is a distribution such that $\Delta u \geqslant 0$. By Example 2.25, Δu is a positive measure μ in $\mathbf{R}^n$. If $v = \mu * E$ is well-defined, then

$$\Delta v = \mu = \Delta u$$

$$\Delta(u - v) = 0$$

By Theorem 4.11 $u - v$ is a harmonic function in $\mathbf{R}^n$, and hence

$$u(x) = \mu * E(x) + h(x)$$

$$= -\frac{n-2}{\omega_{n-1}} \int \frac{\Delta u(\xi)}{|\xi - x|^{n-2}} d\xi + h(x)$$

almost everywhere in $\mathbf{R}^n$, where h is a harmonic function in $\mathbf{R}^n$.

A distribution u in Ω is said to be *subharmonic* if Δu is a positive Radon measure in Ω, i.e., if Δu is a distribution of order 0 and $\Delta u \geqslant 0$. This example, therefore, gives the general representation of a subharmonic functions in $\mathbf{R}^n$, $n \geqslant 3$, which is bounded above by a harmonic function.

The iterated Laplacian operator Δ^2 is only slightly more complicated to deal with, as the next example shows, and points the way to the treatment of higher powers of Δ.

Example 6.8 The equation

$$\Delta^2 u = f \qquad \text{in } \Omega \subset \mathbf{R}^2 \tag{6.23}$$

has to be supplemented with appropriate boundary conditions of the type (6.15) in order to constitute a well-posed boundary value problem. These can be anticipated by replacing equation (6.23) by the pair of equations

$$\Delta u = v \tag{6.24}$$

$$\Delta v = f \tag{6.25}$$

in Ω. In accordance with our previous discussion, equation (6.25) may be solved uniquely for v if either v or $\partial_\eta v$ is specified on $\partial\Omega$, with the appropriate stipulation when $f = 0$ in the Neumann problem. Once v has been determined in Ω, equation (6.24) can also be solved under either the Dirichlet or the Neumann boundary conditions.

Thus equation (6.23) can be solved uniquely for $u \in H^3(\Omega) \cap H^4_{\text{loc}}(\Omega)$, given $f \in H^0(\Omega)$ and, for example, the pair of boundary conditions on $\partial\Omega$

$$u = g_1 \qquad \Delta u = g_2$$

For any pair of functions $u, v \in C^4(\Omega) \cap C^3(\overline{\Omega})$, we obtain the following equation after two successive applications of Green's first identity:

$$\int_\Omega u\, \Delta^2 v = \int_\Omega (\Delta u)\, (\Delta v) + \int_{\partial\Omega} [u\, \partial_\eta\, \Delta v - (\partial_\eta u)\, \Delta v] \tag{6.26}$$

By interchanging u and v, and subtracting, we arrive at

$$\int_\Omega (u\, \Delta^2 v - v\, \Delta^2 u) = \int_{\partial\Omega} [u\, \partial_\eta \Delta v - (\partial_\eta u)\Delta v + (\partial_\eta v)\, \Delta u - v\, \partial_\eta \Delta u]$$

According to equation (4.61), a fundamental solution of Δ^2 in $\mathbf{R}^2$ is given by

$$E(\xi - x) = \frac{1}{8\pi} |\xi - x|^2 \log |\xi - x|$$

Let $G(\xi,x) = E(\xi - x) + h(\xi,x)$, where h satisfies the *biharmonic equation* $\Delta^2 h = 0$ in Ω. Replacing Ω in equation (6.26) by $\Omega_\varepsilon = \{\xi \in \Omega : |\xi - x| > \varepsilon\}$ and $v(\xi)$ by $G(\xi,x)$, and taking the limit as $\varepsilon \to 0$, we obtain

$$u(x) = \int_\Omega G(\xi,x) f(\xi)\, d\xi + \int_{\partial\Omega} [u(\xi)\, \partial_\eta \Delta G(\xi,x) - \partial_\eta u(\xi)\, \Delta G(\xi,x)$$
$$+ \partial_\eta G(\xi,x)\, \Delta u(\xi) - G(\xi,x)\, \partial_\eta \Delta u(\xi)]\, d\sigma \tag{6.27}$$

where u may now be taken in $H^3(\Omega)$.

We see from equation (6.27) that the appropriate homogeneous boundary conditions on G, corresponding to $u = g_1$ and $\Delta u = g_2$, are

$$G = \Delta G = 0$$

Equation (6.27) clearly admits other possibilities for prescribing the boundary conditions on u and, in consequence, their homogeneous counterparts on G.

6.3 THE HEAT EQUATION

In the class of nonelliptic second order differential operators, the two typical examples, which we have already encountered, are

$$\partial_t - \Delta \qquad \text{(the heat operator)} \tag{6.28}$$

$$\partial_t^2 - \Delta \qquad \text{(the wave operator)} \tag{6.29}$$

where t denotes the time variable and Δ is the Laplacian operator in $\mathbf{R}^n$. The Laplacian can actually be replaced by any other elliptic operator in $\mathbf{R}^n$, but that would only add to the technical details and might obscure the important features which we wish to emphasize. So we shall restrict ourselves for the remainder of this chapter to a brief look at the type of boundary value problems which come with these two operators and the properties of their solutions. In the process, we shall establish some uniqueness and smoothness results for these solutions, indicate how they can be constructed for some simple domains, and tie up some loose ends in the examples of Section 3.6.

A fundamental solution of the heat operator has already been derived in Section 4.7, and is given by (4.33) as

$$E(x,t) = (4\pi t)^{-(1/2)n} H(t) \exp(-|x|^2/4t) \tag{6.30}$$

It is a simple matter to verify that this is a C^∞ function in $\mathbf{R}^{n+1} - \{0\}$. The wave operator, on the other hand, has a highly discontinuous fundamental solution. In Section 4.7 it was given for $n = 1$, 2, and 3, respectively, by

$$\tfrac{1}{2}[H(x + t) - H(x - t)]$$

$$\frac{1}{2\pi} H(t - |x|) \frac{1}{2\pi\sqrt{t^2 - |x|^2}}$$

$$\frac{1}{4\pi t} \delta_{|x|-t}$$

It is curious how the one additional time derivative in the wave operator can result in the total loss of all the smoothness properties of the fundamental solution.

The characteristic polynomials of the operators (6.28) and (6.29) are given by $|x|^2$ and $|x|^2 - t^2$, respectively, and since both polynomials vanish at points outside the origin $(t,x) = 0$, we conclude that neither operator is elliptic. However, there is another classification of differential operators, based on the smoothness properties of their solutions, which brings out the difference between $\partial_t - \Delta$ and $\partial_t^2 - \Delta$, and actually places the first in almost the same class as the Laplacian Δ.

Definition The linear partial differential operator L is said to be *hypoelliptic* in $\Omega \subset \mathbf{R}^n$ if, given any open subset ω of Ω, the solution of $Lu = f$ in $\mathcal{D}'(\omega)$ is a $C^\infty(\omega)$ function whenever $f \in C^\infty(\omega)$.

Two immediate conclusions follow from this definition:

(i) From Corollary 1 to the local regularity theorem (5.9), every elliptic operator is hypoelliptic.
(ii) If L is hypoelliptic then for every open set $\Omega \subset \mathbf{R}^n$ the solutions of $Lu = 0$ in $\mathcal{D}'(\Omega)$ are C^∞ functions in Ω. In particular every fundamental solution of L in $\mathbf{R}^n$ is a C^∞ function in $\mathbf{R}^n - \{0\}$.

Example 6.9 Let $L = \sum_{|\alpha| \leq m} c_\alpha \partial^\alpha$ be a differential operator in $\mathbf{R}^n$ with constant coefficients.

(i) If there exists a distribution $T \in \mathscr{E}'$ such that $\delta - LT \in C_0^\infty$, then $Lu \in C^\infty$ implies $u \in C^\infty$; for if $\delta - LT = \psi \in C_0^\infty$, then for any $u \in \mathcal{D}'$ we have

$$\begin{aligned} u &= u * \delta \\ &= u * LT + u * \psi \\ &= Lu * T + u * \psi. \end{aligned}$$

Now if $Lu \in C^\infty$ then $Lu * T \in C^\infty$. Since $u * \psi \in C^\infty$, this implies that $u \in C^\infty$.

(ii) If there exists a distribution $T \in \mathcal{D}' \cap C^\infty(\mathbf{R}^n - \{0\})$ such that $LT - \delta \in C^\infty$, then again $Lu \in C^\infty$ implies that $u \in C^\infty$.

To show this, let $LT - \delta = f \in C^\infty$ and choose $\phi \in \mathcal{D}$ such that $\phi = 1$ on $|x| \leq 1$. By Leibnitz' formula,

$$\begin{aligned} L(\phi T) &= \sum_{|\alpha| \leq m} c_\alpha \sum_{\beta \leq \alpha} \binom{\alpha}{\beta} \partial^\beta \phi \, \partial^{\alpha - \beta} T \\ &= \phi LT + \psi \end{aligned}$$

where ψ is the sum of the terms in which $\beta \neq 0$.

Since T is a C^∞ function in $\mathbf{R}^n - \{0\}$, so is ψ. But since $\phi = 1$ on $\bar{B}(0,1)$, $L(\phi T) = \phi LT = LT$ on $B(0,1)$, and therefore $\psi = 0$ on a neighborhood of 0. Hence $\psi \in C_0^\infty(\mathbf{R}^n)$.

Now

$$\begin{aligned}\phi LT &= \phi(\delta + f)\\ &= \delta + \phi f\\ L(\phi T) &= \delta + \phi f + \psi\end{aligned}$$

Therefore $\delta - L(\phi T) \in C_0^\infty$ and, with supp (ϕT) compact, the operator L satisfies the hypothesis of part (i).

This example implies, in particular, that every linear differential operator with constant coefficients which has a fundamental solution in $C^\infty(\mathbf{R}^n - \{0\})$ has the property that $u \in C^\infty$ whenever $Lu \in C^\infty$. This is not quite the hypoelliptic property, which requires that $u \in C^\infty(\omega)$ whenever $Lu \in C^\infty(\omega)$ for *every* open set ω in $\mathbf{R}^n$.

The remainder of this section is devoted to the heat equation. The wave equation equation will be taken up in Section 6.4.

The Heat Operator: Whereas the solutions of $\Delta u = 0$ in $\mathcal{D}'(\Omega)$ are analytic, being harmonic in the classical sense, this is not the case with $\partial_t - \Delta$. In fact, the fundamental solution (6.30) satisfies $(\partial_t - \Delta)E = 0$ in $\mathbf{R}^{n+1} - \{0\}$, where it is a C^∞ function, but E is clearly not analytic in $\mathbf{R}^{n+1} - \{0\}$ as it vanishes identically in the lower half-space $\mathbf{R}^n \times (-\infty,0)$ but not in $\mathbf{R}^n \times (0,\infty)$. But, as we shall now establish, $\partial_t - \Delta$ is hypoelliptic.

Theorem 6.4 If Ω is an open set in $\mathbf{R}^{n+1}$, $L = \partial_{n+1} - \Delta$, and u is a distribution in Ω which satisfies the homogeneous heat equation $Lu = 0$, then u is a C^∞ function in Ω.

Proof. Let $z_0 = (x_0,t_0)$ be any point in Ω and choose $\varepsilon > 0$ so that $\overline{B}(z_0,4\varepsilon) \subset \Omega$. It is sufficient to show that $u \in C^\infty(B(z_0,\varepsilon))$. Let $\psi \in C_0^\infty(\Omega)$ with $\psi = 1$ on $B(z_0,4\varepsilon)$. Then $\psi u \in \mathcal{E}'(\Omega)$ and its convolution with the fundamental solution E of L exists, so we can write

$$\psi u = \psi u * LE = v * E \tag{6.31}$$

where $v = L(\psi u)$ is also in $\mathcal{E}'(\Omega)$. Since $\psi = 1$ on $B(z_0,4\varepsilon)$, we have $v = Lu$ on $B(z_0, 4\varepsilon)$; and since $Lu = 0$ on Ω by assumption, it follows that $v = 0$ on $B(z_0,4\varepsilon)$ and hence

$$\text{supp } v \subset B^c(z_0,4\varepsilon).$$

Now we choose $\phi \in C_0^\infty(B(0,2\varepsilon))$ such that $\phi = 1$ on $B(0,\varepsilon)$, and rewrite (6.31) as

$$
\begin{aligned}
\psi u &= v * E \\
&= v * (\phi E) + v * (1 - \phi)E.
\end{aligned}
\tag{6.32}
$$

The first term on the right-hand side of (6.32) is zero on $B(z_0,\varepsilon)$ because

$$
\begin{aligned}
\operatorname{supp}(v * \phi E) \cap B(z_0,\varepsilon) &\subset [\operatorname{supp} v + \operatorname{supp}(\phi E)] \cap B(z_0,\varepsilon) \\
&\subset [B^c(z_0,4\varepsilon) + B(0,2\varepsilon)] \cap B(z_0,\varepsilon) \\
&= \emptyset
\end{aligned}
$$

In the second term the product $(1 - \phi)E$ is in $C^\infty(\mathbf{R}^{n+1})$ because E is a C^∞ function in $\mathbf{R}^{n+1} - \{0\}$, while $(1 - \phi)$ is a C^∞ function which vanishes in a neighborhood of 0. Hence $v * (1 - \phi)E \in C^\infty(\mathbf{R}^{n+1})$ by Theorem 3.5 and the fact that $v \in \mathscr{E}'$. But on $B(z_0,\varepsilon)$ we have $u = \psi u = v * (1 - \phi)E$. □

Theorem 6.5 The operator $L = \partial_{n+1} - \Delta$ is hypoelliptic in $\mathbf{R}^{n+1}$.

Proof. Let Ω be an open set in $\mathbf{R}^{n+1}$, $f \in C^\infty(\Omega)$ and $Lu = f$ for some $u \in \mathscr{D}'(\Omega)$. For any $z_0 \in \Omega$ we can choose $\varepsilon > 0$ so that $B(z_0,3\varepsilon) \subset \Omega$, and we shall show that $u \in C^\infty(B(z_0,\varepsilon))$.

Let $\psi \in C_0^\infty(\Omega)$ with $\psi = 1$ on $\bar{B}(z_0,2\varepsilon)$ and define $v = L(\psi u)$. Since $Lu \in C^\infty(\Omega)$, v is a C^∞ function in $B(z_0,2\varepsilon)$. Let $\phi \in C_0^\infty(B(z_0,2\varepsilon))$ and $\phi = 1$ on $B(z_0,\varepsilon)$. Taking E to be the fundamental solution (6.30) of L, and noting that $v \in \mathscr{E}'$,

$$
\begin{aligned}
\psi u &= \psi u * LE \\
&= v * E \\
&= \phi v * E + (1 - \phi)v * E
\end{aligned}
\tag{6.33}
$$

The first term on the right-hand side of (6.33) is in $C^\infty(\mathbf{R}^{n+1})$ because $\phi v \in C_0^\infty(B(z_0,2\varepsilon))$ and $E \in \mathscr{D}'(\mathbf{R}^{n+1})$. The second term satisfies the equation

$$
\begin{aligned}
L[(1 - \phi)v * E] &= (1 - \phi)v \\
&= 0 \qquad \text{on } B(z_0,\varepsilon)
\end{aligned}
$$

By Theorem 6.4 it must be a C^∞ function in $B(z_0,\varepsilon)$. The same is therefore true of ψu, which coincides with u on $B(z_0,\varepsilon)$. □

Having used Ω to denote an open set in $\mathbf{R}^{n+1} = \{(x,t) : x \in \mathbf{R}^n, t \in \mathbf{R}\}$ in our proof of the hypoellipticity of the operator $\partial_{n+1} - \Delta$, we shall find it more convenient in discussing boundary value problems for the heat operator to separate the time coordinate and, for physical reasons, restrict the variable t to $[0,\infty)$. Ω and $\Omega \times (0,\infty)$ will henceforth denote open sets in $\mathbf{R}^n$ and $\mathbf{R}^{n+1}$, respectively. The convolution product, as in equation (6.31) for example, was taken in $\mathbf{R}^{n+1}$. But now, unless otherwise indicated, $*$ will denote convolution in $\mathbf{R}^n$.

The equation

$$(\partial_t - \Delta)u = f \qquad \text{in } \Omega \times (0,\infty) \tag{6.34}$$

describes the flow of heat in a homogeneous isotropic medium which occupies the region Ω. u denotes the temperature at $(x,t) \in \Omega \times (0,\infty)$ and f is the external component of the heat flow due to the presence of a heat source (or sink) at (x,t). f is actually a measure of the volume density of heat sources in Ω at any instant $t > 0$. In the absence of such sources, equation (6.34) takes the form

$$(\partial_t - \Delta)u = 0 \qquad \text{in } \Omega \times (0,\infty) \tag{6.35}$$

and the net heat gain (or loss) can only result from the flow across the boundary $\partial\Omega$. We know, from Theorem 6.4, that equation (6.35) can only have C^∞ solutions in Ω.

From a physical point of view, the temperature distribution in Ω for $t > 0$ is going to depend on the initial temperature at $t = 0$, and therefore the *initial condition*

$$u = g \qquad \text{on } \Omega \times \{0\}, \tag{6.36}$$

where g is a given function on Ω, is necessary for determining the subsequent values of u. Equation (6.36) is really a boundary condition on the surface $\Omega \times \{0\}$ of the domain $\Omega \times (0,\infty) \subset \mathbf{R}^{n+1}$, and is only distinguished by the special role played by the time coordinate. The pair of equations (6.34) and (6.36) is also called a *Cauchy problem* for the heat operator.

The Heat Equation in $\mathbf{R}^n \times (0,\infty)$: In Example 3.22 we found that a solution of the heat equation (6.35) in $\mathbf{R} \times (0,\infty)$ with the initial condition (6.36) is given by the integral (3.23). To justify that result and to generalize it to $\mathbf{R}^n \times (0,\infty)$ let us assume that $g \in L^1(\mathbf{R}^n)$, and that E is the fundamental solution given by (6.30). We already know that the solution u of equation (6.35) is a C^∞ function in $\mathbf{R}^n \times (0,\infty)$ and we now further assume that, as a function of x, it lies in $\mathcal{S}'(\mathbf{R}^n)$. Applying the Fourier transformation to equations (6.35) and (6.36) we obtain the pair of equations

$$\partial_t \hat{u}(\xi,t) + |\xi|^2 \hat{u}(\xi,t) = 0 \qquad (\xi,t) \in \mathbf{R}^n \times (0,\infty)$$
$$\hat{u}(\xi,0) = \hat{g}(\xi) \qquad \xi \in \mathbf{R}^n$$

whose unique solution is

$$\hat{u}(\xi,t) = \hat{g}(\xi)\, e^{-|\xi|^2 t} \tag{6.37}$$

Using equations (4.22) and example (4.2), we can write

$$\begin{aligned} u(x,t) &= \mathcal{F}^{-1}(\hat{g}\, e^{-|\cdot|^2 t})(x) \\ &= [g * \mathcal{F}^{-1}(e^{-|\cdot|^2 t})](x) \\ &= [g * (4\pi t)^{-n/2}\, e^{-|\cdot|^2/4t}](x) \\ &= (g * E)(x,t), \end{aligned}$$

where the convolution of $g \in L^1(\mathbf{R}^n)$ with $E(\cdot,t) \in \mathcal{S}(\mathbf{R}^n)$ for $t > 0$ is a well-defined operation which yields an L^1 function to which formula (4.22) applies (see Example 3.5). Thus

$$u(x,t) = (4\pi t)^{-n/2} \int_{\mathbf{R}^n} g(\xi)\, e^{-|x-\xi|^2/4t}\, d\xi \qquad (x,t) \in \mathbf{R}^n \times (0,\infty) \tag{6.38}$$

and it is left to the reader to verify that this function satisfies equation (6.35) and that $u(x,t) \to g(x)$ as $t \to 0^+$ at every x in $\mathbf{R}^n$.

In this connection, the following points are worth noting:

(i) $|u(x,t)| \to 0$ as $|(x,t)| \to \infty$, which means that the temperature at every point in $\mathbf{R}^n$ tends to 0 as $t \to \infty$. That is because the temperature at ∞ is taken to be 0, and heat always flows from points of higher temperature to points of lower temperature.

(ii) The uniqueness of the solution (6.37) implies that the solution of the initial value problem (6.35), (6.36) is unique in $\mathcal{S}'$.

(iii) Since $E(x,t) \in \mathcal{S}(\mathbf{R}^n)$ for all $t > 0$ it is possible to extend the representation (6.38), which holds for $g \in L^1(\mathbf{R}^n)$, to any g in $\mathcal{S}'(\mathbf{R}^n)$ by the formula $u = g * E$. In that case $g * E$ belongs to $\mathcal{S}'(\mathbf{R}^n)$ for every $t > 0$ (see Exercises 4.13 and 4.14), so we can assume that u is a tempered distribution in $\mathbf{R}^n$ which depends continuously on the parameter $t \in (0,\infty)$, in the sense that $\langle u(t),\phi\rangle$ is a continuous function of t for every $\phi \in \mathcal{S}(\mathbf{R}^n)$. Furthermore, we require that $u(t) \to g$ as $t \to 0^+$.

(iv) For the sake of clarity we shall sometimes write $u_x(t)$ to imply that u is a distribution in the variable $x \in \mathbf{R}^n$ and a function in the variable $t \in$

$(0,\infty)$, or, more precisely, that u is a mapping from $(0,\infty)$ to $\mathcal{S}'(\mathbf{R}^n)$. This is useful when we wish to keep track of the different variables in, say, a Fourier transformation or a convolution operation. For a fixed t_0, $u(t_0)$ naturally denotes the distribution u when $t = t_0$.

As a fundamental solution for the heat operator, E determines a solution in $\mathbf{R}_+^{n+1}$ of the nonhomogeneous equation (6.34) through the convolution product in $\mathbf{R}^{n+1}$,

$$u = f *_{xt} E$$

a formula which makes sense when $f \in \mathcal{E}'(\mathbf{R}_+^{n+1})$. Here $*_{xt}$ denotes convolution with respect to the variables x and t in $\mathbf{R}^{n+1}$, as distinct from the usual convolution $*$ in $\mathbf{R}^n$. We shall also use $*_t$ to denote convolution with respect to t in $\mathbf{R}$. If f, in addition, is locally integrable, this solution takes the form

$$u(x,t) = \int_0^\infty \int_{\mathbf{R}^n} f(\xi,\tau)E(x - \xi, t - \tau)\, d\xi\, d\tau$$

$$= \int_0^t [4\pi(t - \tau)]^{-n/2} \int_{\mathbf{R}^n} f(\xi,\tau) \exp\,[-|x - \xi|^2/4(t - \tau)]\, d\xi\, d\tau \qquad (6.39)$$

The integral representation (6.39) clearly shows that u vanishes on the boundary $t = 0$. If f is taken to be 0 on $\mathbf{R}_-^{n+1} = \mathbf{R}^n \times (-\infty,0)$, u can be extended into the lower half-space as 0. Since E is in $\mathcal{S}(\mathbf{R}^n)$ for every $t > 0$, equation (6.39) also allows f, in its dependence on x, to be extended to $\mathcal{S}'(\mathbf{R}^n)$.

Combining (6.38) and (6.39), we arrive at

$$u = f *_{xt} E + g * E \qquad (6.40)$$

as the solution of equation (6.34) subject to the initial condition (6.36). In equation (6.40) the first convolution is taken with respect to both x and t, while the second is taken with respect to x only.

The Heat Equation in a Bounded Domain: When Ω is bounded the initial condition (6.36) is not sufficient to determine the solution of the heat equation (6.35) uniquely. It then becomes necessary to prescribe a boundary condition on $\partial\Omega \times (0,\infty)$, such as $u = h$, which specifies the temperature on $\partial\Omega$ for all $t \geq 0$, or $\partial_\eta u = h$, which specifies the heat flow across $\partial\Omega$. When the boundary is insulated then the condition $\partial_\eta u = 0$ applies. We recognize these boundary conditions as being of the Dirichlet and Neumann type, respectively, and we can also have a situation where the temperature is prescribed on part of the boundary and the heat flow on the other. When the boundary condition is given on u, the resulting system of equations

$$\begin{aligned} \partial_t u &= \Delta u && \text{in } \Omega \times (0,\infty) \\ u(0) &= g && \text{on } \Omega \\ u &= h && \text{on } \partial\Omega \times [0,\infty) \end{aligned} \tag{6.41}$$

is basically a boundary value problem in the "cylinder" $\Omega \times (0,\infty)$ with u prescribed by the initial condition on the "base" $\Omega \times \{0\}$ and by $u = h$ on the rest of the boundary $\partial\Omega \times [0,\infty)$.

The uniqueness of the solution to the boundary value problem (6.41) follows from a maximum principle similar to the maximum principle for harmonic functions.

Theorem 6.6 Let Ω be a bounded domain in $\mathbf{R}^n$ and t_1 be a finite positive number. If u is a continuous function on $\overline{\Omega} \times [0,t_1]$ which satisfies $(\partial_t - \Delta)u = 0$ in $\Omega \times (0,t_1)$, then u assumes its maximum value on the boundary $\Omega \times \{0\} \cup \partial\Omega \times [0,t_1]$.

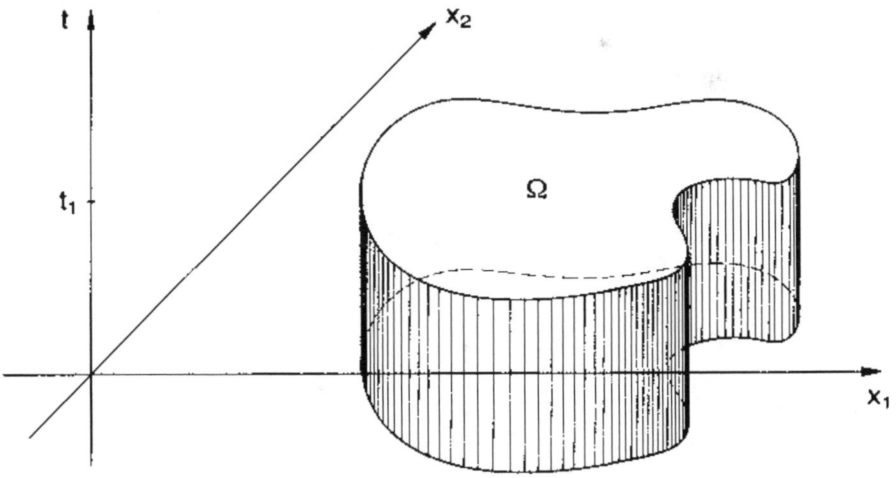

Figure 6.3

Proof. For any $\varepsilon > 0$, let $v(x,t) = u(x,t) + \varepsilon|x|^2$. Since u is a C^∞ function in $\Omega \times (0,t_1)$ by Theorem 6.5, the same is true of v, and

$$(\partial_t - \Delta)v = -2n\varepsilon < 0 \tag{6.42}$$

Suppose v has a maximum value at $z_0 \in \Omega \times (0,t_1)$. Then $\partial_t v = 0$ and $\Delta v \leqslant 0$ at z_0, which contradicts the inequality (6.42). Thus the maximum value of v in $\overline{\Omega} \times [0,t_1]$ can only occur on $\Omega \times \{0\} \cup \partial\Omega \times [0,t_1] \cup \Omega \times \{t_1\}$.

In $\overline{\Omega} \times [0,t_1 - \varepsilon]$, where we now assume that $0 < \varepsilon < t_1$, the maximum value of v can only occur on $\Omega \times \{0\} \cup \partial\Omega \times [0,t_1 - \varepsilon] \cup \Omega \times \{t_1 - \varepsilon\}$. Similarly, if max v in $\overline{\Omega} \times [0,t_1 - \varepsilon]$ occurs at $z_0 \in \Omega \times \{t_1 - \varepsilon\}$, then $\partial_t v \geqslant 0$ and $\Delta v \leqslant 0$ at z_0, which again contradicts the inequality (6.42). Therefore

$$\max_{\overline{\Omega}\times[0,t_1-\varepsilon]} u \leqslant \max_{\overline{\Omega}\times[0,t_1-\varepsilon]} v$$

$$\leqslant \max_{\Omega\times\{0\}\cup(\partial\Omega\times[0,t_1-\varepsilon])} u + \varepsilon \max_{\overline{\Omega}} |x|^2$$

In the limit as $\varepsilon \to 0$, with u continuous on $\overline{\Omega} \times [0,t_1]$, we obtain

$$\max_{\overline{\Omega}\times[0,t_1]} u \leqslant \max_{\Omega\times\{0\}\cup(\partial\Omega\times[0,t_1])} u$$ □

We can also show that the minimum of u can only occur on $\Omega \times \{0\} \cup (\partial\Omega \times [0,t_1])$ by replacing u by $-u$ in the above proof. Thus any solution in $C^0(\overline{\Omega} \times [0,t_1])$ of $(\partial_t - \Delta)\, u = 0$ which vanishes on $\Omega \times \{0\} \cup (\partial\Omega \times [0,t_1])$ must vanish identically in $\overline{\Omega} \times [0,t_1]$.

Corollary If the boundary value problem

$$(\partial_t - \Delta)u = f \qquad \text{in } \Omega \times (0,t_1)$$

$$u = u_0 \qquad \text{on } \Omega \times \{0\} \cup (\partial\Omega \times [0,t_1])$$

has a solution in $C^0(\overline{\Omega} \times [0,t_1])$, it is unique.

A physical interpretation of Theorem 6.6 is provided by the observation that, since heat always flows in a conducting medium from points of higher temperature to points of lower temperature, the highest (and lowest) temperatures can only occur at points where u is prescribed, i.e., at $t = 0$ and on $\partial\Omega$. It is also worth noting that the boundary values of u on $\Omega \times \{t_1\}$ cannot be prescribed, otherwise the problem becomes overdetermined. This is in contrast to the Dirichlet problem for the operator Δ.

Example 6.10 Let us now consider a conducting bar of length l which is insulated except at its ends, and these are held at the same (zero) temperature. If the initial temperature distribution is g, then the boundary value problem which describes the temperature distribution for $t > 0$ is

$$\begin{aligned} \partial_t u &= \partial_x^2 u \qquad x \in (0,l) \quad t > 0 \\ u(0) &= g \qquad x \in (0,l) \\ u &= 0 \qquad x \in \{0,l\} \quad t > 0 \end{aligned}$$

We shall take g in $L^2(0,l)$ and use *separation of variables* to solve the problem. This method is based on the assumption that u, which is a C^∞ function in $(0,l) \times (0,\infty)$, can be expressed as a product $v(x)w(t)$ of a function of x and a function of t, and leads to the differential equation

$$\partial_t u - \partial_x^2 u = vw' - v''w = 0$$

$$\frac{w'}{w} = \frac{v''}{v}$$

Since w'/w is a function of t alone while v''/v is a function of x alone, the above equation can only hold if each side is a (complex) constant, say $-\lambda^2$. The resulting pair of ordinary differential equations,

$$\begin{aligned} w' + \lambda^2 w &= 0 \\ v'' + \lambda^2 v &= 0 \end{aligned}$$

are in fact simple eigenvalue equations whose eigenfunctions are $e^{-\lambda^2 t}$ and $e^{\pm i\lambda x}$, respectively.

Since λ is an arbitrary parameter, any linear combination of $u_\lambda(x,t) = \exp(-\lambda^2 t \pm i\lambda x)$ is a (distributional) solution of the heat equation. We shall choose the particular linear combination which satisfies the initial and the boundary conditions on u.

Recalling the theory of Fourier series in Section 5.5, we take $\lambda = kp$, where p is a positive constant and k is an integer, and set

$$u(x,t) = \sum_{-\infty}^{\infty} c_k \, e^{-k^2p^2t + ikpx} \tag{6.43}$$

From the boundary condition at $x = 0$ we get

$$u(0,t) = \sum_{-\infty}^{\infty} c_k \, e^{-k^2p^2t} = 0 \qquad t > 0$$

which implies that $c_k + c_{-k} = 0$, since the functions $e^{-k^2p^2t}$ are linearly independent for different values of k^2. If u is a real function then $c_{-k} = \bar{c}_k$, and so we conclude that the coefficients c_k are imaginary. Hence (6.43) can be written as

$$u(x,t) = \sum_1^\infty b_k\, e^{-k^2p^2t} \sin(kpx) \tag{6.44}$$

where the coefficients $b_k = 2ic_k$ are now real numbers.

The boundary condition at $x = l$ gives

$$u(l,t) = \sum_1^\infty b_k\, e^{-k^2p^2t} \sin(kpl) = 0 \qquad t > 0$$

Here again the linear independence of the set $\{e^{-k^2p^2t}\}$ implies that $b_k \sin(kpl) = 0$ for all $k \in \mathbf{N}$; and since we cannot allow the coefficients b_k to vanish, as we need them to construct g, we must have $\sin(kpl) = 0$ for all k. This implies that p is an integer multiple of π/l, and hence (6.44) becomes

$$u(x,t) = \sum_1^\infty b_k \exp(-k^2\pi^2t/l^2) \sin(k\pi x/l) \tag{6.45}$$

This sum is an odd periodic function of x of period $2l$. In order to satisfy the initial condition, we should therefore extend g as an odd function into $(-l,l)$ and then as a periodic function, with period $2l$, into $\mathbf{R}$. This extension, which we still denote by g, is now in $\tilde{H}^0$ and coincides with the Fourier series (6.45) at $t = 0$ provided

$$b_k = \frac{2}{l}\int_0^l g(x) \sin(k\pi x/l)\, dx$$

If the bar in this example had been insulated at both ends, the condition $\partial_x u = 0$ at $x = 0$ and $x = l$, applied to the sum (6.43), would result in a cosine series representing the even periodic extension of g.

The method used here can obviously be generalized from the interval $[0,l]$ to the cube $[0,l]^n$ in $\mathbf{R}^n$ by using the representation

$$u(x,t) = \sum c_\alpha \exp(-\langle\alpha,\alpha\rangle p^2t + ip\langle\alpha,x\rangle)$$

For a more general domain $\Omega \subset \mathbf{R}^n$, the solution of the problem by separation of variables will depend on the availability of an orthonormal basis $\{v_k\}$ for

$L^2(\Omega)$ which satisfies $\Delta v_k = \lambda_k v_k$ in Ω and vanishes on $\partial\Omega$. Such a basis, whose elements are eigenfunctions of the Laplacian, can be shown to exist when Ω is bounded and its boundary is smooth [14].

6.4 THE WAVE EQUATION

The wave operator $\partial_{n+1}^2 - \Delta$ in $\mathbf{R}^n \times \mathbf{R}$, sometimes called the *d'Alembertian* and denoted by $\Box$, is the archetype of the hyperbolic operators. It is not hypoelliptic, as may be seen by considering the solution (3.25) in $\mathbf{R} \times \mathbf{R}$ for the initial conditions $u = u_0$, $\partial_t u = 0$. If u_0 is not continuous, then the (weak) solution $u(x,t) = (1/2)[u_0(x - t) + u_0(x + t)]$ will not be continuous. Note also that the two initial conditions on u and $\partial_t u$ are appropriate for the wave equation, in view of the second-order time derivative which appears in the equation.

We start by considering the initial value problem

$$\begin{aligned} &\Box u = f && \text{in } \mathbf{R}^n \times (0,\infty) && (6.46)\\ &u(0) = u_0, \quad \partial_t u(0) = u_1 && \text{on } \mathbf{R}^n \end{aligned}$$

For the purpose of performing convolutions and Fourier transformations, we shall first assume that u_0 and u_1 are in $\mathscr{E}'(\mathbf{R}^n)$. Later on we shall see that this condition can be relaxed. The solution u that we seek is assumed to be a tempered distribution in $\mathbf{R}^n$ which depends on the variable t in $(0,\infty)$ in such a way that $\partial_t^2 u$ is continuous on $(0,\infty)$, with $u \to u_0$ and $\partial_t u \to u_1$ as $t \to 0^+$. This is interpreted, along the lines of Section 2.8, to mean that $\langle u,\phi\rangle$ is a C^2 function of t on $(0,\infty)$ for every $\phi \in \mathscr{S}(\mathbf{R}^n)$, and that $\langle u - u_0,\phi\rangle$ and $\langle \partial_t u - u_1,\phi\rangle \to 0$ as $t \to 0^+$ for all $\phi \in \mathscr{S}(\mathbf{R}^n)$. f is assumed to be in $\mathscr{E}'(\mathbf{R}^n)$ and to depend continuously on t in $(0,\infty)$ with its support in $\mathbf{R}^n \times [0,\infty)$.

With these assumptions we can apply the Fourier transformation to (6.46) to obtain

$$(\partial_t^2 + |\xi|^2)\,\hat{u}_\xi(t) = \hat{f}_\xi(t) \qquad \text{in } \mathbf{R}^n \times (0,\infty) \qquad (6.47)$$

$$\hat{u}(0) = \hat{u}_0, \quad \partial_t\hat{u}(0) = \hat{u}_1 \qquad \text{on } \mathbf{R}^n \qquad (6.48)$$

The general solution of (6.47), which was obtained in Example 3.19, is given by

$$\hat{u} = \hat{f} *_t \hat{E} + c_1 \cos|\xi|t + c_2 \sin|\xi|t$$

where $\hat{E}$ is the fundamental solution of the operator $\partial_t^2 + |\xi|^2$ in $t > 0$, given by

$$\hat{E}_\xi(t) = H(t)\sin|\xi|t/|\xi|$$

and $\hat{f} *_t \hat{E}$ denotes the convolution product (with respect to t) of $\hat{f}$ and $\hat{E}$. Since $f \in \mathscr{E}'(\mathbf{R}^n)$ for every t, its Fourier transform $\hat{f}$ is a C^∞ function whose product with $\hat{E} \in \mathscr{S}'(\mathbf{R}^n)$ is well defined by equation (4.22) and lies in $\mathscr{S}'(\mathbf{R}^n)$. The convolution $\hat{f} *_t \hat{E}$ is then defined by

$$\langle \hat{f} *_t \hat{E}, \phi \rangle(t) = \int_{-\infty}^{\infty} \langle \hat{f}(t - s)\, \hat{E}(s), \phi \rangle\, ds$$

$$= \int_{0}^{t} \langle \hat{f}(t - s)\hat{E}(s), \phi \rangle\, ds \qquad \phi \in \mathscr{S}(\mathbf{R}^n)$$

and lies in $\mathscr{S}'(\mathbf{R}^n)$ with its support in $\mathbf{R}^n \times [0,\infty)$, since both $\hat{f}$ and $\hat{E}$ are supported in $\mathbf{R}^n \times [0,\infty)$. The constants (in time) c_1 and c_2 are evaluated by using equations (6.48) to give

$$\hat{u} = \hat{f} *_t \hat{E} + \hat{u}_0 \cos|\xi|t + \hat{u}_1 \frac{\sin|\xi|t}{|\xi|} \qquad (6.49)$$

When $t > 0$ we can write

$$\cos|\xi|t = \partial_t (\sin|\xi|t/|\xi|)$$

$$= \partial_t(H(t) \sin|\xi|t/|\xi|)$$

Therefore

$$\hat{u} = \hat{f} *_t \hat{E} + \hat{u}_0\, \partial_t\hat{E} + \hat{u}_1\hat{E} \qquad t > 0 \qquad (6.50)$$

Applying the inverse Fourier transform and using Theorem 4.9, we arrive at

$$u = f *_{xt} E + u_0 * \partial_t E + u_1 * E \qquad (6.51)$$

where the first convolution is between f and E as elements of $\mathscr{S}'(\mathbf{R}^{n+1})$. It exists because f has compact support in $\mathbf{R}^n$ and, as functions of t, both f and E are supported in $[0,\infty)$. The other two convolutions also exist since u_0 and u_1 are in $\mathscr{E}'(\mathbf{R}^n)$. The representation (6.51) reduces to (3.25) when $f = 0$ and $n = 1$, and it is left to the reader to check that it satisfies all the equations in (6.46).

Although we shall not do so, it can be shown that the distribution $E_x(t) = \mathscr{F}^{-1}[H(t) \sin|\xi|t/|\xi|]$ is supported in the *forward wave cone*

$$\Gamma_+ = \{(x,t) \in \mathbf{R}^{n+1} : |x| \leq t,\ t \geq 0\}$$

In fact, it turns out that for odd values of $n \geq 3$ the support of E is the boundary of Γ_+, i.e., the set $\partial\Gamma_+ = \{(x,t) : |x| = t,\ t \geq 0\}$; and for all other values of $n \in \mathbf{N}$, supp $E = \Gamma_+$. For a proof of these assertions the reader is referred to Treves [13]. They can also be checked against the expressions given for E when $n = 1, 2$, and 3.

Using the fact that E, as a distribution in x, is supported in the ball $|x| \leq t$ for each $t \geq 0$, we can now allow u_0, u_1 and f in (6.51), as distributions in x, to be any elements in $\mathcal{S}'(\mathbf{R}^n)$. Thus the only restriction on the prescribed data is that f should depend continuously on t.

To investigate the question of uniqueness of the solution (6.51), we shall extend the problem (6.46) to the lower half-space $\mathbf{R}^n \times (-\infty,0)$. Let f be an arbitrary tempered distribution in $\mathbf{R}^n$ which depends continuously on $t \in \mathbf{R}$, and let $f^+ = H(t)f$ and $f^- = H(-t)f$. Since the wave operator $\Box$ is invariant under the transformation $t \mapsto -t$, the solution u of $\Box u = f^-$ in $t < 0$ has the same form as that of $\Box u = f^+$ in $t > 0$. It is easily seen that the fundamental solution of $\partial_t^2 + |\xi|^2$ in $\mathbf{R}^n \times (-\infty,0)$ is $\hat{E}(\xi,-t) = -H(-t) \sin |\xi|t/|\xi|$. If we now denote $\sin |\xi|t/|\xi|$ by $\hat{\Psi}_\xi(t)$, we can define

$$\hat{\Psi}^+ = H(t)\,\hat{\Psi} = \hat{E}$$
$$\hat{\Psi}^- = -H(-t)\hat{\Psi}$$

and their inverse Fourier transforms

$$\Psi^+ = \mathcal{F}^{-1}(\hat{\Psi}^+)$$
$$\Psi^- = \mathcal{F}^{-1}(\hat{\Psi}^-)$$

exist because E exists.

From (6.51) we immediately see that the corresponding solution in $t < 0$ is given by

$$u = f^- *_{xt} \Psi^- - u_0 * \partial_t \Psi^- - u_1 * \Psi^-$$

where the negative signs appear because the time reversal $t \mapsto -t$ transforms the operator ∂_t into $-\partial_t$ both in the initial condition $\partial_t u = u_1$ and in formula (6.51). The convolutions here also make sense since both f^- and Ψ^- are supported in the lower half-space $\mathbf{R}^n \times (-\infty,0]$ and, with Ψ^- supported in the backward wave cone $\Gamma_- = \{(x,t) \in \mathbf{R}^n : |x| \leq |t|, t \leq 0\}$, the prescribed data f^-, u_0 and u_1 can again be any distributions in $\mathcal{S}'(\mathbf{R}^n)$. Consequently,

$$u = f^+ *_{xt} \Psi^+ + f^- *_{xt} \Psi^- + u_0 * \partial_t(\Psi^+ - \Psi^-) + u_1 * (\Psi^+ - \Psi^-) \tag{6.52}$$

which is a C^2 mapping from $\mathbf{R}$ to $\mathcal{S}'(\mathbf{R}^n)$, solves the boundary value problem

$$\begin{aligned} &\Box u = f && \text{in } \mathbf{R}^{n+1} \\ &u(0) = u_0, \quad \partial_t u(0) = u_1 && \text{on } \mathbf{R}^n \end{aligned} \tag{6.53}$$

To prove its uniqueness, let v be a C^2 mapping from $\mathbf{R}$ to $\mathcal{S}'(\mathbf{R}^n)$ which solves the corresponding homogeneous problem

$$\begin{aligned} \Box v &= 0 && \text{in } \mathbf{R}^{n+1} \\ v = \partial_t v &= 0 && \text{on } \mathbf{R}^n \times \{0\} \end{aligned}$$

and suppose $v_x^+(t) = H(t)\, v_x(t)$. Then

$$\begin{aligned} \partial_t^2 v^+ &= \delta' v + 2\delta\, \partial_t v + H\, \partial_t^2\, v \\ &= H\, \partial_t^2\, v \end{aligned}$$

in view of the initial conditions on v. Note that Leibnitz' formula can be used to differentiate Hv with respect to t up to order two, since v is a C^2 function of t. Therefore $\Box v^+ = H\Box v = 0$, and we obtain

$$\begin{aligned} v^+ &= v^+ * (\Box \Psi^+) \\ &= \Box v^+ * \Psi^+ \\ &= 0 \end{aligned}$$

Theorem 6.7 Given u_0, $u_1 \in \mathcal{S}'(\mathbf{R}^n)$ and f a continuous function of t on $[0,\infty)$ with values in $\mathcal{S}'(\mathbf{R}^n)$, the initial value problem (6.46) has a unique solution in the class of C^2 functions of t on $[0,\infty)$ with values in $\mathcal{S}'(\mathbf{R}^n)$, which is given by (6.51).

The representation (6.51) clearly shows that the smoothness of u is determined by the smoothness of the prescribed data, and that, when f, u_0, and u_1 are C^∞ functions, then the same is true of u. That is not to say that the wave operator is hypoelliptic, since we are looking at only a particular class of solutions of $\Box u = f$ in $\mathbf{R}^n \times (0,\infty)$, namely those which satisfy the additional conditions $u(0) = u_0$ and $\partial_t u(0) = u_1$.

But with C^∞ functions we have no way of measuring the smoothness and growth properties of the solution against those of the prescribed data. To be able to do that we resort to Sobolev spaces, and in the process, we shall need the inequality

$$\begin{aligned} |\hat{E}_\xi(t)| &= \frac{|\sin(|\xi| t)|}{|\xi|} \\ &\leq 2(1 + |\xi|^2)^{-(1/2)}(1 + t) \quad \text{for all } (\xi,t) \in \mathbf{R}^n \times [0,\infty) \end{aligned} \tag{6.54}$$

This follows from the observation that, when $|\xi| \leq 1$, we have

$$|\sin(|\xi|t)| \leq |\xi|t \leq \frac{2|\xi|t}{\sqrt{1 + |\xi|^2}}$$

and when $|\xi| > 1$ we have

$$|\sin(|\xi|t)| \leq 1 \leq \frac{2|\xi|}{\sqrt{1 + |\xi|^2}}$$

Let u_0 and u_1 be functions in C_0^∞ and f be a continuous mapping from $[0,\infty)$ to C_0^∞. Now formula (6.49) becomes

$$\hat{u}(t) = \int_0^\infty \hat{E}(t - \tau)\hat{f}(\tau)\, d\tau + \hat{u}_0 \cos|\xi|t + \hat{u}_1 \frac{\sin|\xi|t}{|\xi|}$$

$$= \int_0^t \frac{1}{|\xi|} \sin(|\xi|(t - \tau))\, \hat{f}(\tau)\, d\tau + \hat{u}_0 \cos|\xi|t + \hat{u}_1 \frac{\sin|\xi|t}{|\xi|} \qquad (6.55)$$

After multiplying by $(1 + |\xi|^2)^{(1/2)s}$, using the inequality (6.54), integrating with respect to ξ, and applying Schwartz' inequality, we obtain

$$\|\hat{u}(t)\|_s \leq 2\int_0^t \|\hat{f}(\tau)\|_{s-1}\, (1 + |t - \tau|)d\tau + \|u_0\|_s + 2\|u_1\|_{s-1}\, (1 + t)$$

Since C_0^∞ is dense in H^s, this proves

Theorem 6.8 For any real number s, let $u_0 \in H^s$, $u_1 \in H^{s-1}$ and f be a continuous function of $t \in [0,\infty)$ with values in H^{s-1}. Then the solution u of the initial value problem (6.46) is a continuous function of t with with values in H^s.

When Ω is a bounded domain in $\mathbf{R}^n$ where the wave equation is satisfied, the initial shape and velocity of the waves in Ω are not enough to determine the wave propagation in Ω for all time. In addition, we have to know what happens at the boundary $\partial\Omega$. If the boundary condition on $\partial\Omega$ is independent of t, then the method of separation of variables, which was used for solving the heat equation works equally well here. The only difference is that the time equation is of second (instead of first) order,

$$w'' + \lambda^2 w = 0$$

Consequently the exponential function $e^{-\lambda^2 t}$, which describes heat diffusion in time, is replaced by a linear combination of sin λt and cos λt, with the extra constants determined by the extra initial condition on Ω, namely $\partial_t u = u_1$. Hence the solution, which is periodic in space, is also periodic in time.

For a bounded regular domain $\Omega \subset \mathbf{R}^n$ with a homogeneous boundary condition, the feasibility of using separation of variables to construct u is limited by our ability to find an explicit expression for an orthonormal basis of $H_0^0(\Omega) = L^2(\Omega)$ whose elements satisfy $\Delta v + \lambda^2 v = 0$. These are only known for certain simple geometric shapes, such as the rectangle, cylinder, ball, etc., and they give rise to the various well-known orthonormal systems of special functions.

In Section 6.2 we used Green's function to solve the boundary value problem for Laplace's equation in a bounded domain, but in the treatment of the heat and wave equations we have only considered separation of variables as a tool when the (space) domain is bounded. The choice is really a matter of convenience. We could equally well have used separation of variables in the first instance and Green's function methods in the second. In fact, equation (6.40) shows that the fundamental solution (6.30) is really Green's function for the initial value problem $(\partial_t - \Delta)u = f$ in $\mathbf{R}^n \times (0,\infty)$, $u = g$ on $\mathbf{R}^n \times \{0\}$. Furthermore the Fourier series representation of the (distributional) solution u for heat conduction in a bar, as given by (6.45), can be written as

$$u(x,t) = \sum_{k=1}^{\infty} \left[\frac{2}{l} \int_0^l g(\xi) \sin \frac{k\pi\xi}{l}\, d\xi \right] \left[\exp\left(-\frac{k^2\pi^2 t}{l^2} \right) \sin\left(\frac{k\pi x}{l} \right) \right]$$

$$= \int_0^l \frac{2}{l} \sum_{k=1}^{\infty} \exp\left(-\frac{k^2\pi^2 t}{l^2} \right) \sin\left(\frac{k\pi\xi}{l} \right) \sin\left(\frac{k\pi x}{l} \right) g(\xi)\, d\xi$$

This suggests that

$$G(x,\xi,t) = \frac{2}{l} \sum_{k=1}^{\infty} \exp\left(-\frac{k^2\pi^2 t}{l^2} \right) \sin\left(\frac{k\pi\xi}{l} \right) \sin\left(\frac{k\pi x}{l} \right) \tag{6.56}$$

is the Green's function for the domain $(0,l) \times (0,\infty) \subset \mathbf{R}^2$. Indeed, $G(x,\xi,t - \tau)$ can be shown to satisfy $(\partial_t - \partial_x^2)\, G = \delta_{(\xi,\tau)}$ in this strip (see Exercise 6.20).

Now we look at some representative examples of wave propagation in $\mathbf{R}^n \times [0,\infty)$, with a view to extending some of the observations made in Example 3.23.

Example 6.11 To solve

$$\square u = \delta \otimes f \qquad \text{in } \mathbf{R}^n \times (0,\infty)$$
$$u(0) = \partial_t u(0) = 0 \qquad \text{on } \mathbf{R}^n \tag{6.57}$$

where f is a continuous function of t on $(0,\infty)$ and δ is the Dirac distribution in $\mathbf{R}^n$, we make use of the formula (6.51)

$$u = (\delta \otimes f) *_{xt} E$$
$$= f *_t E$$

Since $\operatorname{supp} f \subset [0,\infty)$ and $\operatorname{supp} E \subset \Gamma_+$, we see that $\operatorname{supp} u \subset \{0\} \times [0,\infty) + \Gamma_+$. Hence

$$\operatorname{supp} u \subset \Gamma_+ = \{(x,t) \in \mathbf{R}^{n+1} : |x| \leq t,\ t \geq 0\}$$

When $n = 1$, $E = (1/2)[H(x + t) - H(x - t)]$, $t \geq 0$, and u is the function

$$u(x,t) = \tfrac{1}{2} \int_{-\infty}^{\infty} H(t - \tau) f(t - \tau)\, H(\tau)\, [H(x + \tau) - H(x - \tau)]\, d\tau$$

$$= \tfrac{1}{2} \int_0^t f(t - \tau)\, [H(x + \tau) - H(x - \tau)]\, d\tau$$

This representation has the following implications:

(i) $u = 0$ when $|x| > t$, which just means that the wave is confined to the cone $\Gamma_{1+} = \{(x,t) \in \mathbf{R}^2 : -t \leq x \leq t,\ t \geq 0\}$

(ii) At any instant $t_0 > 0$ the wave is completely determined by the values of f over the interval from 0 to t_0, and its wave front will have covered a distance along the x-axis equal to t_0 in both the positive and the negative directions. This means that the wave front is traveling with velocities ± 1.

(iii) At any instant $t_0 > 0$, while the wave front is located at the two points $x = \pm t_0$, the rest of the wave fills the space in between. In other words, the wave has a "tail" which trails behind the wave front.

(iv) If the disturbance f has a very short duration so that it is made up entirely of an impulse concentrated at $t = 0$, then we could replace f by δ, as a limiting case, and thereby arrive at

$$u(x,t) = \tfrac{1}{2}\,[H(x + t) - H(x - t)]$$
$$= \begin{cases} \tfrac{1}{2} & \text{when } -t \leqslant x \leqslant t \\ 0 & \text{otherwise} \end{cases}$$

This indicates that the resulting wave u propagates in a diffusive manner with a tail which fills the interior of Γ_{1+}.

Recalling the results of Example 3.23, we see that the wave generated by the disturbance f, which is concentrated at $(x,t) = (0,0)$, travels in a diffusive manner similar to the wave $\frac{1}{2}[H(x + t) - H(x - t)]$ caused by the initial disturbance $\partial_t u = \delta$, but unlike the wave $\frac{1}{2}(\delta_t - \delta_{-t})$ resulting from the initial disturbance $u = \delta$, which travels in a localized wave packet.

Example 6.12 The solution of the initial value problem (6.57) in $\mathbf{R}^3 \times [0,\infty)$ is determined by the fundamental solution of $\Box$ in this region, namely $(1/4\pi t)\ \delta_{|x|-t}$. Formula (6.51) gives

$$u_x(t) = f(t) *_t \left[\frac{1}{4\pi t}\,\delta_{|x|-t}\right]$$
$$= f(t - |x|)/4\pi|x|$$

This wave is also supported in the forward cone Γ_{3+}, so it is concentrated in the ball $\{x \in \mathbf{R}^3 : |x| \leqslant t\}$. As x increases the amplitude of the wave decreases, but the signal is undistorted. In acoustics, this is a mathematical model of a point source of sound located at $x = 0$.

If the disturbance f, on the other hand, has a very short duration, such as a flash of light, which may be idealized by the impulse δ, then u will have its support on the boundary of Γ_{3+}. In other words the signal is then concentrated on the sphere $\{x \in \mathbf{R}^3 : |x| = t\}$, and will be seen by an observer located at x_0 as a single flash which comes on at the instant $t = |x_0|$ and then disappears. This is a manifestation of *Huygens' principle* which applies to wave propagation phenomena in three-dimensional space. In one-dimensional space, the same flash would be seen by the same observer at the instant $t = |x_0|$, but then it would be followed by a tail of light resulting from the contribution of the values of E in the interior of the cone Γ_{1+}.

This difference in the way waves propagate in $\mathbf{R}^n$ is clearly determined by the support of the fundamental solution E, i.e., whether it is Γ_+ or $\partial\Gamma_+$. It turns out that this, in turn, is determined by the analyticity properties of the fundamental solution of the Laplacian operator. The connection comes from the observation that a harmonic function in $\mathbf{R}^{n+1}$ satisfies the wave equation in $\mathbf{R}^n$ when x_{n+1} is replaced by ti. We can use formula (6.19) to represent the harmonic function v in an open subset of $\mathbf{R}^{n+1}$, with E given by (4.40). The function v, being analytic in each of its real variables (Exercise 6.4), may be extended to an open subset of $\mathbf{C}^{n+1}$. Now the function defined by

$$u(x_1, \ldots, x_n, t) = v(x_1, \ldots, x_n, it)$$

satisfies $\Box u = 0$ whenever v satisfies $\Delta v = 0$, and formula (6.19) can be used to represent u. The singularities of E in $\mathbf{R}^{n+1}$ play a crucial role in this representation (see [15] for details). From the formula (4.40) we see that $|x|^{2-(n+1)} = |x|^{1-n}$ is a single-valued analytic function in $\mathbf{R}^{n+1} - \{0\}$ only when n is odd and greater than 1, and this accounts for the fact that $\operatorname{supp} E = \partial\Gamma_+$ for those values only. There is, it seems, a hidden link between analytic function theory and the peculiarities of wave propagation!

EXERCISES

6.1 Use Theorem 6.1 to check the properties of the trace operator as stated in Theorem 6.2.

6.2 Fill in the details of the proof of Theorem 6.3.

6.3 Prove that the boundary value problem for the second order elliptic differential operator L,

$$Lu = f \quad \text{in } \Omega$$

$$\partial_\eta u + \sigma u = g \quad \text{on } \partial\Omega,$$

with $f \in H^0(\Omega)$ and $g \in H^1(\Omega)$, has a unique solution in $H^1(\Omega)$ provided the function σ is bounded and negative on $\partial\Omega$.

6.4 Use the formula (6.19) to show that every harmonic function in an open set $\Omega \in \mathbf{R}^n$ is analytic in Ω.

6.5 Show how Poisson's integral formula (6.21) implies the maximum principle: if u is a harmonic function in an open connected set $\Omega \subset \mathbf{R}^2$, then u cannot assume a maximum value at any point in Ω unless u is a constant.

6.6 Under the hypothesis of Exercise 6.5 prove the *maximum modulus principle:* if u is harmonic in Ω, then $|u|$ cannot assume a maximum in Ω unless u is a constant.

6.7 If u is a harmonic function in the closed ball $\overline{B}(x_0, r) \subset \mathbf{R}^n$, prove the mean-value formula

$$u(x_0) = \frac{1}{\omega_{n-1}} \int_{S_{n-1}} u(x_0 + r\omega)\, d\omega$$

6.8 Prove the maximum principle for harmonic functions in $\Omega \subset \mathbf{R}^n$.

6.9 Let u_1 and u_2 be two solutions of the Poisson equation $\Delta u = f$ in H^s. Show that $u_1 = u_2$ a.e. if $s \geqslant 0$. Conclude from this that $u_1 = u_2$ if $s > (1/2)n$.

6.10 Show that if $u \in \mathcal{S}'$ satisfies the homogeneous *Helmholtz equation* $\Delta u + k^2 u = 0$, $k \in \mathbf{R}$, then $\hat{u} \in \mathcal{E}'$.

6.11 Show that $\Delta u + k^2 u = f$ has at most one solution in $\mathcal{E}'$. Hint: Use Theorem 4.8 and Exercise 6.10.

6.12 Find all solutions of $\Delta u + k^2 u = 0$ in $\mathbf{R}^3$ with spherical symmetry.

6.13 Find the fundamental solution of the operator $-\Delta + k^2$ in $\mathcal{S}'$, and show that the solution of $-\Delta u + k^2 u = f$ is unique in $\mathcal{S}'$.

6.14 Find the Green's function for Δ in the upper half ball $B_+(0,1) = B(0,1) \cap \mathbf{R}^n_+$.

6.15 Discuss the uniqueness of the solution to the Dirichlet problem in the half-plane $\mathbf{R}^2_+$

$$\begin{aligned} &\Delta u = f && \text{in } (-\infty,\infty) \times (0,\infty) \\ &u(x_1,0) = g(x_1) && \text{on } (-\infty,\infty) \end{aligned}$$

6.16 Show that the Cauchy–Riemann operator $\partial_1 + i\partial_2$ is elliptic in $\mathbf{R}^2$, and that the Shrödinger operator $(1/i)\partial_t - \Delta$ is hypoelliptic in $\mathbf{R}^{n+1}$.

6.17 Show that the solution of the initial value problem for the heat equation, as given by (6.38), is analytic in $\mathbf{R}^n$ as a function of x and in $(0,\infty)$ as a function of t.

6.18 Use separation of variables to solve the problem of the conducting bar of length l whose ends are insulated and whose initial temperature distribution is $g \in L^2(0,l)$.

6.19 Solve the boundary value problem for heat flow in a bar of length l, with u = constant at one end and $\partial_x u = 0$ at the other. Assume an initial temperature distribution $g \in L^2(0,l)$ and the presence of a heat source at $x = \frac{1}{2}l$ given by $f = \delta_{(1/2)l} \sin \omega t$.

6.20 Verify that the sum in (6.56) represents Green's function for the heat operator in $\Omega = (0,l) \times (0, \infty)$, in the sense that $(\partial_t - \partial_x^2)\, G = \delta_{(\xi,0)}$ and $G = 0$ on $\partial\Omega$.

6.21 Solve the initial value problem for the free vibrations of an infinite membrane

$$\begin{aligned} &\partial_t^2 u = c^2 \Delta u && \text{in } \mathbf{R}^2 \times (0,\infty) \\ &u(0) = u_0 && \partial_t u(0) = u_1 \end{aligned}$$

where u_0 and u_1 are given in $\mathcal{S}'(\mathbf{R}^2)$.

6.22 Verify that $H(t - |x|)/2\pi(t^2 - |x|^2)^{1/2}$ is a fundamental solution of $\Box$ in $\mathbf{R}^2 \times (0,\infty)$. What is the fundamental solution in $\mathbf{R}^2 \times (-\infty,0)$?

6.23 Prove that any distribution in $\mathcal{D}'(\mathbf{R}^3)$ which satisfies $\Box u = 0$ in $\mathbf{R}^2 \times \mathbf{R}$ and vanishes on either $\mathbf{R}^2 \times (-\infty,0)$ or $\mathbf{R}^2 \times (0,\infty)$ must vanish identically.

6.24 Use separation of variables to determine the free vibrations of a flexible string of length l, which is fixed at its end points:

$$\begin{aligned}
\partial_t^2 u &= \partial_x^2 u && \text{in } (0,l) \times (0,\infty) \\
u &= \tfrac{1}{2}l - |x - \tfrac{1}{2}l| && \text{in } (0,l) \text{ when } t = 0 \\
\partial_t u &= 0 \text{ on } (0,l) && \text{when } t = 0 \\
u = 0 \quad &\text{at} \quad x = 0,l && \text{for } t > 0.
\end{aligned}$$

6.25 Show that the solution obtained in Exercise 6.24 coincides with the solution obtained by using the formula (3.25) with $u_1(x) = 0$ once $u_0(x) = \frac{1}{2}l - |x - (1/2)l|$ is extended as an odd periodic function in $(-\infty,\infty)$.

6.26 Solve the initial value problem (6.57) in $\mathbf{R}^2 \times [0,\infty)$ and show that Huygens' principle does not apply in this dimension.

6.27 Show that the representation (6.52) of the solution to the wave equation in $\mathbf{R}^{n+1}$ is a C^2 mapping of $-\infty < t < \infty$ into $\mathcal{S}'(\mathbf{R}^n)$. Verify that it satisfies the equations (6.53).

6.28 If Ω is an open connected set in $\mathbf{R}^n$, let u_k be an increasing sequence of solutions of the equation $\Delta u_k = 0$, and let $u = \sup u_k$. If u is finite at any point in Ω, show that $\Delta u = 0$ in Ω.

6.29 If Δ is replaced by the operator $\partial_t - \Delta$ in Exercise 6.28, show that the same conclusion may be reached by assuming that u is finite on a dense subset of Ω.

Notation

$\mathbf{Z}$ integers
$\mathbf{N}$ positive integers
$\mathbf{Z}^-$ negative integers
$\mathbf{N}_0$ nonnegative integers
$\mathbf{Z}^n$, $\mathbf{N}^n$, $\mathbf{N}_0^n$ n-tuples of integers in $\mathbf{Z}$, $\mathbf{N}$, $\mathbf{N}_0$
$\mathbf{R}$ real number field
$\mathbf{C}$ complex number field
Φ either $\mathbf{R}$ or $\mathbf{C}$
$\mathbf{R}^n$ n-dimensional real (Euclidean) space
$\mathbf{C}^n$ n-dimensional complex space
$\mathbf{R}^n_+ = \mathbf{R}^{n-1} \times (0,\infty)$
$\mathbf{R}^n_- = \mathbf{R}^{n-1} \times (-\infty,0)$
$\mathring{E}$ interior of the set E, 4
$\overline{E}$ closure of the set E, 4
∂E boundary of the set E, 4
E^c complement of E, 4
$E_1 - E_2 = E_1 \cap E_2^c$
Ω open subset of $\mathbf{R}^n$
$B(x,r)$ open ball in metric space, 6

$B(r) = B(0,r)$
$B_p(r)$, 11
$S_n(r)$ sphere with center 0 and radius r in $\mathbf{R}^{n+1}$, 149
$S_n = S_n(1)$
Γ_+ forward wave cone, 238
Γ_- backward wave cone, 239
$x^\alpha = x_1^{\alpha_1} \cdots x_n^{\alpha_n}$ $x = (x_1, \ldots, x_n) \in \mathbf{R}^n$, $\alpha = (\alpha_1, \ldots, \alpha_n) \in \mathbf{N}_0^n$
$x^j = x_1^j \cdot \cdots \cdot x_n^j$ $j \in \mathbf{N}_0$
$\partial_k^j = \dfrac{\partial^j}{\partial^j x_k}$ $k \in \{1, \ldots, n\}$, 16
$\partial^\alpha = \partial_1^{\alpha_1} \cdots \partial_n^{\alpha_n}$
$\partial^j = \partial_1^j \cdots \partial_n^j$, 95
$D_k = -i\partial_k$, 124
∂_{z_k}, $\partial_{\bar{z}_k}$, 134
L linear partial differential operator, 100
L^* formal adjoint of L, 109
Δ Laplacian operator, 42
$\Box$ d'Alembertian operator, 237
$\bar{\partial} = \frac{1}{2}(\partial_1 + i\partial_2)$ Cauchy–Riemann operator in $\mathbf{R}^2$, 142
$C^m(\Omega)$, $C_K^m(\Omega)$, 16–17
$C_0^\infty(\Omega)$, 24
$C^\infty(\bar{\Omega})$, 190
$C_0^m(\bar{\Omega})$, 205
$\mathcal{D}(\Omega)$, $\mathcal{D}_K$, $\mathcal{D}^m(\Omega)$ test function spaces, 25–27
$\mathcal{D}'(\Omega)$ space of distributions in Ω, 28
$\mathcal{D}^{m\prime}(\Omega)$ distributions of order m, 60
$\mathcal{D}_F'(\Omega)$ distributions of finite order, 61
$\mathcal{D}_+'(\mathbf{R})$ distributions supported in $[0,\infty)$, 89
$\mathcal{E}(\Omega)$, 75
$\mathcal{E}'(\Omega)$ distributions with compact support, 75
$\mathcal{S}$ rapidly decreasing functions, 119
$\mathcal{S}'$ tempered distributions, 121
$L^1(\Omega)$, $L^1_{\mathrm{loc}}(\Omega)$, 28–29
$L^2(\Omega)$, 132
$L^p(\Omega)$, 57
$L^\infty(\Omega)$, 84
$\mathcal{H}$ Hilbert space, 170
$H^m(\Omega)$ Sobolev space, $m \in \mathbf{N}_0$, 172
$H^{-m}(\Omega)$, 177

References

1. J. Dieudonné, "Treatise on Analysis," Vols. I and II, Academic Press, New York, 1970.
2. A. P. Robertson and W. Robertson, "Topological Vector Spaces," Cambridge, 1973.
3. L. Schwartz, "Théorie des Distributions," Vols. I and II, Hermann, Paris, 1957 and 1959.
4. W. Rudin, "Real and Complex Analysis," McGraw-Hill, New York, 1986.
5. M. Spivak, "Calculus on Manifolds," Benjamin, 1965.
6. K. Yosida, "Functional Analysis," Springer-Verlag, Berlin, 1980.
7. J. Lutzen, "The Prehistory of the Theory of Distributions," Springer-Verlag, New York, 1982.
8. L. Hörmander, "The Analysis of Linear Partial Differential Operators" Vols. I and II, Springer-Verlag, Berlin, 1983.
9. S. G. Krantz, "Function Theory of Several Complex Variables," John Wiley, New York, 1982.
10. S. Mizohata, "The Theory of Partial Differential Equations," Cambridge, 1973.
11. I. M. Gel'fand and G. E. Shilov, "Generalized Functions," Vol. I, Academic Press, New York, 1964.

12. J. L. Lions and E. Magenes, "Non-homogeneous Boundary Value Problems and Applications," Vol. I, Springer-Verlag, New York, 1972.
13. F. Treves, "Basic Linear Partial Differential Equations," Academic Press, New York, 1975.
14. L. Bers, F. John and M. Schechter, "Partial Differential Equations," John Wiley, New York, 1964.
15. P. R. Garabedian, "Partial Differential Equations," John Wiley, New York, 1964.

Index